Methods in Molecular Biology

Series Editor
John M. Walker
School of Life and Medical Sciences
University of Hertfordshire
Hatfield, Hertfordshire, AL10 9AB, UK

For further volumes:
http://www.springer.com/series/7651

Heterologous Expression of Membrane Proteins

Methods and Protocols

Second Edition

Edited by

Isabelle Mus-Veteau

Université Côte d'Azur, CNRS, IPMC, France

Editor
Isabelle Mus-Veteau
Université Côte d'Azur
CNRS, IPMC, France

ISSN 1064-3745 ISSN 1940-6029 (electronic)
Methods in Molecular Biology
ISBN 978-1-4939-3635-9 ISBN 978-1-4939-3637-3 (eBook)
DOI 10.1007/978-1-4939-3637-3

Library of Congress Control Number: 2016935861

Printed on acid-free paper

This Humana Press imprint is published by Springer Nature
The registered company is Springer Science+Business Media LLC New York

Preface

Integral membrane proteins (IMPs) account for roughly 30% of all open reading frames in fully sequenced genomes. These proteins are of main importance to living cells. They are involved in fundamental biological processes like ion, water, or solute transport, sensing changes in the cellular environment, signal transduction, and control of cell-cell contacts required to maintain cellular homeostasis and to ensure coordinated cellular activity in all organisms. IMP dysfunctions are responsible for numerous pathologies like cancer, cystic fibrosis, epilepsy, hyperinsulinism, heart failure, hypertension, and Alzheimer's disease. However, studies on these and other disorders are hampered by a lack of information about the involved IMPs. Thus, knowing the structure of IMPs and understanding their molecular mechanism not only is of fundamental biological interest but also holds great potential for enhancing human health. This is of paramount importance in the pharmaceutical industry, which produces many drugs that bind to IMPs and recognizes the potential of many recently identified G-protein-coupled receptors (GPCRs), ion channels, and transporters as targets for future drugs. Fifty percent of all drug targets are GPCRs, which belong to one of the largest and most diverse IMP families encoded by more than 800 genes in the human genome. However, whereas high-resolution structures are available for a myriad of soluble proteins, atomic structures have so far been obtained for only 600 IMPs, with about 300 structures determined in the last 5 years (see: http://blanco.biomol.uci.edu/mpstruc/). Only about 10% of the unique IMP structures are derived from vertebrates. The first IMPs were crystallized due to their natural abundance, circumventing all the difficulties associated with overexpression. However, the majority of medically and pharmaceutically relevant IMPs are present in cells at very low concentrations, making expression of recombinant IMPs in heterologous systems suitable for large-scale production a prerequisite for structural studies. In recent years, the panel of possibilities for the production of IMPs has became larger and larger, from *E. coli* to mammalian cells, passing by other bacteria such as *L. Lactis*, protozoa such as *L. tarentolæ*, yeast, insect cells, frog oocytes, and even acellular systems, in order to create the "right expression system" for each IMP. Indeed, physical-chemical properties of IMPs are very different, and, therefore, it is difficult to predict the best approach. In any case, each system has its pros and its cons, and the choice of the best system allowing the best levels of functional protein production is often empirical. Concurrently with the advances in recombinant IMP production, improvement of the stabilization strategies of IMPs in solution has contributed to the growing number of IMP structures solved. Indeed, purification of IMPs requires the use of detergents to extract IMPs from the membrane and to maintain them in solution. Many IMPs are unstable in detergent solution, and finding suitable detergent and conditions that ensure protein homogeneity, functionality, stability, and crystallization is often a limiting and crucial step. Lots of tools and strategies in the field of heterologous expression systems and stabilization of IMPs for structural analyses are still under development. This volume encompasses chapters from leading experts in the area of membrane proteins who outline step-by-step protocols developed these last few years to improve recombinant IMPs's functional production and stabilization.

Valbonne, France ***Isabelle Mus-Veteau***

Contents

Contributors

FRANÇOIS ANDRÉ • *Institute of Integrative Biology of the Cell (I2BC), CEA, CNRS, Univ Paris-Sud, Université Paris Saclay, Gif-sur-Yvette, France*
JUNI ANDRÉLL • *MRC Laboratory of Molecular Biology, Cambridge, UK*
FEDERICA ANGIUS • *Laboratoire de Biologie Physico-Chimique des Protéines Membranaires, Institut de Biologie Physico-Chimique, CNRS, Univ Paris Diderot, Sorbonne Paris Cité, PSL Research University, Paris, France*
RENATO MANCINI ASTRAY • *Laboratório de Imunologia Viral, Instituto Butantan, Butantã, São Paulo, Brazil*
MARC J. BERGERON • *Division of Cellular and Molecular Neuroscience, Institut Universitaire en Santé Mentale de Québec, Québec, QC, Canada*
FRANK BERNHARD • *Centre for Biomolecular Magnetic Resonance, Institute for Biophysical Chemistry, Goethe-University of Frankfurt/Main, Frankfurt/Main, Germany*
ROSLYN M. BILL • *School of Life & Health Sciences, Aston University, Birmingham, UK*
RAJENDRA BOGGAVARAPU • *Institute of Biochemistry and Molecular Medicine, and Swiss National Centre of Competence in Research (NCCR) TransCure, University of Bern, Bern, Switzerland*
COILIN BOLAND • *Membrane Structural and Functional Biology Group, School of Medicine and School of Biochemistry and Immunology, Trinity College Dublin, Dublin, Ireland*
REBBA C. BOSWELL-CASTEEL • *Department of Biochemistry and Molecular Biology, University of Oklahoma Health Sciences Center, Oklahoma City, OK, USA*
RENATO BRUNI • *New York Consortium on Membrane Protein Structure (NYCOMPS), New York Structural Biology Center, New York, NY, USA*
MARTIN CAFFREY • *Membrane Structural and Functional Biology Group, School of Medicine and School of Biochemistry and Immunology, Trinity College Dublin, Dublin, Ireland*
STEPHANIE P. CARTWRIGHT • *School of Life & Health Sciences, Aston University, Birmingham, UK*
PATRICE CATTY • *Laboratoire de Chimie et Biologie des Métaux, CNRS (UMR-5249)/ CEA/Université Grenoble Alpes, BIG, CEA, Grenoble, France*
VINCENT CHAPTAL • *Drug Resistance and Membrane Proteins Team, UMR CNRS-UCBL1 5086, IBCP, Lyon, France*
SARIKA CHAUDHARY • *Systems Biology Group, CSIR-Institute of Genomics and Integrative Biology, New Delhi, India*
DANIEL O. DALEY • *Department of Biochemistry and Biophysics, Center for Biomembrane Research, Stockholm University, Stockholm, Sweden*
MARIA DALY • *MRC Laboratory of Molecular Biology, Cambridge, UK*
KATIA DUQUESNE • *Aix Marseille Université, Centrale Marseille, CNRS, ISM2 UMR7313, Marseille, France*
PATRICIA C. EDWARDS • *MRC Laboratory of Molecular Biology, Cambridge, UK*
PIERRE FALSON • *Drug Resistance and Membrane Proteins Team, UMR CNRS-UCBL1 5086, IBCP, Lyon, France*
LAURA FIORINI • *Université Côte d'Azur, CNRS, Institut de Pharmacologie Moléculaire et Cellulaire, Valbonne, France*

Eric Forest • *Univ. Grenoble Alpes, IBS, Grenoble, France; CNRS, IBS, Grenoble, France; CEA, IBS, Grenoble, France; Institut de Biologie Structurale, CNRS (UMR 5075)/ CEA/UGA, Grenoble Cedex, France*

Dimitrios Fotiadis • *Institute of Biochemistry and Molecular Medicine, and Swiss National Centre of Competence in Research (NCCR) TransCure, University of Bern, Bern, Switzerland*

Annie Frelet-Barran • *Institute of Integrative Biology of the Cell (I2BC), CEA, CNRS, Univ Paris-Sud, Université Paris Saclay , Gif-sur-Yvette, France; FEMTO-ST Institute, UMR CNRS 6174, University of Bourgogne Franche-Comte, Besançon, France*

Lucia Gonzalez-Lobato • *Drug Resistance and Membrane Proteins Team, UMR CNRS-UCBL1 5086, IBCP, Lyon, France*

Daniel Harder • *Institute of Biochemistry and Molecular Medicine, and Swiss National Centre of Competence in Research (NCCR) TransCure, University of Bern, Bern, Switzerland*

Lucie Hartmann • *Biotechnology and Cell Signalling, IMPReSs Protein Facility, UMR7242 CNRS-University of Strasbourg, Illkirch, France*

Franklin A. Hays • *Department of Biochemistry and Molecular Biology & Stephenson Oklahoma Cancer Center, University of Oklahoma Health Sciences Center, Oklahoma City, OK, USA*

Erik Henrich • *Centre for Biomolecular Magnetic Resonance, Institute for Biophysical Chemistry, Goethe-University of Frankfurt/Main, Frankfurt/Main, Germany*

Stephan Hirschi • *Institute of Biochemistry and Molecular Medicine, and Swiss National Centre of Competence in Research (NCCR) TransCure, University of Bern, Bern, Switzerland*

Oana Ilioaia • *Laboratoire de Biologie Physico-Chimique des Protéines Membranaires, Institut de Biologie Physico-Chimique, CNRS, Univ Paris Diderot, Sorbonne Paris Cité, PSL Research University, Paris, France*

Jean-Michel Jault • *UMR5086 CNRS/Université Claude Bernard Lyon 1, MMSB-IBCP, Lyon, France*

Jennifer M. Johnson • *Department of Biochemistry and Molecular Biology, University of Oklahoma Health Sciences Center, Oklahoma City, OK, USA*

Jennifer L. Johnson • *School of Chemistry & Biochemistry, Georgia Institute of Technology, Atlanta, GA, USA*

Soraia Attie Calil Jorge • *Laboratório de Imunologia Viral, Instituto Butantan, Butantã, São Paulo, Brazil*

Ravi C. Kalathur • *New York Consortium on Membrane Protein Structure (NYCOMPS), New York Structural Biology Center, New York, NY, USA*

Sibel Kalyoncu • *School of Chemistry & Biochemistry, Georgia Institute of Technology, Atlanta, GA, USA*

Martin S. King • *Mitochondrial Biology Unit, Medical Research Council, Cambridge, UK*

Valérie Kugler • *Biotechnology and Cell Signalling, IMPReSs Protein Facility, UMR7242 CNRS-University of Strasbourg, Illkirch, France*

Edmund R.S. Kunji • *Mitochondrial Biology Unit, Medical Research Council, Cambridge, UK*

Raquel L. Lieberman • *School of Chemistry & Biochemistry, Georgia Institute of Technology, Atlanta, GA, USA*

Petr Man • *BioCeV - Institute of Microbiology, Czech Academy of Sciences, Vestec, Czech Republic*

MARCEL MEURY • *Institute of Biochemistry and Molecular Medicine, and Swiss National Centre of Competence in Research (NCCR) TransCure, University of Bern, Bern, Switzerland*
LINA MIKALIUNAITE • *School of Life & Health Sciences, Aston University, Birmingham, UK*
BRUNO MIROUX • *Laboratoire de Biologie Physico-Chimique des Protéines Membranaires, Institut de Biologie Physico-Chimique, CNRS, Univ Paris Diderot, Sorbonne Paris Cité, PSL Research University, Paris, France*
KIAVASH MIRZADEH • *Department of Biochemistry and Biophysics, Center for Biomembrane Research, Stockholm University, Stockholm, Sweden*
JENNIFER MOLLE • *Drug Resistance and Membrane Proteins Team, UMR CNRS-UCBL1 5086, IBCP, Lyon, France*
LUCAS MOYET • *Laboratoire de Physiologie Cellulaire et Végétale, CNRS (UMR-5168)/CEA/INRA (USC1359)/Université Grenoble Alpes, BIG, CEA, Grenoble, France*
ISABELLE MUS-VETEAU • *Université Côte d'Azur, CNRS, Institut de Pharmacologie Moléculaire et Cellulaire, Valbonne, France*
MORTEN H.H. NØRHOLM • *Novo Nordisk Foundation Center for Biosustainability, Technical University of Denmark, Hellerup, Denmark*
MARINELA PANGANIBAN • *New York Consortium on Membrane Protein Structure (NYCOMPS), New York Structural Biology Center, New York, NY, USA*
VALÉRIE PRIMA • *LISM UMR 7255, CNRS and Aix-Marseille Université, Marseille, France*
NORBERT ROLLAND • *Laboratoire de Physiologie Cellulaire et Végétale, CNRS (UMR-5168)/CEA/INRA (USC1359)/Université Grenoble Alpes, BIG, CEA, Grenoble, France*
ALICE J. ROTHNIE • *Life & Health Sciences, Aston University, Birmingham, UK*
RALF-BERNHARDT RUES • *Centre for Biomolecular Magnetic Resonance, Institute for Biophysical Chemistry, Goethe-University of Frankfurt/Main, Frankfurt/Main, Germany*
SUKANYA SAHA • *Systems Biology Group, CSIR-Institute of Genomics and Integrative Biology, New Delhi, India*
EMILINE SAUTRON • *Laboratoire de Physiologie Cellulaire et Végétale, CNRS (UMR-5168)/CEA/INRA (USC1359)/Université Grenoble Alpes, BIG, CEA, Grenoble, France*
DAPHNE SEIGNEURIN-BERNY • *Laboratoire de Physiologie Cellulaire et Végétale, CNRS (UMR-5168)/CEA/INRA (USC1359)/Université Grenoble Alpes, BIG, CEA, Grenoble, France*
ROBERT M. STROUD • *Department of Biochemistry and Biophysics, University of California San Francisco, San Francisco, CA, USA*
JAMES N. STURGIS • *LISM UMR 7255, CNRS and Aix-Marseille Université, Marseille, France*
SANDRA FERNANDA SUÁREZ-PATIÑO • *Laboratório de Imunologia Viral, Instituto Butantan, Butantã, São Paulo, Brazil*
CHRISTOPHER G. TATE • *MRC Laboratory of Molecular Biology, Cambridge, UK*
SOBRAHANI THAMMINANA • *Department of Biochemistry and Biophysics, University of California at San Francisco, San Francisco, CA, USA*
STEPHEN TODDO • *Department of Biochemistry and Biophysics, Center for Biomembrane Research, Stockholm University, Stockholm, Sweden*
ZÖHRE UCURUM • *Institute of Biochemistry and Molecular Medicine, and Swiss National Centre of Competence in Research (NCCR) TransCure, University of Bern, Bern, Switzerland*

MARC UZAN • *Laboratoire de Biologie Physico-Chimique des Protéines Membranaires, Institut de Biologie Physico-Chimique, CNRS, Univ Paris Diderot, Sorbonne Paris Cité, PSL Research University, Paris, France*
RENAUD WAGNER • *Biotechnology and Cell Signalling, IMPReSs Protein Facility, UMR7242 CNRS-University of Strasbourg, Illkirch, France*
BENJAMIN WISEMAN • *UMR5086 CNRS/Université Claude Bernard Lyon 1, MMSB-IBCP, Lyon, France*
FAN ZHANG • *MRC Laboratory of Molecular Biology, Cambridge, UK*

Chapter 1

Cell-Free Production of Membrane Proteins in *Escherichia coli* Lysates for Functional and Structural Studies

Ralf-Bernhardt Rues, Erik Henrich, Coilin Boland, Martin Caffrey, and Frank Bernhard

Abstract

The complexity of membrane protein synthesis is largely reduced in cell-free systems and it results into high success rates of target expression. Protocols for the preparation of bacterial lysates have been optimized in order to ensure reliable efficiencies in membrane protein production that are even sufficient for structural applications. The open accessibility of the semisynthetic cell-free expression reactions allows to adjust membrane protein solubilization conditions according to the optimal folding requirements of individual targets. Two basic strategies will be exemplified. The post-translational solubilization of membrane proteins in detergent micelles is most straightforward for crystallization approaches. The co-translational integration of membrane proteins into preformed nanodiscs will enable their functional characterization in a variety of natural lipid environments.

Key words G-protein-coupled receptors, Nanodiscs, Synthetic biology, Membranes, Membrane protein crystallization, Lipid screening

1 Introduction

Cell-free (CF) expression in lysates of certain *Escherichia coli* strains has become a standard tool for the preparative-scale production of a wide variety of membrane proteins [1]. In particular advantageous is the adaption and fine-tuning of the expression environment by supplied compounds according to the requirements of individual membrane proteins and their intended applications [2]. CF expression can be considered as a core technology in the emerging field of synthetic biology as it combines natural biosynthetic pathways with artificial folding environments. The continuously growing number of CF expression conditions for membrane protein production can generally be classified into three basic modes [3]. In the precipitate forming P-CF mode, the synthesized membrane proteins precipitate after translation due to the lack of any supplied hydrophobic environment. The proteins need then to be

Isabelle Mus-Veteau (ed.), *Heterologous Expression of Membrane Proteins: Methods and Protocols*, Methods in Molecular Biology, vol. 1432, DOI 10.1007/978-1-4939-3637-3_1, © Springer Science+Business Media New York 2016

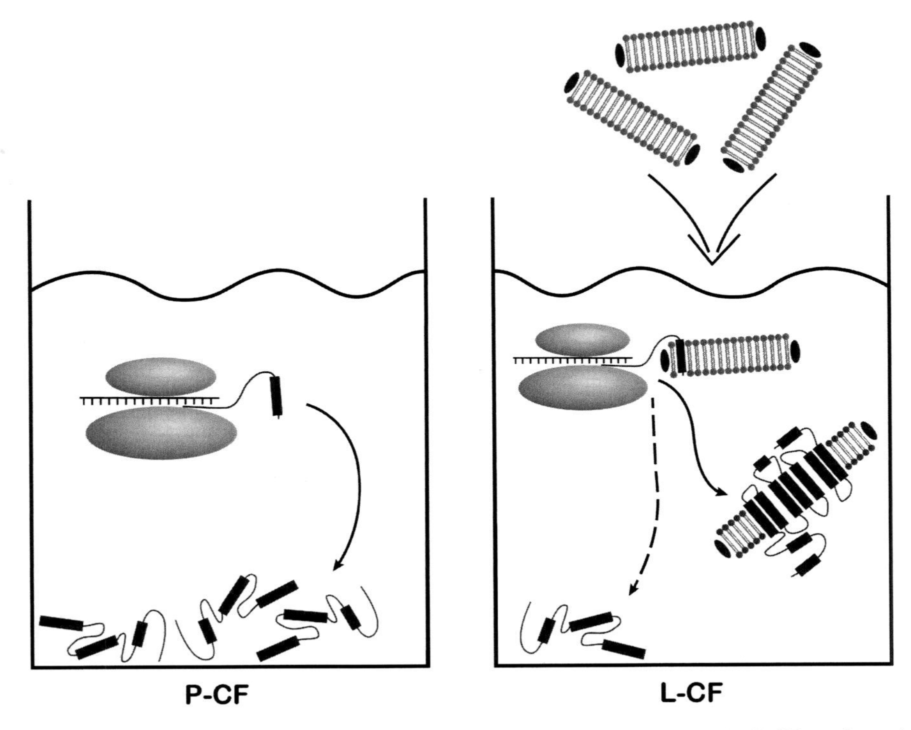

Fig. 1 CF expression modes exemplified in the protocols. DgkA is synthesized in the P-CF mode as initial precipitate and post-translationally solubilized in detergent for its subsequent crystallization. The ETB receptor is synthesized in the L-CF mode and co-translationally inserted into supplied NDs

solubilized post-translationally (Fig. 1). In the detergent-based D-CF or membrane-based L-CF modes, the membrane proteins will be co-translationally solubilized by supplied detergents or membranes, respectively. The selected expression modes can have strong impact on efficiencies, costs, and final sample quality and should be carefully and extensively screened before preparative-scale production is approached.

The structural as well as the functional characterization of membrane proteins synthesized in all three CF modes has already been accomplished [3]. The development of customized reaction protocols and screening of additives are mostly indispensable for obtaining suitable and efficient CF production protocols. However, based on the currently accumulated knowledge, some preliminary guidelines start to appear that could help to focus on the screening of the most promising reaction compounds and expression conditions. In this chapter, we present most recent optimizations in protocol and reaction design that streamline membrane protein expression projects and reduce costs as well as workload. We further highlight new technical details that could be valuable upon establishing CF expression technologies.

CF expression in *E. coli* lysates is excellent for prokaryotic as well as for eukaryotic membrane proteins. We describe applications of CF-synthesized membrane protein samples from two core expression modes in the most efficient continuous-exchange cell-free (CECF) configuration [4]. The P-CF production of the membrane-integrated enzyme DgkA results into sufficient quality for its subsequent high-diffraction crystallization [5]. The L-CF production of G-protein-coupled receptors (GPCRs) in the presence of supplied nanodiscs (NDs) allows their subsequent characterization of ligand-binding properties [6]. This production strategy is in particular suitable for critical membrane proteins that are sensitive against contact with detergents. Both described protocols might serve as guidelines for similar work with related proteins.

2 Materials

All stock solutions should be prepared with ultrapure water and stored at –20 °C if not otherwise stated.

2.1 General Materials

1. Fermenter for bacterial cultures, e.g., 5–10 L volume.
2. French Press.
3. Photometer.
4. Standard centrifuges and set of rotors.
5. Thermo shaker for incubation.
6. Chromatographic system (e.g., Äkta purifier, GE Healthcare).
7. Q-Sepharose column (GE Healthcare).
8. Immobilized Metal Affinity Chromatography (IMAC) material or column (Cube Biotech).
9. Sonicator.
10. Centriprep filter devices, 10 kDa MWCO (Millipore).
11. Mini-extruder (Avanti Polar Lipids).

2.2 *E. coli* Lysate Preparation

1. *E. coli* strains A19, BL21, or C43.
2. 2× TPG medium: 10 g/L yeast extract, 16 g/L tryptone, 5 g/L NaCl, 100 mM glucose, 22 mM KH_2PO_4, 40 mM K_2HPO_4.
3. Antifoam (Sigma).
4. 40× LY-A/B buffer: 400 mM Tris-acetate pH 8.2, 560 mM $Mg(OAc)_2$, 2.4 M KCl.
5. 1× LY-A buffer (washing buffer) diluted from the 40× LY-A/B stock, supplemented with 6 mM ß-mercaptoethanol.

6. 1× LY-B buffer (lysis buffer) diluted from the 40× LY-A/B stock, supplemented with 1 mM DTT and 1 mM phenylmethanesulfonylfluoride (PMSF).
7. 40× LY-C buffer: 400 mM Tris-acetate pH 8.2, 560 mM $Mg(OAc)_2$, 2.4 M KOAc.
8. 1× LY-C + DTT buffer (dialysis buffer): Diluted from the 40× LY-C stock, supplemented with 0.5 mM DTT.
9. 5 M NaCl.

2.3 T7-RNA Polymerase Preparation

1. *E. coli* BL21 (DE3) Star × pAR1219 [7].
2. LB medium: 10 g/L peptone, 5 g/L yeast extract, 5 g/L NaCl.
3. 1 M isopropyl-β-D-1-thiogalactopyranoside (IPTG).
4. 30% (w/v) streptomycin sulfate in H_2O.
5. Buffer-T7RNAP-A (equilibration buffer): 30 mM Tris–HCl pH 8.0, 50 mM NaCl, 1 mM EDTA, 10 mM ß-mercaptoethanol, 5% glycerol.
6. Buffer-T7RNAP-B (dialysis buffer): 10 mM K_2HPO_4/KH_2PO_4 pH 8.0, 10 mM NaCl, 0.5 mM EDTA, 1 mM DTT, 5% glycerol.
7. Resuspension buffer: 30 mM Tris–HCl, pH 8.0, 10 mM EDTA, 50 mM NaCl, 5% glycerol, and 10 mM β-mercaptoethanol.

2.4 DNA Template Preparation

1. Specific primers designed for the target DNA.
2. Vent polymerase (New England Biolabs).
3. PCR purification kit (Qiagen).
4. Restriction enzymes and ligase for template preparation.
5. Plasmid DNA purification kit (Machery-Nagel/Qiagen).
6. Agarose (Rotigarose, Roth).

2.5 CECF Expression Reactions

1. MD100 dialysis cartridges as reaction mix containers (Scienova).
2. 96-Deep-well microplates as feeding mix containers (Ritter riplate PP, 2 mL).
3. Dialysis tubes, 12–14 kDa MWCO (Spectrum).
4. Slide-A-Lyzer devices, 10 kDa MWCO (Thermo Scientific).
5. Optional: High-yield *E. coli* lysates including T7RNAP (Cube Biotech) as controls.
6. Stock solutions required for CECF reactions are listed in Table 1.

Table 1
Reagent example for CECF expression reaction with 1 mL RM and 14 mL FM

Compound	Stock conc.	Final conc.	Mix	µL
Preparation of master mix:				*µL/16.5 mL*[a]
Mix of 20 amino acids	25 mM[b]	1 mM	RM + FM	660 (44/616)
Acetyl phosphate (Li^+, K^+), pH 7.0[c]	1 M	20 mM	RM + FM	330 (22/308)
Phospho(enol)pyruvic acid (K^+), pH 7.0[c]	1 M	20 mM	RM + FM	330 (22/308)
75 × NTP mix, pH 7.0[c]	90 mM ATP	1.2 mM		
	60 mM G/C/UTP	0.8 mM	RM + FM	221 (15/206)
HEPES/KOH, pH 8.0[c]	2.4 M	100 mM	RM + FM	688 (46/642)
50 × Salt mix	50x	1x	RM + FM	330 (22/308)
DTT	500 mM	2 mM	RM + FM	66 (4/62)
Preparation of 1 mL RM:				*µL/1 mL*
MilliQ water				final 1 mL[d]
Master-mix				154
E. coli lysate + T7RNAP	1×	0.33–0.4×	RM	330–400
DNA template	0.2–0.5 µg/µL	2–20 ng/µL	RM	4–100
t-RNA *(E. coli)*	40 mg/mL	0.5 mg/mL	RM	12.5
Pyruvate kinase	10 mg/mL	0.04 mg/mL	RM	4
RiboLock	40 U/µL	0.3 U/µL	RM	7.5
Preparation of 14 mL FM:				*mL/14 mL*
MilliQ water				final 14 mL[d]
Master mix				2.15
LY-C buffer[e]				4.62–5.60
Optional compounds[f]				
Nanodiscs	0.5–1.2 mM	30–100 µM	RM	Variable[g]
DTT[h]	500 mM	2–10 mM	RM + FM	Variable[g]
Complete cocktail (Roche)[i]	50×	1×	RM + FM	20 + 280
PEG 8,000	40 %	2 %	RM + FM	50 + 700
Glutathione reduced/oxidized	200 mM each	0.2–5 mM	RM + FM	Variable[g]
Mix of 20 amino acids[j]	25 mM[b]	+1–2 mM	FM	Variable[g]

(continued)

Table 1 (continued)

Compound	Stock conc.	Final conc.	Mix	µL
Preparation of 50 × salt mix[k]	*Stock conc.*	*g/20 mL*		
Folinic acid (Ca^{2+})	5 mg/mL	0.1		
$Mg(OAc)_2 \times 4\ H_2O$[l]	355 mM	1.523		
KOAc	6.5 M	12.759		

[a]A 10 % excess volume is added in order to compensate for the loss of volume effect upon mixing of the individual compounds in the master mix (i.e., 1.1 mL RM + 15.4 mL FM). Nevertheless, the calculation is for a final volume of 1 mL RM and 14 mL FM. Volumes are rounded. Given are the total volumes with the individual volumes for RM and FM in parenthesis.
[b]Stock stays turbid, mix thoroughly before pipetting.
[c]Adjusted with KOH.
[d]Fill up to a final volume of 1 mL (RM) or 14 mL (FM).
[e]LY-C buffer is added in order to compensate for the Mg^{2+} and K^{+} ions brought into the RM by the *E. coli* lysate (=4.9 mM Mg^{2+} with 0.35×). The volume of LY-C buffer in the FM corresponds therefore with the lysate volume in the RM.
[f]If optional compounds are added, the MilliQ water volumes of the RM and/or FM have to be adjusted accordingly.
[g]Volume depends on concentration of stocks and/or on desired final concentration.
[h]Concentration of DTT might be increased for improving target protein quality.
[i]Dissolve one tablet in 1 mL MilliQ water.
[j]An additional supply of amino acids in the FM can improve expression efficiencies.
[k]Dissolve in water at 40 °C.
[l]Final Mg^{2+} concentration will be 12 mM in the reaction (7.1 mM from salt mix and 4.9 mM from lysate/LY-buffer).

2.6 MSP1E3D1 Preparation and ND Formation

NDs are prepared by using the membrane scaffold protein derivative MSP1E3D1 and DMPC, DMPG, DOPC, or DOPG lipid in appropriate molar ratios.

1. pET-28-MSP1E3D1 vector [8].
2. BL21(DE3) Star cells.
3. LB-medium: 10 g/L peptone, 5 g/L yeast extract, 5 g/L NaCl.
4. 10 % (w/v) glucose stock solution.
5. 1 M IPTG stock solution.
6. Complete EDTA-free protease inhibitor (Roche).
7. 10 % (v/v) Triton X-100 stock solution.
8. MSP-A buffer: 40 mM Tris/HCl, pH 8.0, 300 mM NaCl, 1 % (v/v) Triton X-100.
9. MSP-B buffer: 40 mM Tris/HCl, pH 8.9, 300 mM NaCl, 50 mM cholic acid.
10. MSP-C buffer: 40 mM Tris/HCl, pH 8.0, 300 mM NaCl.
11. MSP-D buffer: 40 mM Tris/HCl, pH 8.0, 300 mM NaCl, 50 mM imidazole.

12. MSP-E buffer: 40 mM Tris/HCl, pH 8.0, 300 mM NaCl, 300 mM imidazole.
13. MSP-F (dialysis) buffer: 40 mM Tris/HCl, pH 8.0, 300 mM NaCl, 10% (v/v) glycerol.
14. Lipid-cholate stock solutions: 50 mM Lipid, 100 mM sodium cholate.
15. 10% DPC stock solution (for complete solubilization ultrasonic water bath is required).
16. ND-A buffer: 10 mM Tris/HCl, pH 8.0, 100 mM NaCl.
17. Bio-beads (Bio-Rad).
18. Ni-NTA resin (Cube Biotech).

2.7 P-CF Production of DgkA

1. Empigen BB dissolved.
2. DgkA-buffer-1: 1 mM TCEP, 300 mM NaCl, and 20 mM HEPES, pH 7.5.
3. DgkA-buffer-2: 1 mM TCEP, 150 mM NaCl, and 20 mM HEPES, pH 7.5.
4. DgkA-buffer-3: 1 mM TCEP, 150 mM NaCl, and 10 mM HEPES, pH 7.5.
5. DgkA-buffer-4: 1 mM TCEP, 100 mM NaCl, and 10 mM Tris/HCl, pH 7.4.

3 Methods

3.1 *E. coli* Lysate Preparation

Depending on the nature or intended application of the synthesized protein, a particular *E. coli* strain may be selected for lysate preparation (*see* **Note 1**). Some preferred *E. coli* strains are A19, BL21, or C43 and the resulting efficiencies of the corresponding lysates in CF membrane protein production are comparable. Strain A19 is low in endogenous RNase and BL21 derivatives generally contain different endogenous pools of proteases if compared with *E. coli* K strains such as A19. The strains C43 and C41 are well known as standard for the cellular production of membrane proteins and copy numbers of beneficial proteins may be increased. In the suggested 2x TPG medium, the final yield of lysate will be in between 4 and 7 mL per 1 L of fermenter broth. The processing of the crude cell lysate depends on the intended application of the CF-synthesized proteins and several options will be discussed. The following protocol exemplifies lysate preparation out of a culture volume of 10 L:

1. For the pre-culture, inoculate 120 mL 2× TPG medium with freshly grown overnight cultures of the selected *E. coli* strain and shake overnight at 37 °C (*see* **Note 2**).

2. Inoculate a fermenter with 2× TPG medium with the preculture in a ratio of 1:100 and incubate at 37 °C with vigorous stirring (500–700 rpm) and aeration until mid-log phase (OD_{600} approximately 3.5–4.5, *see* **Note 3**).
3. Cool down the fermenter broth from 37 °C to approximately 20 °C within 20–40 min.
4. Harvest the cells by centrifugation at 6800 × *g* for 15 min.
5. Suspend and wash the pellet with 300 mL LY-A buffer, and centrifuge at 8000 × *g* for 10 min.
6. Repeat this step two more times.
7. Weigh the pellet and suspend it in 110 % (w/v) LY-B buffer.
8. Disrupt cells with French Press or a similar device; the solution should become grayish and viscous. Centrifuge the solution at 30,000 × *g* for 30 min.
9. Transfer supernatant into a fresh tube (*see* **Note 4**). Centrifuge the supernatant one more time at 30,000 × *g* for 30 min.
10. Transfer supernatant to a fresh tube and adjust to a final concentration of 400 mM NaCl with a 5 M NaCl stock solution. Incubate at 42 °C in a water bath for 45 min (*see* **Note 5**).
11. Dialyze the turbid solution overnight against approx. 100 times volume of LY-C+ DTT buffer using a 12–14 kDa cutoff membrane. Apply two changes of LY-C + DTT buffer.
12. Fill the turbid solution into a centrifuge tube and centrifuge at 30,000 × *g* for 30 min.
13. Transfer supernatant into a fresh tube and repeat centrifugation once (= S30 lysate).

    Option: Lysates can be centrifuged for an additional 1 h at 60,000 × *g* (= S60 lysate) in an ultracentrifuge in order to reduce residual background of endogenous porins such as OmpF (*see* **Note 6**). Centrifugation at higher g-forces could be applied in order to further reduce backgrounds [9] but protein expression efficiencies of the lysate might be reduced as well.

14. Remove supernatant and mix shortly. The final total protein concentration of the lysate should be in between 20 and 50 mg/mL.
15. Adjust the lysate with appropriate concentration of T7RNAP (*see* Subheading 3.2). The optimal concentration should be determined with a pilot screen after T7RNAP purification (*see* **Note 7**).
16. Aliquot into suitable volumes (*see* **Note 8**). Shock-freeze in liquid nitrogen and store at −80 °C. Frozen extract is stable for many months. Aliquots can be thawed on ice and left on ice for few hours during setup of reactions.

17. Check efficiency of each new lysate batch with the expression of GFP or any other suitable protein standard and determine the basic Mg^{2+} ion optimum. The Mg^{2+} ion optimum is usually within the range of 12–24 mM.

3.2 T7RNAP Preparation

T7RNAP is produced from the *E. coli* strain BL21 (DE3) Star × pAR1219 by conventional cultivation in Erlenmeyer flasks with LB medium:

1. Inoculate 1 L LB medium 1:100 with a fresh pre-culture of BL21 (DE3) Star × pAR1219.
2. Incubate the culture at 37 °C on a shaker until OD_{600} = 0.6–0.8.
3. Induce T7RNAP expression with 1 mM IPTG.
4. Incubate for further 3–5 h at 37 °C.
5. Harvest the cells by centrifugation at 6800 × *g* for 15 min at 4 °C.
6. Resuspend the pellet in 30 mL resuspension buffer.
7. Disrupt the cells with French Press, centrifuge at 20,000 × *g* for 30 min, and transfer supernatant into a fresh tube.
8. Precipitate nucleic acids in the supernatant with a final concentration of 3% streptomycin sulfate. Add dropwise the required amount of stock solution, mix, and incubate for 5 min on ice.
9. Centrifuge the turbid solution at 20,000 × *g* for 30 min.
10. Load the supernatant on a 40 mL Q-Sepharose column equilibrated with 2 column volumes (CV) equilibration buffer.
11. Wash the column with equilibration buffer at a flow rate of 4 mL/min until A_{280} of the elution is stable.
12. Elute bound proteins with a gradient from 50 to 500 mM NaCl at a flow rate of approximately 3 mL/min.
13. Check for a prominent band at approximately 90 kDa by SDS-PAGE analysis and Coomassie Blue staining and pool the fractions with highest T7RNAP content.
14. Dialyze pooled fractions against dialysis buffer.
15. Concentrate T7RNAP to 4–8 mg/mL by ultrafiltration (*see* **Note 9**).
16. Adjust T7RNAP solution to final concentration of 50% glycerol and store at −80 °C. Stored aliquots are stable for many months. Working aliquots could be stored at −20 °C.

3.3 DNA Template Design and Preparation

In general, two different transcription systems can routinely be employed for CF expression. Most efficient is expression under control of T7 regulatory elements and with supplied T7RNAP. DNA templates for CF expression can be prepared either by amplification of linear transcriptional units by polymerase chain reactions (PCR), or by cloning the gene of interest into suitable vectors such as plasmids from the pET (Merck Biosciences) or pIVEX (Roche Diagnostic) series. An alternative option is to use the endogenous bacterial RNA polymerase already present in the lysate in combination with corresponding regulatory regions such as the *tac* promoter. The T7RNAP and the *E. coli* RNA polymerase have different characteristics in view of processivity and initiation of transcription events. It could therefore be helpful to evaluate both transcription systems in particular for the production of very large transcripts. High-quality and -purity standards of the DNA template are essential for CF expression. Final concentrations of DNA template within CF reactions are in between 0.1 and approx. 20 ng/μL of reaction mixture.

Plasmid templates should be prepared with commercial standard kits such as "Midi" or "Maxi" DNA purification kits or PCR purification kits (Qiagen, Macherey Nagel) according to the manufacturers' instructions. The finally purified DNA should best be dissolved in pure MilliQ water or in low-molarity buffers without EDTA. The final concentrations of template stocks should be in between 0.2 and 0.5 mg/mL.

Templates may be engineered with several modifications in order to improve expression, purification, or monitoring. (i) Small N-terminal expression tags comprising up to six codons could be valuable in order to reduce secondary structure formation of the mRNA and to facilitate interactions of the mRNA ribosomal binding site with the ribosomal subunits. We recommend the H-tag sequence (AAA CCA TAC GAT GGT CCA) immediately placed behind the ATG start codon as a first choice [10]. (ii) C-terminal standard purification tags such as a poly(His)$_{10}$-tag or the StrepII-tag will streamline the purification of the synthesized membrane protein out of the reaction mixture. In order to improve accessibility of the purification tag and thus the recovery of the protein from affinity purification, a small linker of 4–10 amino acids should be placed in between target protein and purification tag. (iii) Derivatives of green fluorescent protein (GFP) such as superfolder GFP could be attached to the C-terminus of the target protein. GFP-tags will help to monitor and to quantify solubilization efficiencies of membrane proteins CF synthesized in the presence of supplied hydrophobic additives directly in the RM. Monitor tags could significantly accelerate the screening of compounds such as different sets of NDs as well as the determination of specific activities of the resulting membrane protein samples.

3.4 Basic CECF Expression Reactions

With *E. coli* lysates and T7RNAP transcription, routine GFP expression yields of 4–5 mg/mL of RM can be obtained (*see* **Note 10**). The RM volume is separated from a 14 to 20 times larger feeding mix (FM) volume via a semipermeable membrane with a molecular weight cutoff (MWCO) of 12–14 kDa. The FM supplies a reservoir of small-molecule precursors and additionally acts as a dilution reservoir for inhibitory by-products formed during the expression reaction in the RM. The choice of reaction containers is important in order to ensure efficient reagent exchange in between the two compartments. We recommend commercial dialysis cartridges (Xpress Micro Dialyzer MD100, 12–14 kDa MWCO, Scienova) as convenient reaction containers for analytical or semi-preparative-scale CECF reactions. The cartridges hold the RM volume and are placed into cavities of 96-deep-well plates (Ritter) holding appropriate volumes of FM (Fig. 2). Analytical scale CECF expressions necessary for protocol optimization or compound screening are best performed in RM volumes of 25–100 μL (*see* **Note 11**).

CF expression protocols established in analytical scale reactions can be scaled up to many mL RM volumes without any loss of efficiencies as long as sufficient exchange surface in between RM and FM is provided. For preparative-scale CECF reactions with higher RM volumes, commercial Slide-A-Lyzer devices with an MWCO of 10 kDa (Thermo Scientific) are suitable. Slide-A-Lyzers are available in sizes for 0.5–3 mL or 3–12 mL RM volumes. The

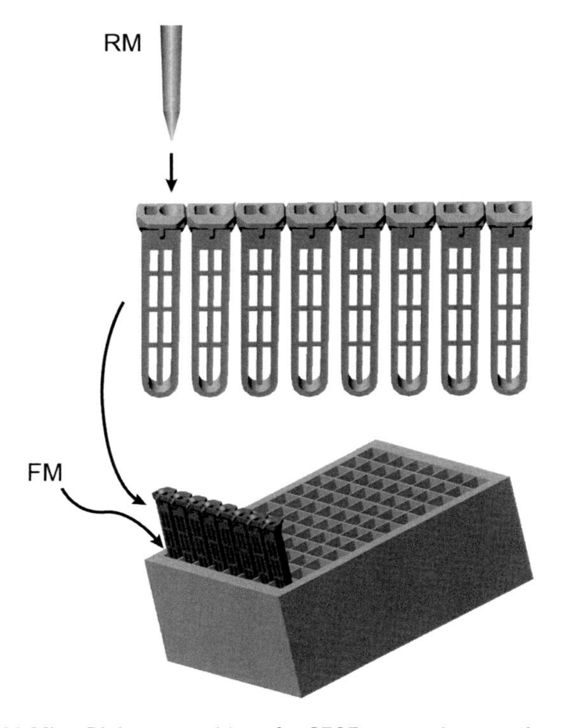

Fig. 2 MD100 MicroDialyzer cartridges for CECF expression reactions. The RM is filled into the round opening of the cartridge and the FM is filled into the cavities of 96-deep-well plates

Slide-A-Lyzers are placed into suitable containers holding appropriate volumes of FM such as plastic or glass trays or custom-made Plexiglas containers [11]. Alternatively, commercial d-tube dialyzers (Novagen) with screw cups or standard dialysis tubes closed with knots or clips can be used as RM container (*see* **Note 12**). Suitable 15–50 mL plastic tubes (Falcon) could serve as FM container.

1. Calculate the desired total RM and FM volumes. Depending on the selected reaction containers, the total FM volume is approximately 14–20 times the volume of the RM.
2. Calculate total volumes of the individual compound stocks (*see* Table 1). For the preparation of master mixes, the calculation of a 10 % excess volume of the individual compounds is recommended in order to compensate for the loss of volume effect upon mixing of the individual reagent volumes.
3. Prepare the common master mix for RM and FM (*see* Table 1).
4. Option: Additional low-molecular-weight compounds such as DTT may now be added to the master mix if desired.
5. Mix the master mix thoroughly and remove appropriate volume for the preparation of the FM.
6. Add appropriate volume of LY-C buffer to the FM (*see* **Note 13**).
7. Fill up with MilliQ water to the final FM volume.
8. Remove appropriate volume from the master mix for the preparation of the RM.
9. Complete RM with the high-molecular-weight compounds (*see* Table 1).
10. Option: Additional high-molecular-weight compounds such as ND may now be added to the RM.
11. Fill up with MilliQ water to the final RM volume.
12. Transfer FM into a suitable container (e.g., cavities of 96-deep-well plates).
13. Transfer RM into a suitable container (e.g., dialysis cartridges).
14. Combine the RM and FM containers (e.g., place filled dialysis cartridge into the corresponding cavity of a 96-deep-well plate).
15. Incubate reaction for 6–9 h or overnight at 25–30 °C (*see* **Note** 14). The reaction should be slightly agitated while incubating in order to promote reagent exchange through the membrane.
16. After incubation, mix the RM thoroughly in order to suspend potentially formed precipitates. Then remove the RM from the container with a pipette and analyze protein expression.

3.5 Troubleshooting and CF Expression Optimization

Problems are caused either by low performance of the expression system or by bad sample quality. The complexity of protein expression is largely reduced in CF reactions as issues with membrane translocation, proteolytic degradation, or toxicity are less relevant. Optimization of system performance is therefore usually fast and initial problems with insufficient expression efficiencies can be addressed by systematic and standardized screening procedures as listed below. In contrast, sample quality optimization is a very individual process that depends on characteristic properties of the target proteins and on the availability of quality assays. However, the open nature of CF reactions offers a unique versatility in order to optimize membrane protein sample quality by testing different expression modes and by screening individually designed hydrophobic environments [2].

The performance of CF membrane protein production is best optimized in the P-CF expression mode. Turbidity of the RM and precipitate formation will already serve as preliminary indicator for expression success. Analytical scale screening reactions should always be performed in duplicate. In addition, control reactions with expression of GFP should be included in order to rule out technical problems.

1. Quality and purity of the DNA template are important and should be checked by spectroscopy or agarose gel electrophoreses. Templates should be purified by standard affinity chromatography (*see* Subheading 3.3).
2. The concentration of Mg^{2+} ions is crucial for efficient expression and must be screened for each new template or even new reagent stock. Screening should be performed within the range of 10–24 mM and appropriate amounts of additional $Mg(OAc)_2$ must be added into the FM and RM.
3. Inefficient initiation of translation is a frequent problem and can be addressed by addition of small expression tags to the 5-prime end of the coding region (*see* Subheading 3.3). Alternatively, the natural nucleotide sequence within the first six to ten codons could be AT enriched by silent mutagenesis in order to suppress secondary structure formation.
4. Codon usage in particular of larger reading frames of eukaryotic origin might be adjusted to the *E. coli* preferences. In addition, an overall high GC content of the coding region should be avoided.

3.6 Preparation of NDs

NDs are assembled in vitro out of purified membrane scaffold proteins (MSP) and detergent-solubilized lipids. Thereby the use of defined MSP-to-lipid ratios is crucial for optimal ND assembly performance. Several engineered MSP derivatives can be used and they direct the final diameter of the NDs ranging from some 5 nm to approximately 15 nm [8]. We exemplify production and assembly with the MSP derivative MSP1E3D1 resulting into NDs with a diameter of approximately 12 nm. As controls, preassembled NDs with a variety of MSP derivatives and lipid combination and suitable for L-CF expression approaches can be obtained from commercial sources (Cube Biotech).

1. For MSP1E3D1 production, 4 × 600 mL LB supplemented with a final concentration of 0.5 % (w/v) glucose and 30 μg/mL kanamycin in 2 L buffled Erlenmeyer flasks are inoculated 1:12 with a fresh overnight culture of strain BL21 (DE3) Star × pET-28-MSP1E3D1.
2. Incubate at 37 °C and at 180–220 rpm until OD_{600} = 1.0 is reached.
3. Induce MSP1E3D1 expression with 1 mM final IPTG concentration.
4. Continue incubation for 1 h at 37 °C.
5. Reduce temperature to 28 °C and continue incubation for further 4 h.
6. Harvest cells by centrifugation at 6000 × *g* for 10 min at 4 °C and discard supernatants.
7. Combine the pellets and wash once with MSP-C buffer; store at −20 °C or continue with preparation.
8. Suspend the pellets in 50 mL MSP-C buffer supplemented with one tablet of complete protease inhibitor (Roche).
9. Adjust to a final concentration of 1 % (v/v) Triton X-100.
10. Disrupt cells by ultra-sonication (3 × 60 s and 3 × 45 s), with a rest period of at least 60 s on ice between each cycle. Gently mix after each sonication cycle.
11. Centrifuge the suspension for 20 min at 30.000 × *g* at 4 °C and filter supernatant through a 0.45 μm filter.
12. Equilibrate a Ni^{2+}-immobilized IMAC column (10 mL bed volume) with 5 CV MSP-A buffer and load filtered supernatant on the column. Flow rate should be 1–2 mL/min.
13. Wash the column with each 5 CV MSP-A, -B, -C, and -D buffer and with a flow rate of 2 mL/min.
14. Elute protein in 1 mL fractions with MSP-E buffer and pool MSP1E3D1 containing fractions. Adjust to 10% (v/v) glycerol.
15. Dialyze for 3 h at 4 °C against 5 L MSP-F buffer, change to fresh 5 L MSP-F buffer, and continue dialysis overnight.
16. Remove precipitated protein by centrifugation at 18,000 × *g* for 30 min at 4 °C. The protein concentration in the supernatant should be in between 80 and 100 μM. Aliquot the supernatant, shock-freeze in liquid nitrogen, and store frozen at −80 °C.

17. For ND assembly, mix purified MSP1E3D1, selected lipid-cholate stock, DPC, and ND-A buffer in a tube. The final concentration of DPC is 0.1 % (w/v), and the MSP1E3D1:lipid ratio is variable, e.g., 1:115 (DMPC), 1:110 (DMPG), 1:80 (DOPC), and 1:90 (DOPG).
18. Incubate the mixture at room temperature for 1 h.
19. Add 0.5 g Bio-beads (equilibrated with ND-A buffer) per mL solution and incubate for further 4 h on a shaker at room temperature in order to remove the detergent. Alternatively, dialysis against ND-A buffer might also be used for the removal of detergent.
20. Centrifuge the solution at 18,000 × *g* for 30 min in order to completely remove Bio-beads from the solution.
21. Fill the supernatant into a Centriprep concentrating unit (MWCO 10 kDa) equilibrated with ND-A buffer and centrifuge at 2000 × *g* for 20 min several times until the final MSP1E3D1 concentration is approximately 2.4 mM, corresponding to 1.2 mM ND concentration. The homogeneity of ND samples could be checked by size-exclusion chromatography using a Superdex 200 3.2/30 column.
22. Aliquot and freeze ND samples in liquid nitrogen and store the aliquots at −80 °C until further use.

3.7 Application I: P-CF Expression of Membrane Proteins for Crystallization (Example of DgkA)

Diacylglycerol kinase DgkA is a membrane-integrated enzyme catalyzing the ATP/Mg^{2+}-dependent conversion of diacylglycerol to phosphatidic acid. Engineering resulted into the stabilized but still functionally active derivative Δ7 DgkA [5]. The protein is modified by a C-terminal poly(His)-tag for purification. The protein is P-CF synthesized from a pIVEX vector under control of a T7 promoter (Fig. 3). Reactions are incubated at 30 °C with slight shaking for 16 h. Upon P-CF expression, the reactions will become cloudy during incubation due to the precipitation of the synthesized membrane protein.

3.7.1 Expression Protocol Development

1. Set up reactions according to Table 1 and with a final plasmid template concentration of 2–10 ng/μL RM.
2. Determine optimal Mg^{2+} concentration in 50 μL analytical scale reactions with 825 μL feeding mixture in a first screen covering a Mg^{2+} range of 12–24 mM.
3. After incubation, mix the reaction in order to suspend the pelleted fraction and transfer the suspension into a fresh 1.5 mL Eppendorf tube.
4. Centrifuge at 12,000 × *g* for 10 min at 4 °C.
5. Remove the supernatant and wash the pellet twice with 500 μL of LY-C buffer.

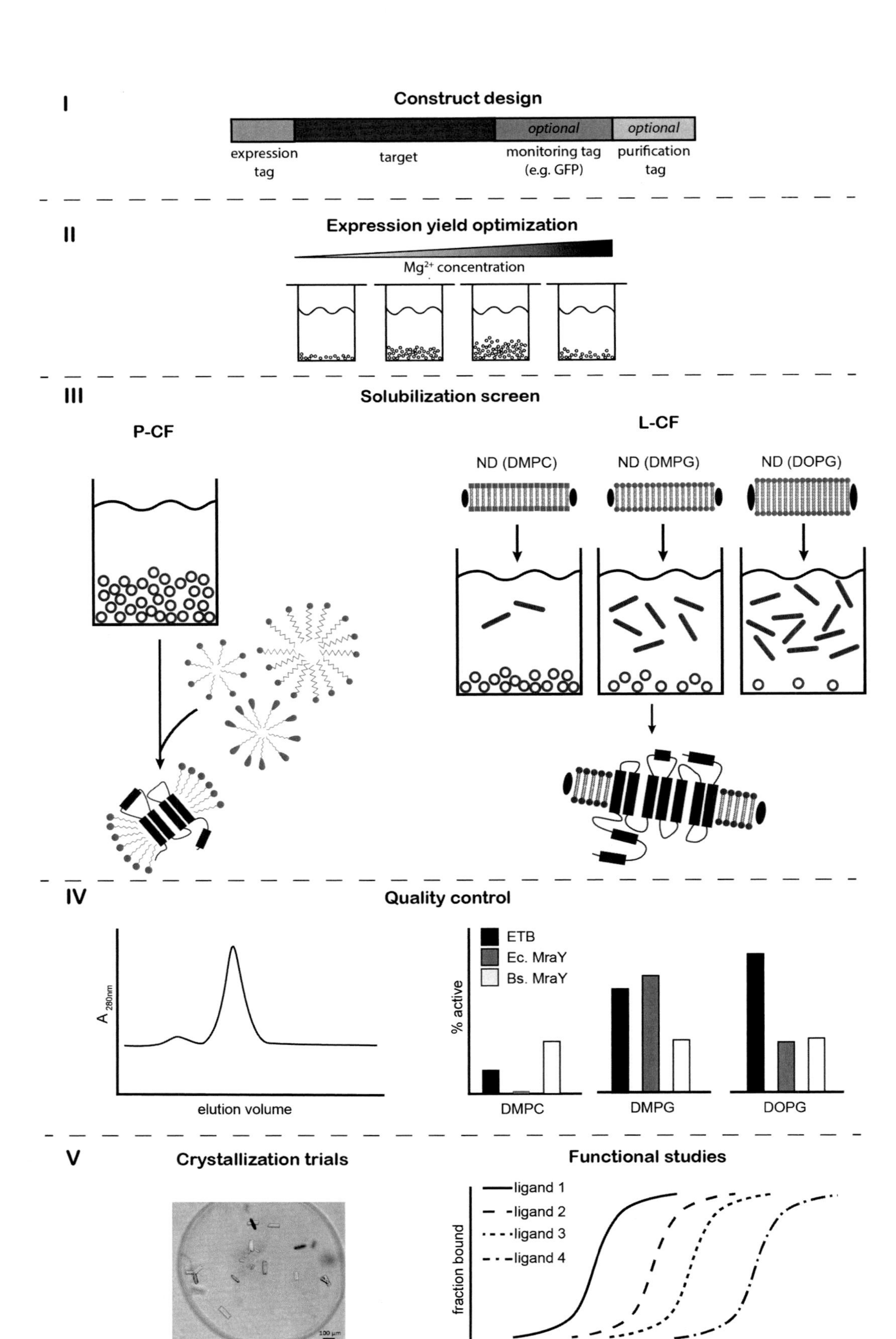
I
Construct design
optional
optional
expression tag
target
monitoring tag (e.g. GFP)
purification tag
II
Expression yield optimization
Mg2+ concentration
III
Solubilization screen
P-CF
L-CF
ND (DMPC)
ND (DMPG)
ND (DOPG)
IV
Quality control
A 280nm
elution volume
ETB
Ec. MraY
Bs. MraY
% active
DMPC
DMPG
DOPG
V
Crystallization trials
100 µm
Functional studies
ligand 1
ligand 2
ligand 3
ligand 4
fraction bound
ligand concentration

6. Suspend the final washed pellet in 50 μL (= initial reaction mix volume) of LY-C buffer.
7. Analyze suitable aliquots (usually in between 1 and 10 μL) of the suspension by 12% SDS-PAGE and estimate the DgkA expression according to a standard on the gel (*see* **Note 15**).

3.7.2 Solubilization Screen

1. If expression efficiency is sufficient, express 100 μL of DgkA, make 5 × 20 μL aliquots of the resulting suspensions, and centrifuge and wash the pellet as described above.
2. With the five pellet aliquots, screen the post-translational solubilization by addition of detergent solutions at final concentrations of 1–3%. A comprehensive solubilization screen should include mild as well as some harsh detergents and we would recommend to start with a set composed out of the detergents DDM, DH_7PC, DPC, LMPG, and SDS. Detergents could be dissolved in LY-C buffer but other buffer compositions could be used as well. P-CF-synthesized DgkA is solubilized in a solution of 3% (w/v) Empigen BB dissolved in DgkA-buffer-1 within 30 min by incubation at 4 °C.
3. Suspend the pellets in 100 μL of detergent solution and incubate with slight shaking for 0.5–2 h (*see* **Note 16**).
4. Centrifuge, evaluate the solubilization efficiency by the residual pellet size, and analyze the supernatants by SDS-PAGE.

3.7.3 Preparative-Scale Production and Purification

1. Prepare and incubate a preparative-scale reaction with several mL of RM and an RM:FM ratio of 1:14 (e.g., 3 mL RM and 42 mL FM).
2. Suspend the formed precipitate in the reaction container thoroughly by pipetting, and then transfer the RM into a fresh tube. Perform all the following steps at 4 °C or on ice.
3. Pellet the precipitate by centrifugation and wash the pellet twice in LY-C buffer.
4. Solubilize DgkA in a solution of 3% (w/v) Empigen BB dissolved in DgkA-buffer-1 within 30 min by incubation at 4 °C.
5. Centrifuge at 20,000 × *g* for 10 min to remove insoluble material.
6. Transfer the clear supernatant in a fresh tube and add 5 mL Ni-NTA resin pre-equilibrated with 0.5% (w/v) Empigen BB in DgkA-buffer-1.

Fig. 3 Flow charts for the CF production of DgkA in the P-CF mode and the ETB receptor in the L-CF mode. The production process is systematically optimized in individual steps. I: Template design, e.g., tag optimization; II: yield optimization, e.g., Mg^{2+} optimization; III: protein production in the corresponding CF expression modes; IV: purification and quality control of the protein samples; V: application, e.g., crystallization or biochemical characterization

7. Incubate for 30 min by slight shaking and then pack the resin into a gravity column (1 × 15 cm) and wash with 10 CV of 3% (w/v) Empigen BB in DgkA-buffer-1.
8. Perform a second wash with 15 CV 3% (w/v) Empigen BB in DgkA-buffer-1 supplemented with 40 mM imidazole.
9. Exchange the detergent Empigen BB with the detergent DM on column by washing with 12 CV with 0.25% (w/v) DM in DgkA-buffer-2.
10. Elute the protein with 0.25% (w/v) DM and 250 mM imidazole in DgkA-buffer-3 (*see* **Note 17**).
11. Concentrate the protein in the elution fraction to 12 mg/mL in an Amicon 50 kDa concentrator.
12. Load the concentrated DgkA sample on a gel-filtration column Superdex S200 16/60 equilibrated with 0.25% (w/v) DM in DgkA-buffer-4.
13. Peak fractions elute at approx. 4.5 mL and are concentrated to 12 mg/mL.
14. Samples can be flash frozen in liquid nitrogen and stored at −80 °C up to 6 months or immediately used for crystallization or functional characterization [5].

3.8 Application II: L-CF Expression of GPCR Samples in the Presence of NDs for Biochemical Studies

The co-translational insertion and functional folding of GPCRs into pre-assembled NDs strongly depend on the lipid properties including head group composition, fluidity, as well as matching of the bilayer thickness with the hydrophobic area of the membrane protein [12]. Screening for optimal lipid composition of the NDs is therefore essential for the production of high-quality samples. The final ND concentration in the reaction is furthermore determined by the expression efficiency of the GPCR. An initial concentration screen should therefore subsequently be performed with each selected ND/lipid combination. Membrane protein-GFP fusions will accelerate the lipid screening by monitoring the fluorescence in the supernatant of the reaction. The following protocol exemplifies the production of ND complexes of the human full-length endothelin-B (ETB) receptor, a key modulator in blood pressure regulation (Fig. 3). The resulting ETB/ND complexes are suitable for the characterization of ligand-binding properties, e.g., by surface plasmon resonance techniques [6]. The protocol might be suitable for other GPCRs as well.

1. Prepare a set of pre-formed NDs containing membranes with various lipid compositions (*see* Subheading 3.6). Commonly used lipids for initial screens are DMPC, DMPG, DOPC, and DOPG. The ND stocks should be at 0.5 mM to 1 mM.
2. For each pre-formed ND type, set up three analytical scale (25–100 μL) CECF reactions for expression of the ETB-

superfolderGFP construct. Prepare a master mix for a corresponding number of reactions according to Table 1.

3. Split master mix into RM and FM and complete FM (*see* Table 1).
4. Supplement RM according to Table 1 but leave out the water.
5. Divide RM into aliquots according to the number of analyzed ND types and complete with corresponding volumes of ND stock and water. Final ND concentrations in the RM should be equal for each reaction and within the range of 60–80 μM. The high ND concentration should ensure a sufficiently high ratio of NDs to the synthesized GPCR. The solubilization of the GPCR is then determined by its association with the lipids.
6. Fill RM and FM into the containers, assemble CECF reactions, and incubate overnight at 30 °C with continuous shaking.
7. Remove RMs into fresh tube, centrifuge at 18,000 × *g* for 10 min, and analyze supernatants by superfolderGFP fluorescence (*see* **Note 18**).
8. Select the ND types giving highest fluorescence in the reaction supernatant and refine the optimal ND concentration. Prepare a master mix for ten reactions for the screening of five ND concentrations (e.g., 0, 20, 40, 80, 100 μM) in triplicates. Proceed according to **step 3**.
9. GPCR/ND complexes may be purified from the reaction supernatant by affinity chromatography by taking advantage of terminal purification tags attached to the GPCR (*see* **Note 19**) or by affinity to immobilized ligands.

4 Notes

1. Strains selected for lysate preparation may further contain specific mutations in order to reduce non-desired background activities or they may contain additional gene copies encoding for helper proteins such as chaperones. Although not tested, it appears very likely that many *E. coli* strains can be used successfully for lysate preparation according to the described protocol. However, a growth curve of new strains should always be recorded in order to determine the optimal time points for cooling and harvesting.
2. Using freshly grown overnight cultures for inoculation is essential in order to obtain reliable growth curves.
3. The time of harvesting at mid-log phase is most important for lysate preparation. The indicated OD_{600} values are only examples and may vary with each individual fermentation setup. Please also consider that cells continue to grow during cooling.
4. Carefully transfer the supernatant by pipetting.

5. The salt and temperature step is essential for high-quality lysate and it will cause a significant precipitate.
6. S30 lysate still contains small vesicles originating from the cell membrane in amounts of approximately 100 μg/mL lysate. In particular porins such as OmpG and OmpF are still detectable in these vesicles. The porin background is almost completely removed in S60 lysates.
7. From a T7RNAP stock of 4 mg/mL, approximately 20–30 μL in 1 mL of lysate is usually appropriate.
8. Repeated freezing and thawing cycles may reduce lysate efficiencies. Efficiency of lysate batches is usually evaluated by GFP expression. The synthesis of membrane proteins in particular upon co-translational solubilization in the D-CF and L-CF modes might be lower due to feedback mechanisms on the translation process. Highest membrane protein production can usually be obtained in the P-CF mode.
9. T7RNAP may already start to precipitate at these concentrations and ultrafiltration should be stopped as soon as first precipitates are formed.
10. Commercial lysates with pre-adjusted T7RNAP concentrations (Cube Biotech) may be used as controls.
11. MD100 MicroDialyzers fit volumes from 10 to 100 μL while the cavities of 96-deep-well plates can fit up to 2 mL. However, we do not recommend using volumes below 25 μL for RM and 500 μL for FM. The MicroDialyzers should be loaded and unloaded from the *round* opening (Fig. 2). We recommend usage of autoclaved MicroDialyzers for optimal expression efficiency.
12. Standard dialysis tubes should be washed before usage. Boil the membranes for 1–2 min in 5–20 mM NaH_2PO_4, wash 2–3 times with MilliQ water, and boil again for 1–2 min in 10 mM EDTA (pH 8.0). Afterwards wash the membrane extensively with MilliQ water, including one boiling step in MilliQ water for 1 min in order to completely remove the EDTA.
13. Addition of LY-C buffer is necessary in order to compensate for the Mg^{2+} and K^{+} ions present in the lysate in the RM.
14. Incubation time and temperature may be subject of optimization and can depend on the target protein. Lower incubation temperature may reduce the expression yield but the resulting sample quality could be improved.
15. Membrane proteins may migrate faster in SDS gels as expected according to their molecular mass due to incomplete denaturation.
16. The solubilization procedure may be refined by adjusting temperature, incubation time, and volume of the added detergent solution.

17. The yield of purified DgkA out of 1 mL RM is approximately 1 mg.
18. GFP fluorescence is only a preliminary monitor for solubilization efficiency and does not necessarily correlate with the functional folding of the GPCR. Specific functional assays such as radioligand-binding assays or measurements by surface plasmon resonance are necessary in order to define sample quality [6].
19. The MSP protein usually contains a poly(His)$_6$-tag. Purification of the target protein via metal chelate affinity chromatography is therefore not recommendable if empty NDs need to be removed. The GPCR should therefore be modified with alternative purification tags like the StrepII-tag. Alternatively, the poly(His)$_6$-tag of the MSP may be removed via TEV-protease cleavage.

Acknowledgments

This work was funded by the Collaborative Research Center (SFB) 807 of the German Research Foundation (DFG) and by the German Ministry of Education and Science (BMBF). Support was further obtained by Instruct, part of the European Strategy Forum on Research Infrastructures (ESFRI).

References

1. Henrich E, Hein C, Dötsch V et al (2015) Membrane protein production in Escherichia coli cell-free lysates. FEBS Lett 589:1713–1722
2. Hein C, Henrich E, Orbán E et al (2014) Hydrophobic supplements in cell-free systems: designing artificial environments for membrane proteins. Eng Life Sci 14:365–379
3. Junge F, Haberstock S, Roos C et al (2011) Advances in cell-free protein synthesis for the functional and structural analysis of membrane proteins. N Biotechnol 28:262–271
4. Kigawa T, Yabuki T, Yoshida Y et al (1999) Cell-free production and stable-isotope labeling of milligram quantities of proteins. FEBS Lett 442:15–19
5. Boland C, Li D, Shah ST et al (2014) Cell-free expression and *in meso* crystallisation of an integral membrane kinase for structure determination. Cell Mol Life Sci 71:4895–4910
6. Proverbio D, Roos C, Beyermann M et al (2013) Functional properties of cell-free expressed human endothelin A and endothelin B receptors in artificial membrane environments. Biochim Biophys Acta 1828:2182–2192
7. Li Y, Wang E, Wang Y (1999) A modified procedure for fast purification of T7 RNA polymerase. Protein Expr Purif 16:355–358
8. Denisov IG, Grinkova YV, Lazarides AA et al (2004) Directed self-assembly of monodisperse phospholipid bilayer nanodiscs with controlled size. J Am Chem Soc 126:3477–3487
9. Berrier C, Guilvout I, Bayan N et al (2011) Coupled cell-free synthesis and lipid vesicle insertion of a functional oligomeric channel MscL: MscL does not need the insertase YidC for insertion *in vitro*. Biochim Biophys Acta 1808:41–46
10. Haberstock S, Roos C, Hoevels Y et al (2012) A systematic approach to increase the efficiency of membrane protein production in cell-free expression systems. Protein Expr Purif 82:308–316
11. Schneider B, Junge F, Shirokov VA et al (2010) Membrane protein expression in cell-free systems. Methods Mol Biol 601:165–186
12. Soubias O, Gawrisch K (2012) The role of the lipid matrix for structure and function of the GPCR rhodopsin. Biochim Biophys Acta 1818:234–240

Chapter 2

Membrane Protein Production in the Yeast, *S. cerevisiae*

Stephanie P. Cartwright, Lina Mikaliunaite, and Roslyn M. Bill

Abstract

The first crystal structures of recombinant mammalian membrane proteins were solved in 2005 using protein that had been produced in yeast cells. One of these, the rabbit Ca^{2+}-ATPase SERCA1a, was synthesized in *Saccharomyces cerevisiae*. All host systems have their specific advantages and disadvantages, but yeast has remained a consistently popular choice in the eukaryotic membrane protein field because it is quick, easy and cheap to culture, whilst being able to post-translationally process eukaryotic membrane proteins. Very recent structures of recombinant membrane proteins produced in *S. cerevisiae* include those of the *Arabidopsis thaliana* NRT1.1 nitrate transporter and the fungal plant pathogen lipid scramblase, TMEM16. This chapter provides an overview of the methodological approaches underpinning these successes.

Key words Membrane protein, Recombinant, *S. cerevisiae*, P_{GAL} promoter

Abbreviations

BCA	Bicinchoninic acid
BSA	Bovine serum albumin
CCD	Charge-coupled device
DoE	Design of experiments
GFP	Green fluorescent protein
GOI	Gene of interest
GPCR	G protein-coupled receptor
h	Hour
LioAc	Lithium acetate
P_{GAL1}	*GAL1* promoter
PEG	Polyethylene glycol
s	Second
T4L	T4 lysozyme

Isabelle Mus-Veteau (ed.), *Heterologous Expression of Membrane Proteins: Methods and Protocols*, Methods in Molecular Biology, vol. 1432, DOI 10.1007/978-1-4939-3637-3_2, © Springer Science+Business Media New York 2016

1 Introduction

Over 1,500 species of yeast are known, but only a small minority of them have been employed as host organisms for the production of recombinant membrane proteins [1]. The two most important are *Saccharomyces cerevisiae* and *Pichia pastoris*; these eukaryotic microbes grow quickly in complex or defined media in a range of convenient formats (from multi-well plates to shake flasks and bioreactors) of various sizes [1].

In 2005, the first crystal structures of mammalian membrane proteins derived from recombinant sources were solved using protein that had been produced in yeast: the rabbit Ca^{2+}-ATPase SERCA1a was produced in *S. cerevisiae* [2] and the rat voltage-dependent potassium ion channel Kv1.2 was produced in *P. pastoris* [3]. Several other host cells have been used since then for eukaryotic membrane protein production [4], all with their own specific advantages and disadvantages. However, yeasts have remained a consistently popular choice [5, 6] because they are quick, easy, and cheap to culture whilst still being able to post-translationally process eukaryotic membrane proteins. Recent structures of recombinant membrane proteins produced in *S. cerevisiae* include those of the *Arabidopsis thaliana* NRT1.1 nitrate transporter and the fungal plant pathogen lipid scramblase, TMEM16.

S. cerevisiae has several advantages over the other commonly-used yeast species, *P. pastoris*: its genetics are better understood (http://www.yeastgenome.org/); it is supported by a more extensive body of literature; and there is a wider range of tools and strains available from both commercial and academic sources. In our laboratory, we often start with *P. pastoris* and, if the production is not straightforward, turn to *S. cerevisiae* to troubleshoot, thereby benefitting from the best attributes of the two hosts [1]. Notably, the structure of the human histamine H_1 receptor was obtained in this way: initial screening to define the best expression construct was performed in *S. cerevisiae* [7] followed by protein production in *P. pastoris* [8].

1.1 Designing a Yeast Expression Plasmid

Yeast expression plasmids used for recombinant protein production typically contain a 2 μ origin of replication and have a copy number of approximately 20 per cell [9]. Critical elements of such expression plasmids are the gene sequence encoding the target membrane protein, the promoter and terminator sequences, and any tags that might aid functional gene expression and protein purification.

In 2013 and 2014, all eight α-helical transmembrane protein structures derived from yeast (structures with PDB codes 4CL4, 4WIS, 4NEF, 4RDQ, 4M1M, 4JCZ, 3WME and 4WFF) were synthesized under the control of a strong, inducible promoter. For the two structures solved using protein produced in *S. cerevisiae* (4CL4 and 4WIS), the promoter was P_{GAL1}, which is induced with galactose. This promoter is the basis of the commercially-available pYES2 plasmid (Life Technologies V825-20, Fig. 1) as well as the plasmid, pRS426GAL1 [10]; both are suitable plasmids for

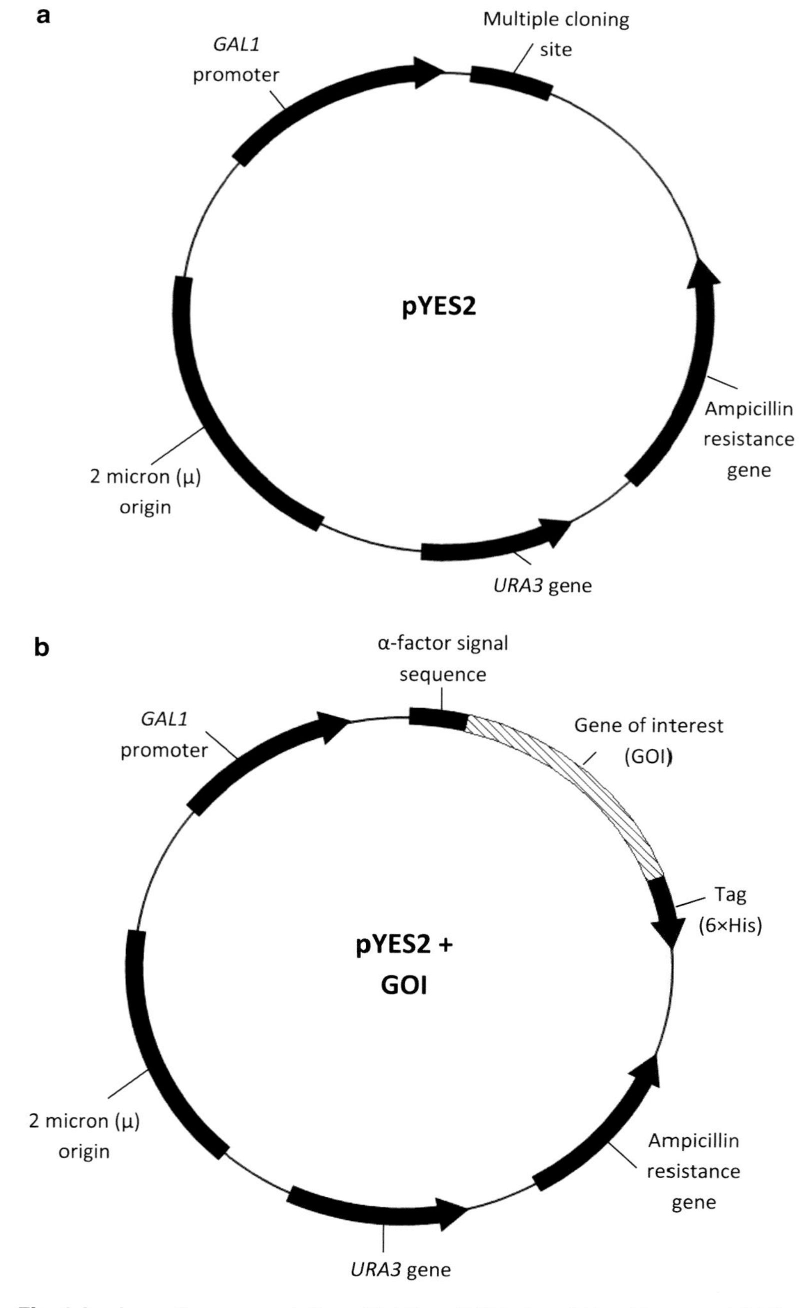

Fig. 1 A schematic representation of (**a**) the pYES2 plasmid backbone and (**b**) the pYES2 plasmid containing the gene of interest (GOI)

initiating expression trials in *S. cerevisiae*. Notably, neither plasmid contains the *S. cerevisiae* α-mating factor sequence signal, which is believed to correctly target recombinant membrane proteins to the yeast membrane. For example, its presence had a positive impact on the yield of the mouse 5-HT_{5A} serotonin receptor [11] but dramatically reduced expression of the human histamine H_1 receptor [8]. The following sequence (containing the Kex2/Ste13 processing sites) may therefore be added by PCR or gene synthesis as an optional element when designing the expression plasmid: 5′ATGAGATTTCCTTCAATTTTTACTGCAGTTT TATTCGCAGCATCCTCCGCATTAGCTGCTCCAGTCAAC ACTACAACAGAAGATGAAACGGCACAAATTCCGGCT GAAGCTGTCATCGGTTACTTAGATTTAGAAGGGGAT TTCGATGTTGCTGTTTTGCCATTTTCCAACAGCACAAAT AACGGGTTATTGTTTATAAATACTACTATTGCCAGCATTG CTGCTAAAGAAGAAGGGGTATCTTTGGATAAAAGAGAGG CTGAAGCT 3′

Other commonly-used sequences in *S. cerevisiae* expression plasmids include those that encode polyhistidine (hexa-, octa- (present in pRS426GAL1), and decahistine tags are all common), green fluorescent protein (GFP; present in pRS426GAL1), and T4 lysozyme (T4L). These and others have been reviewed extensively elsewhere [12, 13]. In summary, polyhistidine tags are routinely fused to recombinantly-produced membrane proteins to facilitate rapid purification by metal chelate chromatography using Ni^{2+}-affinity resins. In many cases, the tag is not removed prior to crystallization trials, although protease cleavage sites can be engineered into the expression plasmid if this is desired [6]. GFP tags are used differently, typically to assess functional yield or homogeneity of the purified recombinant protein prior to crystallization trials. However, GFP tags remain fluorescent in yeast (and other eukaryotic) cells irrespective of whether the partner membrane protein is correctly folded in the plasma membrane [14]. GFP is therefore an inappropriate marker to assess the folding status of recombinant membrane proteins produced in yeast, although it is still useful in analyzing the stability of a membrane protein by fluorescence size-exclusion chromatography [15]. Finally, most G protein-coupled receptor (GPCR) crystal structures have been obtained using a fusion protein strategy where the flexible third intracellular loop is replaced by T4L; recently modified T4L variants having been developed to optimize crystal quality or promote alternative packing interactions [16]. Overall, the precise combination and location (at either terminus or within the protein sequence) of any tags needs to be decided based upon their proposed use (for targetting, as an epitope, to promote stability, for purification or as a tool to assess protein quality) and the biochemistry of the target recombinant membrane protein. Once the final, preferred sequence has been designed, it is possible to codon optimize it for expression in *S. cerevisiae*; recent data suggest that the codon sequence around the translation start site has a bigger influence on membrane protein yields than codon choice in the rest of the open reading frame when recombinant proteins are produced in *E. coli* [17] or *P. pastoris* [18]. The use of degenerate PCR primers to screen for the optimal codon sequence around the start codon may therefore be worth considering [19].

1.2 Choosing an S. cerevisiae Strain

A popular expression strain for structural applications is the *pep4* deletion strain, FGY217 (*MATa, ura3-52, lys2Δ201, pep4Δ*) (34) in which the gene for proteinase A has been deleted to reduce protease-mediated protein degradation. In addition, the yeast deletion collections comprise over 21,000 mutant strains with deletions of the approximately 6,000 *S. cerevisiae* ORFs [20] available as both *MAT*a and *MAT*α mating types. These strains can be obtained from Euroscarf (http://web.uni-frankfurt.de/fb15/mikro/euroscarf/) or the American Type Culture Collection (http://www.atcc.org/). Complementing this, Dharmacon sells the Yeast Tet-Promoters Hughes Collection (yTHC) with 800 essential yeast genes under the control of a tetracycline-regulated promoter that permits their experimental regulation. These strain resources are supported by a wealth of information in the *Saccharomyces* Genome Database (http://www.yeastgenome.org/). Use of specific strains from these collections offers the potential to gain mechanistic insight into the molecular bottlenecks that preclude high recombinant protein yields [21].

1.3 A Note on the Yeast Membrane

The yeast membrane has a different composition from that of mammalian membranes which may be important for some membrane protein targets. Yeast strains have therefore been developed that contain cholesterol rather than the native yeast sterol, ergosterol. This was achieved by replacing the *ERG5* and *ERG6* genes of the ergosterol biosynthetic pathway with the higher eukaryotic (e.g., zebrafish and human) genes of the cholesterol biosynthesis pathway, *DHRC24* and *DHRC7* [22–24], respectively. Cell viability does not appear to be impaired in these "humanized" yeast cells, although growth rates and densities are somewhat affected. However, this may be an acceptable trade-off in return for higher yields of functional recombinant membrane protein. Since a relatively small number of heterologous membrane proteins have been produced in cholesterol-producing yeast strains to date, potential exists to optimize recombinant protein production by using them.

1.4 Culturing Yeast Cells to Maximize Functional Recombinant Protein Yields

During a recombinant membrane protein production experiment, understanding the relative importance of the different experimental variables and their influence on protein yield and quality is an essential part of its optimization. Common approaches to increase functional yields are to lower the growth temperature of the expressing culture, alter the pH or composition of the growth medium, or to change the culture aeration strategy [25]. The addition of molecules such as dimethyl sulfoxide (DMSO), histidine, glycerol, or specific ligands has also been used to increase yields of GPCRs in *P. pastoris* [26] and transporters in *S. cerevisiae* [27]. Often these variables are optimized in a stepwise manner, one factor at a time. A more effective method is to implement a statistical design of experiments (DoE) approach because the influence of numerous factors and their interactions, which may be nonlinear in nature, can be determined [28]. Irrespective of the approach taken, it is important to systematically investigate the effects of all input parameters in order to maximize membrane protein yields from recombinant yeast cultures.

2 Materials

2.1 Yeast Strains and Plasmids

1. Yeast expression strain, e.g., the *pep4* deletion strain, FGY217 (*MATa, ura3-52, lys2Δ201, pep4Δ*) [29].
2. Expression plasmid, e.g., pYES2 (Life Technologies, V825-20) or pRS426GAL1 [10] containing the gene and other sequences of interest.

2.2 Yeast Transformation and Culture Conditions

1. YPD medium (stable at room temperature): 1% yeast extract, 2% bacto-peptone, 2% glucose. For plates, add 2 % bacto-agar.
2. 40% glucose stock solution in water, filter sterilized (0.2 mm pore size).
3. Carrier DNA: 7 mg/mL sonicated salmon testes DNA; store at −20 °C.
4. 100 mM lithium acetate (LiOAc).
5. Polyethylene glycol (PEG) 3350 (50% w/v).
6. 1 M LiOAc.

2.3 Membrane Preparation and Immunoblotting

1. Growth/induction medium lacking uracil with either 2% glucose (for growth) or 2% galactose (for induction). For plates, add 2 % bacto-agar.
2. Breaking buffer: 50 mM Na_2HPO_4, 50 mM NaH_2PO_4, 2 mM EDTA pH 7.4, 100 mM NaCl, 5% glycerol.
3. Protease inhibitor cocktail IV (Merck Millipore).
4. Acid-washed glass beads, 500 μm.
5. 2 mL screw-cap breaking tubes.
6. Buffer A: 20 mM HEPES, 50 mM NaCl, 10% glycerol, pH 7.
7. Bicinchoninic acid (BCA) protein assay kit.
8. Tris-glycine SDS gels.
9. Powdered milk or BSA.

10. Primary anti-His_6 monoclonal antibody (Clontech).
11. Secondary antibody peroxidase conjugate.
12. PBS (phosphate buffered saline) (for 1 L): 1.44 g Na_2HPO_4-$2H_2O$ (8.1 mM phosphate), 0.25 g KH_2HPO_4 (1.9 mM phosphate), 8 g NaCl, 0.2 g KCl, adjust pH to 7.4 using 1 M NaOH or HCl.
13. PBS-Tween 20 (PBST): 1 L PBS, 2 mL Tween 20 (0.2 %).
14. 5× Laemmli sample buffer: 0.08 M Tris–HCl pH 6.8, 12.5 % glycerol, 2.5 % SDS, 6.25 % β-mercaptoethanol, 0.01 % bromophenol blue.
15. Pre-stained protein standard.
16. ECL detection kit.
17. Coomassie brilliant blue R-250.

2.4 Equipment

1. 50 mL, 500 mL, and 2.5 L baffled shake flasks.
2. Water bath or heat block.
3. TissueLyser (Qiagen) for small-scale membrane preparation.
4. Cell disruptor (e.g., Avestin C3) for large-scale membrane preparation.
5. Floor-standing centrifuge such as Beckman Coulter Avanti J-20 and rotors such as JLA 8.1000 and JA 25.50 (Beckman).
6. Benchtop ultracentrifuge, Beckman Coulter Optima MAX series with TLA-55 and TLA-120.1 rotors (Beckman).
7. 1 mL, 10.4 mL and 50 mL polycarbonate ultracentrifuge tubes (Beckman).
8. LAS-1000–3000 charge-coupled device (CCD) imaging system.

3 Methods

3.1 Generation of Yeast Transformants

1. Aseptically pick a single colony from a freshly streaked YPD agar plate (*see* **Note 1**) into 5 mL YPD medium and culture overnight at 30 °C with 220 rpm agitation.
2. Dilute the overnight culture into 5 mL YPD (to an OD_{600} ~ 0.25) and culture to an OD_{600} of 1.0 (*see* **Note 2**).
3. Harvest the cells by centrifugation at 5,300 × *g* for 3 min and remove the supernatant.
4. Wash the cell pellet in 500 μL 100 mM lithium acetate (LiOAc) and transfer to a 1.5 mL microcentrifuge tube.
5. Harvest the cells by centrifugation at 13,000 × *g* for 15 s and remove the supernatant.
6. Repeat **steps 4** and **5**.

7. Boil salmon testes DNA for 5 min and chill on ice for 2 min.
8. Add the individual components of the transformation solution to the yeast pellet in the following order (*see* **Note 3**):
   (a) 240 μL PEG 3350 (50% w/v).
   (b) 35 μL 1 M LiOAc.
   (c) 25 μL boiled and chilled salmon testes DNA.
   (d) 0.5 μg plasmid DNA made up to 50 μL with water.
9. Vortex the mixture vigorously until the pellet has been completely suspended, which can take up to a minute.
10. Incubate the cells with the transformation mixture at 30 °C for 30 min.
11. Shock the cells at 42 °C for 20 min.
12. Pellet the cells at 6,000 × g for 15 s and remove the transformation mixture.
13. Add 0.5 mL sterile water to the cells and gently suspend them with a pipette (*see* **Note 4**).
14. Plate 100 μL of the cell suspension on a selective agar plate lacking uracil (this is the appropriate selection for use with pYES2 or pRS426GAL1).
15. Incubate plates for 2–3 days at 30 °C; if colonies do not form continue to incubate the plate for up to a week.

3.2 Screening for High-Yielding Transformants

1. Inoculate 10 mL growth medium (lacking uracil plus 2% glucose) with a single colony in a 50 mL baffled shake flask. Typical screens involve assaying 10–20 single colonies.
2. Incubate the cultures overnight in an orbital shaker at 30 °C, 220 rpm.
3. Spot 10 μL from each overnight culture onto a selective plate and allow the spots to dry. Clearly label each spot on the plate then incubate it at 30 °C for 2–3 days.
4. Measure the OD_{600} of each overnight culture. Dilute the cultures to OD_{600} = 0.12 in 10 mL of growth medium (lacking uracil plus 2% glucose) and culture them to an OD_{600} of 0.6 (*see* **Note 5**).
5. Harvest the cultures by centrifugation in 15 mL Falcon tubes at 1,500 × *g* for 5 min. Remove the supernatants and suspend each pellet in 10 mL of induction medium (lacking uracil plus 2% galactose). Incubate the cultures for 22 h at 30 °C, 220 rpm.
6. Harvest cells at 5,300 × g, 4 °C for 3 min; remove supernatants and keep the cell pellet on ice.
7. Wash cells once in 2 mL ice-cold breaking buffer then harvest by centrifugation at 5300 × *g*, 4 °C for 3 min.

8. Add 1 mL glass beads to a breaking tube and place on ice; repeat this so there is one breaking tube for each colony being screened.
9. Suspend the cells (from **step 7**) in 1 mL breaking buffer (supplemented with protease inhibitor cocktail IV at a dilution of 1:500) and add to the breaking tube that contains the glass beads. Repeat for all harvested cell samples and keep the tubes on ice.
10. Place a TissueLyser breaking tube holder at −80 °C (or −20 °C) for 10 min.
11. Place the breaking tubes into the chilled TissueLyser breaking tube holder and lyse the cells in a TissueLyser at 50 Hz for 10 min.
12. Remove the supernatants from the glass beads using a pipette and transfer them to clean microcentrifuge tubes.
13. Remove cell debris by centrifugation at 17,000 × *g* for 15 min at 4 °C and transfer the supernatants to clean ultracentrifuge tubes (*see* **Note 6**).
14. Harvest total membrane pellets by centrifugation at 190,000 × *g* for 1 h. Remove the supernatants and suspend each membrane pellet in 100 μL buffer A. Membrane suspensions can be stored at −20 °C.
15. Assay the amount of total membrane protein in each membrane suspension using a BCA assay kit according to the kit instructions.
16. Confirm the presence and relative amount of target protein in each membrane suspension by immunoblot:
    (a) Each sample to be loaded on an SDS-PAGE gel should contain approximately 50 μg total membrane protein, although this will vary according to the expression level of the protein of interest.
    (b) In preparing each sample, mix 4 volumes membrane suspension and 1 volume 5× Laemmli sample buffer. Incubate this mixture for 10 min (*see* **Note 7**).
    (c) Load samples on an SDS-PAGE gel remembering to include both a protein ladder and a standard to allow comparisons between blots. Follow the “Bio-Rad General Protocol for Western Blotting” (http://www.bio-rad.com/webroot/web/pdf/lsr/literature/Bulletin_6376.pdf).
    (d) Depending on the availability of protein-specific antibodies or the presence of tags (such as polyhistidine), incubate the blot with appropriate primary and secondary antibodies.

(e) Visualize the blot using an ECL kit and a CCD imaging system.

(f) Analyze the blot to identify high-yielding colonies (using ImageJ for example).

17. Pick high-yielding colonies from the spot plate (**step 3**) and prepare a glycerol stock for long-term storage (*see* **Note 8**) and subsequent scale up to produce large quantities of the target membrane protein, as described in Subheading 3.3.

3.3 Scaling-Up Recombinant Membrane Protein Production

1. Isolate a single colony on a selective plate from a glycerol stock (Subheading 3.2, **step 17**) or use a fresh transformant (isolated on a spot plate as in Subheading 3.2, **step 3**).
2. Inoculate 10 mL growth medium (lacking uracil plus 2 % glucose) with the single colony and incubate overnight in an orbital shaker at 220 rpm and 30 °C.
3. The following day, transfer the 10 mL overnight culture to a 500 mL shake flask containing 150 mL growth medium (lacking uracil plus 2 % glucose) and incubate overnight as in **step 1**.
4. On the third day, dilute the 150 mL overnight culture to $OD_{600} = 0.12$ in 1 L medium lacking uracil and containing 0.1 % glucose and grow in a 2.5 L baffled shake flask at 220 rpm and 30 °C. Induce the culture with 2 % galactose (although this can be optimized further) when the OD_{600} has reached 0.6.
5. Harvest the cells (*see* **Note 9**) by centrifugation at 5,300 × *g*, 4 °C for 10 min in a floor-standing centrifuge such as a Beckman Coulter Avanti J-20; the JLA 8.1000 rotor holds bottles of 1 L capacity that can be used for this step. Discard the supernatant. The pellet should ideally weigh between 10 and 20 g for efficient large-scale membrane preparation. If the weight is less than 10 g, the protocol in Subheading 3.2 (from **step 7**) can be used instead.
6. Suspend the cell pellet in 25 mL breaking buffer for every 1 L of original culture. Add protease inhibitor cocktail IV at a 1:500 dilution.
7. Break the cells using a high pressure homogenizer, such as an Avestin C3, according to the manufacturer's instructions. It is important to keep the sample at low temperature (~4 °C) to prevent protein degradation. The low temperature should be maintained from this step onwards.
8. Separate the cell lysate from the cell debris and unbroken cells by centrifugation at 8,000 × *g*, 4 °C for 30 min.
9. Collect the membrane fraction by centrifugation of the supernatant at 100,000 × *g*, 4 °C for 60 min (allow extra time for acceleration and deceleration) in an ultracentrifuge such as a

Beckman Coulter Optima L-80 XP; the 70.1 Ti rotor holds 12 tubes. Discard the supernatant and suspend the pellets in 6 mL per original 1 L culture. To ease the resuspension of membrane pellet, the membranes can be soaked overnight at 4 °C in 1 mL buffer A per tube; pellets should then be homogenized and made up to the required volume with buffer A. Measure total membrane protein concentration using a BCA protein assay kit according to the manufacturer's instructions prior to extraction and purification of the target protein.

4 Notes

1. The plate should be no older than 5 days; if the colonies are older, this can lead to a reduction in the competence of the cells.
2. Transformation is more likely to be successful when cells are growing logarithmically.
3. It is important to add the PEG 3350 first to protect the yeast cells from the high concentration of LiOAc.
4. At this point the cells are fragile and need to be suspended gently with a pipette; do not vortex.
5. This ensures that the cells are in the logarithmic growth phase during induction.
6. An alternative method for separating cell lysates from glass beads is to collect them into 15 mL Falcon tubes. To do that, cut a round hole in the cap of a Falcon tube, pierce the bottom of the breaking tube with a needle, and insert it into the cut cap. Place the cap onto the Falcon tube and collect the lysate by centrifugation at 5,300 × *g* for 3 min; the glass beads and cell debris are retained in the breaking tube. Transfer the supernatants to clean ultracentrifuge tubes.
7. Incubate the mixture between 4 °C (on ice) and 70 °C; the best temperature for the particular protein of interest must be determined empirically by examining the immunoblot to ensure the protein has entered the gel and has not aggregated or degraded. Lower temperatures may require longer incubation times.
8. Transformants can often be stored as glycerol stocks at −80 °C, but their stability should be assessed to confirm this on a case-by-case basis. For unstable transformants, it will be necessary to do a fresh transformation prior to each scale-up experiment.
9. The incubation period may need to be optimized; recombinant protein may be detected 4 h post-induction, but a 22 h culture period is often used for convenience. To optimize the

post-induction incubation period, collect samples at several time intervals and analyze by immunoblot as detailed in Subheading 3.2, **step 16**.

Acknowledgments

RMB acknowledges funding from the Biotechnology and Biological Sciences Research Council (BBSRC; via grants BB/I019960/1 and BB/L502194/1) and the Innovative Medicines Joint Undertaking under Grant Agreement number 115583 to the ND4BB ENABLE Consortium.

References

1. Darby RA, Cartwright SP, Dilworth MV, Bill RM (2012) Which yeast species shall I choose? *Saccharomyces cerevisiae* versus *Pichia pastoris* (review). Methods Mol Biol 866:11–23. doi:10.1007/978-1-61779-770-5_2
2. Jidenko M, Nielsen RC, Sorensen TL, Moller JV, le Maire M, Nissen P, Jaxel C (2005) Crystallization of a mammalian membrane protein overexpressed in *Saccharomyces cerevisiae*. Proc Natl Acad Sci U S A 102(33):11687–11691. doi:10.1073/pnas.0503986102
3. Long SB, Campbell EB, Mackinnon R (2005) Crystal structure of a mammalian voltage-dependent Shaker family K^+ channel. Science 309(5736):897–903. doi:10.1126/science.1116269
4. Bill RM, von der Haar T (2015) Hijacked then lost in translation: the plight of the recombinant host cell in membrane protein structural biology projects. Curr Opin Struct Biol 32:147–155. doi:10.1016/j.sbi.2015.04.003
5. Bill RM (2014) Playing catch-up with *Escherichia coli*: using yeast to increase success rates in recombinant protein production experiments. Front Microbiol 5:85. doi:10.3389/fmicb.2014.00085
6. Byrne B (2015) *Pichia pastoris* as an expression host for membrane protein structural biology. Curr Opin Struct Biol 32:9–17. doi:10.1016/j.sbi.2015.01.005
7. Shiroishi M, Tsujimoto H, Makyio H, Asada H, Yurugi-Kobayashi T, Shimamura T, Murata T, Nomura N, Haga T, Iwata S, Kobayashi T (2012) Platform for the rapid construction and evaluation of GPCRs for crystallography in *Saccharomyces cerevisiae*. Microb Cell Factories 11:78. doi:10.1186/1475-2859-11-78
8. Shiroishi M, Kobayashi T, Ogasawara S, Tsujimoto H, Ikeda-Suno C, Iwata S, Shimamura T (2011) Production of the stable human histamine H(1) receptor in *Pichia pastoris* for structural determination. Methods 55(4):281–286. doi:10.1016/j.ymeth.2011.08.015
9. Christianson TW, Sikorski RS, Dante M, Shero JH, Hieter P (1992) Multifunctional yeast high-copy-number shuttle vectors. Gene 110(1):119–122. doi:0378-1119(92)90454-W
10. Drew D, Kim H (2012) Preparation of *Saccharomyces cerevisiae* expression plasmids. Methods Mol Biol 866:41–46. doi:10.1007/978-1-61779-770-5_4
11. Weiss HM, Haase W, Michel H, Reilander H (1995) Expression of functional mouse 5-HT5A serotonin receptor in the methylotrophic yeast *Pichia pastoris*: pharmacological characterization and localization. FEBS Lett 377(3):451–456. doi:10.1016/0014-5793(95)01389-X
12. Terpe K (2003) Overview of tag protein fusions: from molecular and biochemical fundamentals to commercial systems. Appl Microbiol Biotechnol 60(5):523–533. doi:10.1007/s00253-002-1158-6
13. Manjasetty BA, Turnbull AP, Panjikar S, Bussow K, Chance MR (2008) Automated technologies and novel techniques to accelerate protein crystallography for structural genomics. Proteomics 8(4):612–625. doi:10.1002/pmic.200700687
14. Thomas J, Tate CG (2014) Quality control in eukaryotic membrane protein overproduction. J Mol Biol 426(24):4139–4154. doi:10.1016/j.jmb.2014.10.012
15. Drew D, Newstead S, Sonoda Y, Kim H, von Heijne G, Iwata S (2008) GFP-based optimization scheme for the overexpression and purification of eukaryotic membrane proteins in *Saccharomyces cerevisiae*. Nat Protoc 3(5):784–798. doi:10.1038/nprot.2008.44

16. Thorsen TS, Matt R, Weis WI, Kobilka BK (2014) Modified T4 lysozyme fusion proteins facilitate G protein-coupled receptor crystallogenesis. Structure 22(11):1657–1664. doi:10.1016/j.str.2014.08.022
17. Norholm MH, Toddo S, Virkki MT, Light S, von Heijne G, Daley DO (2013) Improved production of membrane proteins in *Escherichia coli* by selective codon substitutions. FEBS Lett 587(15):2352–2358. doi:10.1016/j.febslet.2013.05.063
18. Oberg F, Ekvall M, Nyblom M, Backmark A, Neutze R, Hedfalk K (2009) Insight into factors directing high production of eukaryotic membrane proteins; production of 13 human AQPs in *Pichia pastoris*. Mol Membr Biol 26(4): 215–227. doi:10.1080/09687680902862085
19. Mirzadeh K, Martinez V, Toddo S, Guntur S, Herrgard MJ, Elofsson A, Norholm MH, Daley DO (2015) Enhanced protein production in *Escherichia coli* by optimization of cloning scars at the vector-coding sequence junction. ACS Synth Biol 4(9):959–965. doi:10.1021/acssynbio.5b00033
20. Giaever G, Nislow C (2014) The yeast deletion collection: a decade of functional genomics. Genetics 197(2):451–465. doi:10.1534/genetics.114.161620
21. Bonander N, Hedfalk K, Larsson C, Mostad P, Chang C, Gustafsson L, Bill RM (2005) Design of improved membrane protein production experiments: quantitation of the host response. Protein Sci 14(7):1729–1740. doi:10.1110/ps.051435705
22. Souza CM, Schwabe TM, Pichler H, Ploier B, Leitner E, Guan XL, Wenk MR, Riezman I, Riezman H (2011) A stable yeast strain efficiently producing cholesterol instead of ergosterol is functional for tryptophan uptake, but not weak organic acid resistance. Metab Eng 13(5):555–569. doi:10.1016/j.ymben.2011.06.006
23. Morioka S, Shigemori T, Hara K, Morisaka H, Kuroda K, Ueda M (2013) Effect of sterol composition on the activity of the yeast G-protein-coupled receptor Ste2. Appl Microbiol Biotechnol 97(9):4013–4020. doi:10.1007/s00253-012-4470-9
24. Hirz M, Richter G, Leitner E, Wriessnegger T, Pichler H (2013) A novel cholesterol-producing Pichia pastoris strain is an ideal host for functional expression of human Na, K-ATPase alpha3beta1 isoform. Appl Microbiol Biotechnol 97(21):9465–9478. doi:10.1007/s00253-013-5156-7
25. Holmes WJ, Darby RA, Wilks MD, Smith R, Bill RM (2009) Developing a scalable model of recombinant protein yield from *Pichia pastoris*: the influence of culture conditions, biomass and induction regime. Microb Cell Factories 8:35. doi:10.1186/1475-2859-8-35
26. Andre N, Cherouati N, Prual C, Steffan T, Zeder-Lutz G, Magnin T, Pattus F, Michel H, Wagner R, Reinhart C (2006) Enhancing functional production of G protein-coupled receptors in *Pichia pastoris* to levels required for structural studies via a single expression screen. Protein Sci 15(5):1115–1126. doi:10.1110/ps.062098206, ps.062098206
27. Figler RA, Omote H, Nakamoto RK, Al-Shawi MK (2000) Use of chemical chaperones in the yeast *Saccharomyces cerevisiae* to enhance heterologous membrane protein expression: high-yield expression and purification of human P-glycoprotein. Arch Biochem Biophys 376(1):34–46. doi:10.1006/abbi.2000.1712, S0003-9861(00)91712-0
28. Bora N, Bawa Z, Bill RM, Wilks MD (2012) The implementation of a design of experiments strategy to increase recombinant protein yields in yeast (review). Methods Mol Biol866:115–127.doi:10.1007/978-1-61779-770-5_11
29. Kota J, Gilstring CF, Ljungdahl PO (2007) Membrane chaperone Shr3 assists in folding amino acid permeases preventing precocious ERAD. J Cell Biol 176(5):617–628. doi:10.1083/jcb.200612100

Chapter 3

Membrane Protein Production in *Escherichia coli*: Protocols and Rules

Federica Angius, Oana Ilioaia, Marc Uzan, and Bruno Miroux

Abstract

Functional and structural studies on membrane proteins are limited by the difficulty to produce them in large amount and in a functional state. In this review, we provide protocols to achieve high-level expression of membrane proteins in *Escherichia coli*. The T7 RNA polymerase-based expression system is presented in detail and protocols to assess and improve its efficiency are discussed. Protocols to isolate either membrane or inclusion bodies and to perform an initial qualitative test to assess the solubility of the recombinant protein are also included.

Key words Production of recombinant proteins, *E. coli*, T7 RNA polymerase

1 Introduction

Membrane protein (MP) production is still a challenge for biochemists and biophysicists. Over the last decade, eukaryotic expression systems have emerged and have proven to be very useful for structural studies of eukaryotic MP such as G-protein-coupled receptors [1]. However bacterial expression systems remain widely used. We have recently conducted a global survey of the protein data bank (PDB) and found that half of unique MP structures deposited in the PDB have been produced in *E. coli* [2]. Provided that the recombinant MP is well folded within the membrane of the host, bacteria can produce, at very low cost, sufficient amount of the target MP for X-ray crystallization or NMR studies. *E. coli* is also the most versatile host for specific isotopic labeling of proteins required for NMR studies. In this review, we focus on the T7 RNA polymerase (T7 RNAP) bacterial expression system which is, so far, the most efficient in producing large amount of membrane proteins for structural studies [2]. Figure 1 provides an overview of how the expression system works in the bacterial host BL21λ(DE3). The gene encoding the T7 RNAP is inserted in the lambda DE3 under the control of the *lacUV5* promoter. Upon addition of

Isabelle Mus-Veteau (ed.), *Heterologous Expression of Membrane Proteins: Methods and Protocols*, Methods in Molecular Biology, vol. 1432, DOI 10.1007/978-1-4939-3637-3_3, © Springer Science+Business Media New York 2016

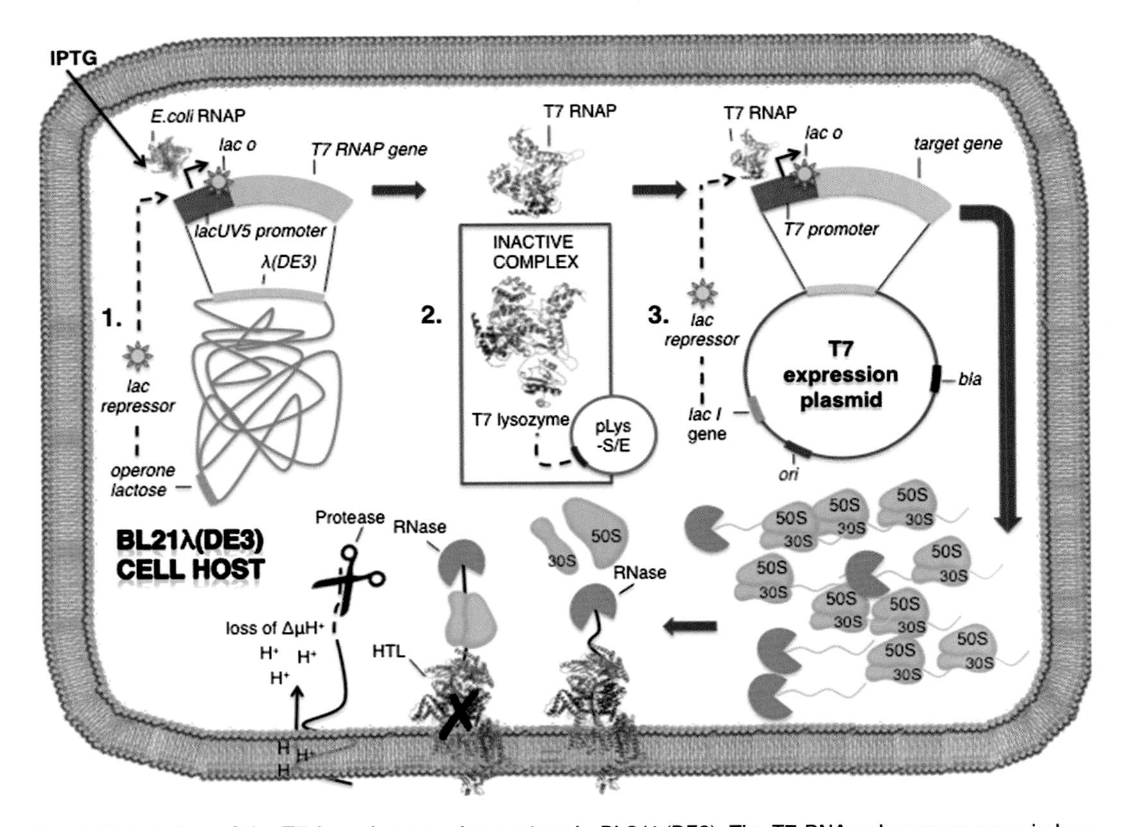

Fig. 1 Global view of the T7-based expression system in BL21λ(DE3). The T7 RNA polymerase gene is lysogenic in the genome and its expression is under the control of the IPTG-inducible *lacUV5* promoter. Upon addition of IPTG, the T7RNAP will specifically transcribe the target gene inserted in the T7 expression plasmid and the target MP might be produced at very high levels. However, overexpression of the target mRNA is, most of the time, toxic to the cell because it overloads the translation machinery and uncouple transcription from translation. The newly synthetized membrane protein might also overload the folding and secretion machineries causing mistargeting of the overproduced MP, protein aggregation, and ultimately proton leak and loss of energy homeostasis. To circumvent these difficulties, the expression system can be regulated by several ways: *1.* Repressing the *lacUV5* promoter using a *lac* repressor; *2.* expressing the T7 lysozyme from a pLys-S/E plasmid, which will inhibit its activity; *3.* inserting the *lac* repressor in the T7 multi-copy expression plasmid

IPTG, isopropyl β-D-1-thiogalactopyranoside, a non-metabolized derivative of lactose, the T7 RNAP is produced and will specifically transcribe the target gene inserted in a T7 expression vector downstream of the T7 promoter. The mRNA of the target MP is highly expressed because the T7 RNAP transcriptional elongation rate is ten times faster than the *E. coli* enzyme. In addition, the T7 expression vector is present in multiple copies. In many cases, the target mRNA overloads the translation machinery triggering ribosome destruction and growth arrest [3]. Naked un-translated mRNA are rapidly degraded by RNases and, in some cases, the RNA degradation is faster than the transcriptional activity of the T7 RNAP leading to lower yield of the target than expected [4]. For this

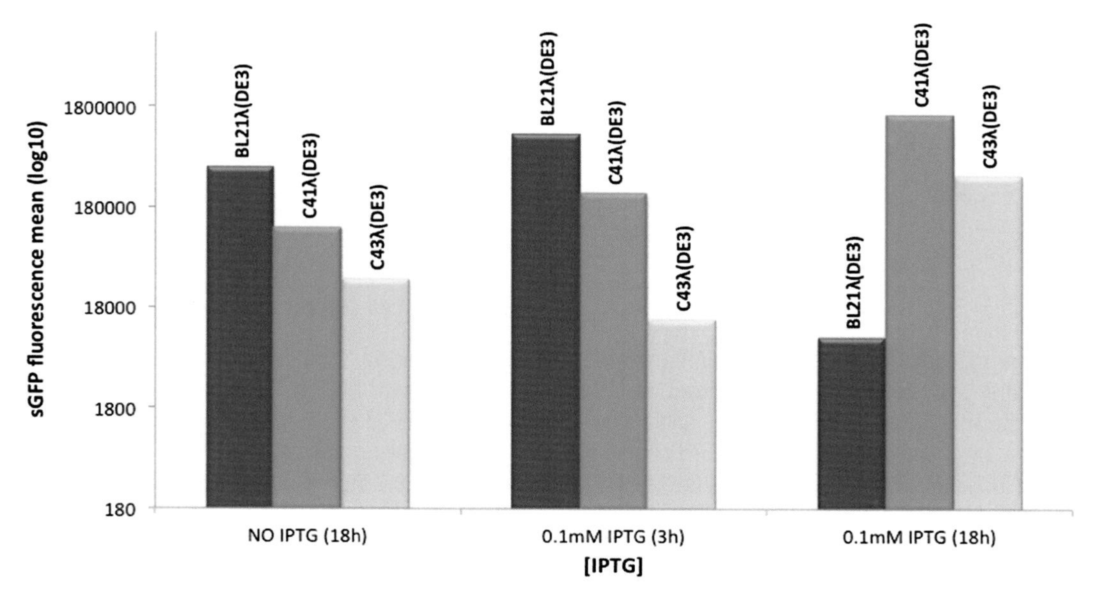

Fig. 2 Analysis of GFP fluorescence by flow cytometry. The pHis17-sGFP T7 expression plasmid has been transformed in the BL21λ(DE3) (*green*), C41λ(DE3) (*light green*), and C43λ(DE3) (*yellow*) bacterial hosts. The fluorescence has been recorded with the flow cytometer Accuri C6 3 h and 18 h after induction with 0.1 mM of IPTG. To assess the basal level of expression of sGFP in each host, cells were grown for 18 h with no addition of IPTG

reason, we provide here a rapid protocol to assess the levels of your target mRNA when the yield of the corresponding MP is low. Insertion of the target MP at the *E. coli* membrane can also overload the translocation and secretion machineries. The recombinant MP will then be not only misfolded and produced at low level but proton permeability of the bacterial membrane might be compromised leading to cell death. Over the last 20 years, the T7 expression system has been optimized to improve its regulation and extend its ability to produce large amount of MP. For instance the use of lysozyme has been shown to strongly inhibit the activity of the T7 RNAP, thus providing a means to decrease the basal activity of the LacUV5 promoter and to tune the activity of the T7-RNAP upon induction [5, 6]. Other groups have isolated mutant hosts from the parental strain BL21λ(DE3) [7, 8]. Some of them namely C41λ(DE3) and C43λ(DE3) have proven to be extremely useful for structural biologists; these mutant hosts contributed to 28 % of non-*E. coli* unique MP structures and 19 % of *E. coli* unique MP structures deposited into the PDB [2]. To illustrate that the mutant hosts are better regulated than the parental strain, we used the green fluorescent protein (superfold version, sGFP) as gene reporter. After transformation with the pRSET-sGFP expression plasmid, cells have been induced at $OD_{600nm} = 0.6$ with 0.1 mM IPTG. Figure 2 shows the mean green fluorescent intensity analyzed by flow cytometry. In all three bacterial hosts, the basal level of sGFP fluorescence after an overnight culture in 2*TY-rich

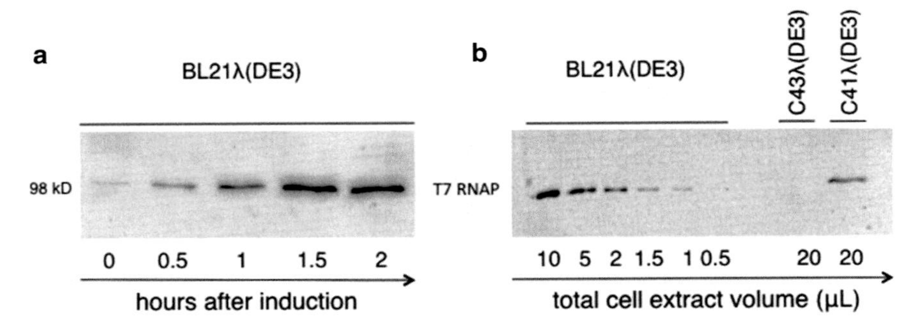

Fig. 3 Immunodetection of the T7 RNAP enzyme in T7 expression hosts. Total cell extracts were loaded on SDS-PAGE and proteins were transferred on nitrocellulose membrane. The T7 RNAP protein was revealed using the anti-T7RNAP from Novagene and a second antibody coupled to peroxidase. Peroxidase activity was detected by chemiluminescence. (**a**) Time course of the T7 RNAP protein expression in BL21λ(DE3) upon addition of 0.7 mM IPTG. (**b**) Expression levels of T7 RNAP 2 h after 0.7 mM IPTG induction in BL21λ(DE3), C43λ(DE3), C41λ(DE3). For BL21λ(DE3) host, decreasing amounts of total cell extract have been loaded to compare the intensity of the signal with the mutant hosts

medium is very high showing that the expression system is leaky. However, basal fluorescence intensities are 4 and 13 times decreased in C41λ(DE3) and C43λ(DE3), respectively. Upon addition of IPTG, fluorescence intensity increased twice in BL21λ(DE3) host 3 h after induction and decreased strongly after overnight induction. This is due to loss of the expression plasmid, cell death, and lysis. In contrast, sGFP production reached a maximal value after overnight induction in both C41λ(DE3) and C43λ(DE3) hosts (13- and 11-fold induction, respectively). At the molecular level, Wagner et al. have shown that, in C41λ(DE3) and C43λ(DE3) hosts, the strong *lacUV5* promoter recombined with the wild-type genomic copy of the *lac* promoter. Consequently, the amount of T7 RNAP enzyme produced upon addition of IPTG is ten times reduced in C41λ(DE3) and undetectable in C43λ(DE3) using the commercially available anti-T7 RNAP antibody from Novagene (Fig. 3). In this chapter we provide protocols to design your construct and choose the appropriate host/vector combination, isolate new bacterial hosts, set up growth conditions, assess your expression system by flow cytometry, fractionate bacterial cells, and perform a first biochemical analysis.

2 Materials

2.1 Materials for RNA Isolation and Sucrose Gradient

1. Tips, plastic tubes, glass, gloves, water must be RNase free.
2. A dry bath to warm up the samples to 65 °C.
3. A laboratory fume hood.
4. A spectrophotometer.

5. QIAGEN RNase-Free DNase I Set.
6. Lysis solution for 700 μL of culture: 35.5 μL 20% SDS + 7 μL of 200 mM Na-EDTA + 500 μL water-saturated phenol.
7. Water-saturated phenol.
8. Phenol/chloroform solution v/v 1:1. The chloroform should contain isoamyl alcohol in a proportion v/v 24:1.
9. 3 M Na-acetate pH 5.
10. Ethanol 100% RNase free.
11. Ethanol 70% RNase free.
12. Gradient maker.
13. 10 mM Tris–HCl, pH 8.
14. Sucrose solutions: 50% and 5% (w/v) prepared in 10 mM Tris–HCl, pH 8.
15. Beckman Coulter Ultra-Clear™ centrifuge tubes.

2.2 Media, Buffers, and Chemicals

1. 1*LB Medium: 10 g Bacto Tryptone, 5 g Bacto Yeast Extract, and 5 g NaCl. Add ultrapure water to 900 mL. Adjust the pH to 7.2 with NaOH. Add water at a final volume of 1 L and autoclave for 20 min at 121 °C.
2. 2*TY Medium: 16 g Bacto Tryptone, 10 g Bacto Yeast Extract, and 5 g NaCl. Add ultrapure water to 800 mL. Adjust the pH to 7.2 with NaOH, adjust the final volume to 900 mL, and autoclave for 20 min at 121 °C.
3. 2*TY with glucose: 2 g Glucose in 100 mL of final volume of water, filter sterilize. Add the glucose solution in the autoclaved medium. Adding glucose could be useful if you wish to repress further the expression vector before induction.
4. Isopropyl-beta-D-galactoside (IPTG): Prepare 100 mM, 500 mM, 700 mM, and 1 M stock solutions in ultrapure water, sterilize with 0.22 μm filter, aliquot, and store at −20 °C.
5. Antibiotics: Prepare 1000 times stock solutions of antibiotics. Ampicillin (100 mg/mL) can be prepared in ultrapure water and stored at −20 °C. Tetracycline (12,5 mg/mL) and kanamycin (30 mg/mL) are freely soluble in water but in time the solutions can turn turbid due to precipitation. It is thus recommended to prepare it in 95% ethanol. Dilute 1000 times the stock solution in medium prior to use.
6. Phosphate-buffered saline (PBS): 10 mM Phosphate, 150 mM NaCl, pH 7.4 (tablets are commercially available).
7. TEP buffer: 10 mM Tris–HCl, pH 8, 1 mM EDTA, and 0.001% PMSF.
8. Triton X-100.
9. Dodecyl-maltoside (DDM).

10. Phosphododecylcholine (Fc12).
11. Sodium dodecyl sulfate (SDS).

2.3 Web Resources

2.3.1 Sequence Analysis and Molecular Biology Tools

1. **SMART** (protein domain database): http://smart.embl-heidelberg.de/
2. **Jpred** (secondary structure prediction): http://www.compbio.dundee.ac.uk/jpred/index.html
3. **ExPasy**: http://www.expasy.ch/
4. **Amplify** (PCR simulation, oligonucleotide design) http://engels.genetics.wisc.edu/amplify/

Vector design:

5. **Dna20:** https://www.dna20.com/resources/bioinformatics-tools
6. **Serial-Cloner**: http://serialbasics.free.fr/Serial_Cloner-Download.html
7. **APE**: http://biologylabs.utah.edu/jorgensen/wayned/ape/

Molecular and structural biology websites

8. Steewe White: http://blanco.biomol.uci.edu/mpstruc/
9. Dror Warschawski: http://www.drorlist.com/nmr/MPNMR.html

Academic expression plasmid resources

10. Protein Science Initiative. http://psimr.asu.edu/about.html

3 Methods

3.1 Designing Constructs for Expression

1. Before starting molecular cloning experiments, check if your target MP is already available in an expression vector (*see* Protein Science Initiative Web site) and search the literature to see if your MP target or related proteins have been produced in recombinant systems (*see* **Note 1**).
2. Be aware that *E. coli* cannot produce at high levels proteins larger than 90 kDa. Ribosomes drop off very long mRNA leading to incomplete synthesis products. If possible break up your protein into smaller fragments. Use SMART (protein domain identification) or Jpred (secondary structure prediction) to define the boundaries carefully.
3. Addition of purification Tag: For N-terminal constructs start protein synthesis with three amino acids before the Tag. In pRSET vector (Invitrogen), the N-terminal sequence is **MRGS**-(His)6 which gives a very good yield of recombinant protein. Consider adding more than 6 histidines (up to 12 but then preferentially in C-terminal position) to achieve a stronger

binding on Nickel column. There is no generic rule regarding cleavage sequences after the Tag but TEV cleavage sequence is widely used for MP as the TEV protease is still active in the presence of the most commonly used detergents [9].

4. Your target MP might not spontaneously go to the inner membrane of the bacteria. Consider making a C-terminal fusion with periplasmic maltose-binding protein (MBP), which contains a periplasmic signal sequence, to target your MP to the *E. coli* membrane [10, 11].
5. If possible engineer dua-ribosome-binding site (RBS) expression vectors like pET-Duet (Novagen) so that you can clone a fluorescent protein (FP) gene downstream of your target MP gene. This allows you to follow the cell population of bacteria by flow cytometry, to assess the stability and toxicity of your expression vector quickly, and to establish the optimal induction conditions (*see* Subheading 3.5.1). FP fusion with your target MP is also an option developed successfully by several laboratories [12].

3.2 Selecting the Optimal Expression Vector/Bacterial Host

Selecting the right combination of vector/bacterial host is an essential step to achieve the optimal production of your MP (*see* **Note 2**). The following rules apply to the T7 RNAP-based expression system:

1. In combination with C41λ(DE3) and C43λ(DE3) bacterial hosts use high-copy-number plasmids like those containing the pMB1 origin of replication (200–600 copies/cell). A non-exhaustive list is pMW7 and derivatives (pHis and pRun) [13, 14], pGEM (Promega), pRSET and pDEST (Invitrogen), pIVEX (5prime), and pPR-IBA (IBA). Avoid *lacI* and *lacO* sequences in the plasmid.
2. In combination with BL21λ(DE3), use preferentially medium-copy-number vectors and those containing *lacI* and *lacO* sequences like pET (3, 9, 14, 17, 20, 23 from Novagen), to reduce the amount of T7 RNAP before induction. Consider using the companion plasmid pLyS to inhibit the T7 RNAP after induction.
3. The BL21AI host, which contains the T7RNA polymerase gene under the control of the arabinose promoter, and the Lemo21 host [6], which contains a companion plasmid expressing the lysozyme under the rhamnose promoter, may also be useful to titrate the amount or activity of T7 RNA polymerase (*see* **Note 3**).

3.3 Viability Test on Agar Plate

1. Preparing the agar plates: Melt 500 mL (or less) 2*YT agar medium in water bath at 100 °C. When the solution is clear switch the temperature to 55 °C. Add 500 μL 0.7 M sterile IPTG and mix vigorously. Add the required antibiotic.

Pour the plates and wait for 1 h till the agar is solid. Incubate the plate upside down O/N at 37 °C.

2. Cell transformation: Take out of the −80 °C freezer a stock of competent cells and thaw them on ice. Add 10 ng of plasmid to 50 μL of cells and leave it on ice for 20 min. Place the microcentrifuge tube on a water bath set at 42 °C for 90 s. Replace the tube on ice for 5 min. Add 500–700 μL of SOC or LB media (without antibiotic) and allow the culture to grow for 45 min at 37 °C. Plate 100 μL of transformation mix on 2*TY agar plate containing only the antibiotic and 100 μL on agar plate with both antibiotic and IPTG. Incubate O/N at 37 °C.
3. Analysis of the cell population: Count the number of colonies on both plates. If you see no colony on IPTG-containing plates then expression of your target MP compromises cell growth. Transform the cells with the empty plasmid to verify if the plasmid is also toxic for the cell. If you have the same number of colonies in both conditions, then expression of your target MP is not toxic. Usually the size of the colonies is reduced in the presence of IPTG.

3.4 Selecting a Host Strain Adapted to the Expression of Your Target MP

The protocol below allows you to genetically isolate low-frequency mutants when your bacterial host/vector expression system compromises cell growth upon induction (*see* **Note 4**). The protocol uses the green fluorescent protein from Aequora Victoria [15] but can be performed without FP marker [7].

3.4.1 Selection Procedure

1. It is essential to work in a sterile environment. Pre-warm five 250 mL flasks containing 50 mL of 2*TY media. Add antibiotic prior to use. Autoclave 40 microcentrifuge tubes and fill them with 900 μL of sterile water or 2*TY media.
2. Transform the parental strain with your target MP expression vector, ideally in co-expression with a fluorescent protein. Inoculate a freshly transformed colony in 50 mL 2*TY medium. Grow cells until OD_{600nm} has reached 0.4–0.6. Add IPTG to 0.7 mM final concentration. One, two, and three hours after induction, take 1 mL of culture. After a low-speed centrifugation (300 × *g* for 2 min), gently resuspend the pellet in 1 mL of sterile water.
3. Perform serial dilutions from 1/10 to $1/10^4$ and immediately plate 100 μL of all dilutions on the IPTG/antibiotic-containing plates. Incubate the plates O/N at 37 °C.
4. Check the presence of green fluorescent colonies (Fig. 4) under UV light (above 300 nm to avoid the mutagenic effect of UV). Select ten small fluorescent colonies (*see* **Note 5**) for each selection experiment and make over-day cultures in tubes with 1 mL 2*TY containing ampicillin. After 2–3 h (when the

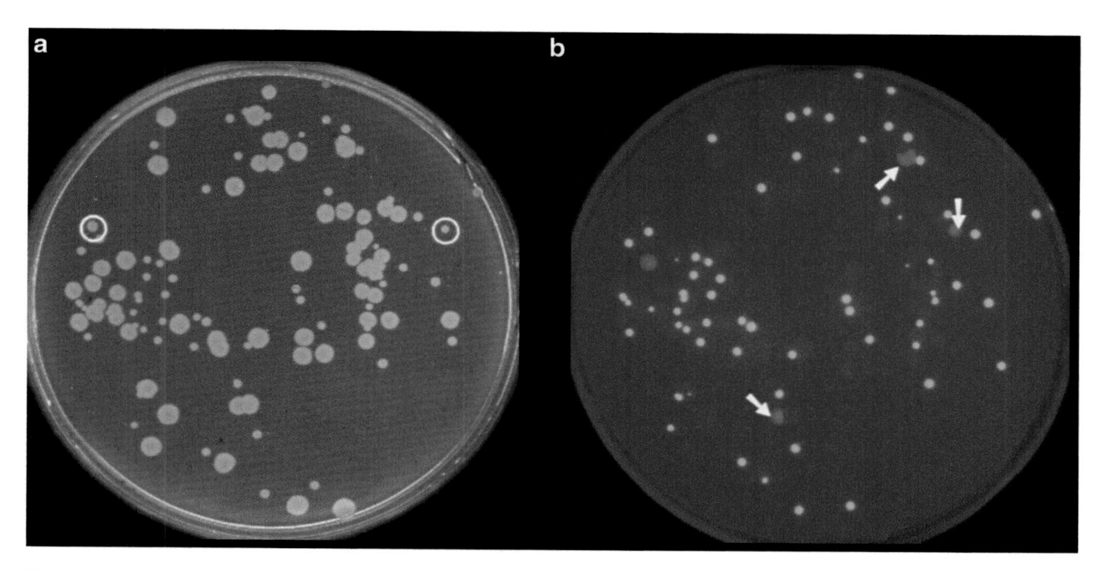

Fig. 4 Selection of bacterial hosts using GFP as gene reporter. After transformation of the pMW7-GFP expression plasmid, cells are grown in 2*TY medium and induced at $OD_{600\ nm} = 0.6$. Two hours after induction, serial dilutions of the culture are plated on IPTG-containing plates. The next day, plates were illuminated under a normal light (**a**) or UV light (**b**)

culture becomes turbid) save the mutants on 2*TY ampicillin agar plates with and without IPTG. Incubate the plates O/N at 37 °C.

5. Check that all mutants now grow on IPTG-containing plate. The size of the colonies in the presence of IPTG can be considered as a characteristic feature for a couple of vector/host.

3.4.2 Localization of the Mutation

You need to check if the mutation is in the bacterial genome or in the plasmid.

1. Plasmid rescue: Purify the plasmid DNA from each clone and transform the initial host, i.e., BL21λ(DE3). Plate 100 μL of the transformation mix on 2*TY plates with antibiotic and with or without IPTG. Incubate O/N at 37 °C. Check the presence of colonies on IPTG-containing agar plates. No colony means that the mutation is not in the expression vector and consequently most likely in the bacterial host genome.
2. Curing the bacterial host from the plasmid: Grow the bacterial mutant host in a 250 mL flask containing 50 mL of 2*TY without antibiotic. Make daily serial 10 times dilutions of the culture and plate 100 μL of the $1/10^8$ and $1/10^7$ dilutions on 2*TY agar plates containing IPTG but no antibiotic. After O/N incubation at 37 °C, check the fluorescence under UV light. Usually, after 3–5 days of culture in the absence of antibiotic, large colonies that have lost their green fluorescence and therefore the expression plasmid will appear.

3. Isolate a large nonfluorescent colony, make a glycerol stock, and prepare calcium-competent cells. Transform the cured host with the original expression vector (not the cured one) and verify that the "colony size phenotype" on an IPTG-containing plate is restored.

3.5 Optimization of Growth Conditions

1. General rules: Choose expression hosts where the T7 RNAP expression of activity is tightly controlled. Systematically test induction temperature below 25 °C. Check the optimal IPTG concentration for your target MP (*see* Subheading 3.5.1).
2. Specific rules for BL21λ(DE3) host: Do not induce the target MP with IPTG or follow Alfasi's protocol [8] by adding extremely low concentration of IPTG (10 μM). Use pLysS as companion plasmid to downregulate the activity of the T7RNAP.
3. Specific condition for C41λ(DE3) and C43λ(DE3) bacterial mutant hosts: Test cell viability on IPTG plates. If the expression of your target MP is not toxic in C41λ(DE3) then keep this mutant host and induce with 0.1 and 0.7 mM IPTG. Test 3-h and O/N induction. Use C43λ(DE3) when expression of your target gene is toxic for C41λ(DE3). Add IPTG at 0.7 mM O/N.

3.5.1 Exploring Induction Conditions and Mutant Hosts by Flow Cytometry

If you co-express a fluorescent protein, superfold GFP for instance, then you can detect GFP fluorescence on FL1 graph (*see* **Note 6**) and also check the size (FSC) and the granularity (SSC) of your cells by using a density plot. As illustrated in Fig. 2, this can be extremely useful to compare rapid expression hosts as well as expression conditions.

1. Native conditions: Collect cells by centrifugation (300 × *g* for 2 min) and remove the supernatant. Resuspend cells in 0.5–1 mL PBS. Repeat the washing step three times. Dilute 1/1000 the cells in PBS before loading the sample on the cytometer.
2. Analysis on fixed cells: Collect cells by centrifugation and aspirate the supernatant. Resuspend cells in 0.5–1 mL PBS. Add formaldehyde to 4%. Fix for 10 min at 37 °C. Wash the cells three times with PBS. Dilute 1/1000 the cells in PBS before loading the sample on the cytometer.

3.5.2 Testing the mRNA Stability: Phenol/Chloroform RNA Extraction from E. coli

If your target membrane protein is not produced in several vector/host combinations then you should check the mRNA stability of the target gene either by quantitative real-time PCR or by "Northern blot" analysis. Obtaining high-quality RNA is the first and often most critical step. Phenol/chloroform extraction is an easy way to remove proteins from nucleic acid samples: nucleic acids remain in the aqueous phase while proteins separate into the organic phase or lie at the phase interface. This protocol can be used also to extract RNA from bacteria grown in a rich medium. Phenol is a dangerous poison that burns the skin and the lungs upon inhalation. You must wear gloves and manipulate carefully under fume hood. A solution of PEG400 is recommended for first aid.

1. First phenol extraction: Add 700 μL of cell culture into the lysis solution maintained at 65 °C. Keep at 65 °C and vortex vigorously intermittently about ten times for 10 s. Cool the tubes on ice and centrifuge for 2 min, 15,000 × *g*, at 4 °C. Transfer the aqueous phase in a new Eppendorf tube, being careful not to contaminate with the interface phase.
2. Second phenol extraction: Add an equal volume of water-saturated phenol. Place the tubes at 65 °C and vortex vigorously intermittently about ten times for 10 s. Cool the tubes on ice and centrifuge for 2 min, 15,000 × *g*, at 4 °C. Transfer the aqueous phase in a new Eppendorf tube, being careful not to contaminate with the interface phase.
3. First phenol/chloroform extraction: Add an equal volume of phenol/chloroform and vortex vigorously intermittently about ten times for 10 s; this step can be done at room temperature. Centrifuge for 2 min, 15,000 × *g*, at 4 °C. Transfer the aqueous phase in a new Eppendorf tube, being careful not to contaminate with the interface phase.
4. First RNA precipitation: Add 1/10 the volume of 3 M Na-acetate, pH 5 (or 5 M NaCl) and 2.5 volume ethanol. Mix and place at −20 °C for 1–2 h (*see* **Note 7**). Centrifuge at 15,000 × *g* for 30–60 min at 4 °C. Remove carefully the supernatant and wash the pellet with 1 mL of 70% ethanol. Centrifuge at 15,000 × *g* for 30 min at 4 °C. Dry pellet in air and resuspend in 400 μL of sterile water.
5. Treatment with RNase-free DNase I (*see* **Note 8**; protocol provided from QIAGEN).
6. Second phenol/chloroform extraction as described in **step 3**.
7. Second RNA precipitation as described in **step 4**.
8. Measure the RNA concentration with a spectrophotometer: 1 A_{260} corresponds to 40 μg/mL RNA. Check the purity of the RNA by estimating the ratio 260/280 nm which must be 2.0.

3.6 Collecting Membranes or Inclusion Bodies from E. coli

This section provides protocols for isolation of inclusion bodies and bacterial membranes when internal membrane proliferation occurs within the cell (*see* **Note 10**).

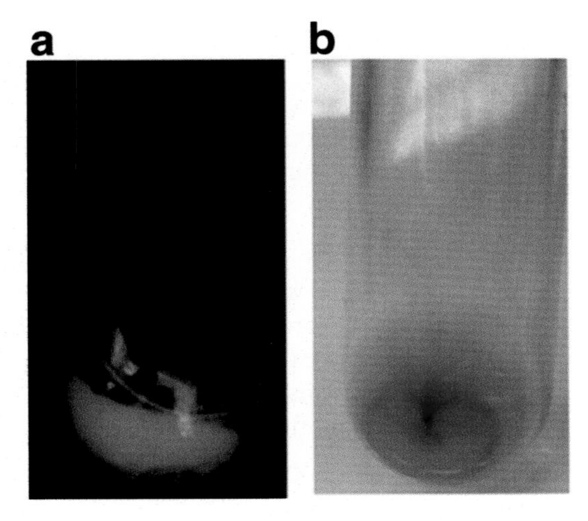

Fig. 5 Pictures of inclusion bodies and intracellular membrane pellets. (**a**) Inclusion bodies of OmpF protein 10,000 × *g* pellet, (**b**) intracellular membrane 100,000 × *g* pellet containing the b subunit of the ATP-synthase

3.6.1 Check the Presence of Inclusion Bodies

1. Breaking the cells: Harvest the culture by centrifugation at 7000 × *g*, 10 min, at 4 °C. Resuspend the pellet in 25 mL of TEP buffer. Disrupt the bacteria by passing the suspension twice in a French Press or cell disruptor.
2. Differential centrifugation: Pellet the cell debris at 600 × *g* for 10 min. Keep the supernatant. Collect the putative inclusion bodies by centrifuging the supernatant at 10,000 × *g* for 15 min at 4 °C. You should see a white brawny pellet. Collect the bacterial membranes by centrifugation of the 10,000 × *g* supernatant at 100,000 × *g* for 1 h.
3. Wash of the inclusion bodies: Wash the first pellet obtained at 10,000 × *g* with 25 mL TEP buffer supplemented with 2% Triton X-100. Centrifuge at 10,000 × *g* for 30 min. Inclusion body pellet is usually white (*see* Fig. 5a). Repeat the wash. Resuspend the pellet in 25 mL TEP buffer without detergent and centrifuge at 10,000 × *g*. Repeat the wash in order to remove all traces of detergent.
4. Resuspend the pellet in 2 mL TEP buffer and proceed to protein assay.

3.6.2 Collecting E. coli Membranes in the Absence of Inclusion Bodies

1. Breaking the cells: Harvest the culture by centrifugation at 7000 × *g*, 10 min, at 4 °C. Resuspend the pellet in 25 mL of TEP buffer. Disrupt the bacteria by passing the suspension twice in a French Press or cell disruptor.
2. Differential centrifugation: Collect the P1 pellet of internal membranes by low-speed centrifugation: 2500 × *g* for 10 min (*see* **Note 11**). Centrifuge the supernatant (S1) at 100,000 × *g* for 1 h to recover the inner and outer membranes.

3. Wash P1 with 25 mL of TEP buffer and centrifuge at 2500 × *g* for 10 min at 4 °C to remove unbroken cells (P2).
4. The supernatant (S2) contains the washed internal membranes.
5. Centrifuge for 1 h at 100,000 × *g* in order to pellet the internal membranes. You should see a brown pellet (*see* Fig. 5b).
6. Resuspend the pellet in 2 mL of buffer and assay the protein concentration.

3.6.3 Sucrose Gradient Protocol

The purpose of sucrose gradient is to concentrate and separate membrane vesicles according to their specific density. For high-purity requirements, continuous gradients are used. If you do not have access to gradient maker, then use step gradients.

1. Setting up the gradient maker: Attach capillary tubes to the end of the tubing emerging from the gradient maker. Gradient maker tubing must be clean; otherwise the sucrose gradient will not flow correctly. Close the mixer between the two compartments of the maker. Add the higher percentage sucrose solution to the outlet side of the maker. Start the stirring and add the lower percentage sucrose solution to the other compartment. Place the capillary tube on the top of the ultracentrifuge tube. Switch on the peristaltic pump and open the mixer between the two sucrose solutions. Check the flow rate of the pump to ensure that the gradient is poured drop by drop. When the gradient is completed, stop the stirrer and carefully remove the capillary tube.
2. Sample loading and centrifugation: Gently load 1 mL of 2 mg/mL protein sample on the top of the gradient paying attention not to mix the sample with the gradient. Make sure to fill up the ultracentrifuge tube (12 mL) and balance tubes with 10 mM Tris–HCl, pH 8. Centrifuge for at least 18 h at 100,000 × *g* and 4 °C. Collect 1 mL fraction in Eppendorf tubes from top to bottom.
3. Run an SDS-PAGE gel with all the fractions to detect your target MP in the different types of membranes.

3.6.4 Testing the Solubility of Your Target MP

Usually, folded MP in native membranes can be solubilized with detergent. However after production in heterologous membranes, it is frequently occurring that the target MP is difficult to solubilize. Although it is associated to the membrane fraction it might be misfolded and therefore behaves like inclusion bodies. A simple test is to compare the solubility of your target MP in three different detergents: dodecyl-maltoside (DDM), phosphododecylcholine (Fc12), and SDS.

1. Solubilization of the target MP: Prepare three Eppendorf tubes with 100 μg of your target MP in TEP buffer supplemented with 150 mM NaCl. In each tube add separately one

detergent to 1% final concentration for DDM and Fc12 and 2% for SDS (at least ten times above the critical micelle concentration). Adjust the final volume to 100 μL with buffer to perform the solubilization at 1 mg/mL. Incubate for 1 h at 4 °C and ultracentrifuge at 100,000 × *g* for 30 min.

2. Run an SDS-PAGE to check if the target MP is in the supernatant (solubilized) or in the pellet.
3. If the target MP is solubilized only by SDS, then you have inclusion bodies. If it is solubilized by all three detergents then it is likely to be well folded. If DDM cannot solubilize your target MP then try other detergents but it is likely that your target MP is misfolded.

4 Notes

1. Expression protocols are usually poorly described and you will have to go to several previous publications to find out the exact expression vector or host that was used. Try to find out the exact final yield of purified MP target per liter of culture. Below 1 mg/L you may spend 90% of your time growing cells to perform a single biophysical analysis.
2. There is a plethora of vectors and expression systems commercially available. A systematic analysis of expression protocols in bacteria [2] showed that for 80% of membrane protein structures, the two main expression systems use the T7 and arabinose promoter-based expression plasmids. The distribution of secondary structures among the different expression system is asymmetrical. For instance beta-barrel membrane protein structures were preferentially obtained using the arabinose promoter-based expression system or the T7 system with BL21λ(DE3) as expression host. In contrast, alpha helical integral membrane proteins (IMP) were almost all produced in the T7 system. The C41λ(DE3) and C43λ(DE3) bacterial hosts succeeded in producing 50% of heterologous IMP.
3. Lysozyme is a natural inhibitor of T7 RNA polymerase. If you chose to downregulate your expression system by using companion plasmids that express lysozyme (pLyS/E), take into consideration that it requires the addition of a second antibiotic which could affect considerably the cell growth.
4. When the production of the target MP is toxic, the bacteria are unable to form colonies on plates containing the inducer. The selection of new bacterial hosts (mutation in the expression vector is rare) is based upon their ability to form colonies on plate in the presence of inducer, here IPTG. By analyzing their presence, number, and size you can determine the degree of toxicity of the expression of the target protein.

5. In order to isolate and select the new mutant hosts, it is critical to have no more than 200 colonies on the plate. The frequency of occurrence of mutant hosts varies from 10^{-4} to 10^{-6}, hence the importance of diluting the culture. Figure 4 shows a fairly good correlation between the size of the colonies and the intensity of fluorescence. Most of the normal-size colonies do not exhibit fluorescence; they have lost the ability to express the gene. The smaller colonies on the other hand are almost all highly fluorescent.
6. Flow cytometers measure a variety of cellular characteristics such as relative cell size, internal complexity/granularity, cell surface properties/refractive indices, levels of autofluorescence, presence or absence of an exogenous fluorescent probe, and relative fluorescence intensities. Here we used the BD Accuri™ C6 Cytometer. The C6 cytometer is equipped with a blue and a red laser (488 and 640 nm, respectively) and four fluorescent detectors. Standard optical filters are FL1 533/30 nm (e.g., FITC/GFP), FL2 585/40 nm (e.g., PE/PI), FL3 > 670 nm (e.g., PerCP, PerCP-Cy™5.5, PE-Cy7), and FL4 675/25 nm (e.g., APC). Follow the wash procedure of the cytometer before and after the analysis to avoid cells aggregating within the cytometer.
7. If you are in hurry, you can stop the protocol at this step and store the samples at −20 °C. You can also shorten the protocol by omitting the first phenol/chloroform extraction but this is possible only if you have not grown your cells in a rich medium.
8. The treatment with DNase I-RNase free is necessary because samples from high-density *E. coli* culture may be contaminated with DNA.
9. After the 4 °C centrifugation, leave the samples at room temperature for few seconds. This clarifies the solution and helps to transfer the aqueous phase in a new clean tube.
10. On some occasions, upon overexpression of a membrane protein in *E. coli* membrane proliferation has been observed [16–20]. For instance, the overproduction of the *E. coli* b-subunit of the $F1F_o$ ATP synthase resulted in the development of a large network of intracytoplasmic membranes (ICM, Fig. 5b). The bacterial host responds to the overproduction of a membrane protein by synthesizing lipids and by converting phosphatidyl glycerol into cardiolipids at the stationary phase [18, 19].
11. It is highly unusual to collect membranes at 2500 × *g*. It might be due to the high number of cells (1 L culture at $OD_{600nm} = 8$ concentrated in 25 mL), or the high density of the membranes and their association with DNA and cell debris. However, after washing of the 2500 × *g* pellet, membranes will not anymore pellet at 2500 × *g* but, as expected, at 100,000 × *g*.

Acknowledgments

This work was supported by the Centre National de la Recherche Scientifique, INSERM, and by the "Initiative d'Excellence" program from the French State (Grant "DYNAMO," ANR-11-LABEX-0011-01). FA is supported by a DYNAMO PhD fellowship.

References

1. Tate CG (2012) A crystal clear solution for determining G-protein-coupled receptor structures. Trends Biochem Sci 37:343–352. doi:10.1016/j.tibs.2012.06.003
2. Hattab G, Warschawski DE, Moncoq K, Miroux B (2015) Escherichia coli as host for membrane protein structure determination: a global analysis. Sci Rep 5:12097
3. Dong H, Nilsson L, Kurland CG (1995) Gratuitous overexpression of genes in Escherichia coli leads to growth inhibition and ribosome destruction. J Bacteriol 177:1497–1504
4. Arechaga I, Miroux B, Runswick MJ, Walker JE (2003) Over-expression of Escherichia coli F1F(o)-ATPase subunit a is inhibited by instability of the uncB gene transcript. FEBS Lett 547:97–100
5. Moffatt BA, Studier FW (1987) T7 lysozyme inhibits transcription by T7 RNA polymerase. Cell 49:221–227
6. Wagner S, Klepsch MM, Schlegel S et al (2008) Tuning Escherichia coli for membrane protein overexpression. Proc Natl Acad Sci 105:14371–14376. doi:10.1073/pnas.0804090105
7. Miroux B, Walker JE (1996) Over-production of proteins in Escherichia coli: mutant hosts that allow synthesis of some membrane proteins and globular proteins at high levels. J Mol Biol 260:289–298. doi:10.1006/jmbi.1996.0399
8. Alfasi S, Sevastsyanovich Y, Zaffaroni L et al (2011) Use of GFP fusions for the isolation of Escherichia coli strains for improved production of different target recombinant proteins. J Biotechnol 156:11–21. doi:10.1016/j.jbiotec.2011.08.016
9. Gräslund S, Nordlund P, Weigelt J et al (2008) Protein production and purification. Nat Methods 5:135–146. doi:10.1038/nmeth.f.202
10. Shibata Y, White JF, Serrano-Vega MJ et al (2009) Thermostabilization of the neurotensin receptor NTS1. J Mol Biol 390:262–277. doi:10.1016/j.jmb.2009.04.068
11. Egloff P, Hillenbrand M, Klenk C et al (2014) Structure of signaling-competent neurotensin receptor 1 obtained by directed evolution in Escherichia coli. Proc Natl Acad Sci U S A 111: E655–E662. doi:10.1073/pnas.1317903111
12. Drew D, Lerch M, Kunji E et al (2006) Optimization of membrane protein overexpression and purification using GFP fusions. Nat Methods 3:303–313. doi:10.1038/nmeth0406-303
13. Way M, Pope B, Gooch J et al (1990) Identification of a region in segment 1 of gelsolin critical for actin binding. EMBO J 9:4103–4109
14. Orriss GL, Runswick MJ, Collinson IR et al (1996) The delta- and epsilon-subunits of bovine F1-ATPase interact to form a heterodimeric subcomplex. Biochem J 314(Pt 2): 695–700
15. Walker J, Miroux B (1999) Selection of Escherichia coli hosts that are optimized for the overexpression of proteins. Man. Ind. Microbiol. Biotechnol. 2nd Ed. MIMB2.
16. von Meyenburg K, Jørgensen BB, van Deurs B (1984) Physiological and morphological effects of overproduction of membrane-bound ATP synthase in Escherichia coli K-12. EMBO J 3:1791–1797
17. Wilkison WO, Walsh JP, Corless JM, Bell RM (1986) Crystalline arrays of the Escherichia coli sn-glycerol-3-phosphate acyltransferase, an integral membrane protein. J Biol Chem 261: 9951–9958
18. Weiner JH, Lemire BD, Elmes ML et al (1984) Overproduction of fumarate reductase in Escherichia coli induces a novel intracellular lipid-protein organelle. J Bacteriol 158: 590–596
19. Arechaga I, Miroux B, Karrasch S et al (2000) Characterisation of new intracellular membranes in Escherichia coli accompanying large scale over-production of the b subunit of F(1) F(o) ATP synthase. FEBS Lett 482:215–219
20. Eriksson HM, Wessman P, Ge C et al (2009) Massive formation of intracellular membrane vesicles in Escherichia coli by a monotopic membrane-bound lipid glycosyltransferase. J Biol Chem 284:33904–33914. doi:10.1074/jbc.M109.021618

Chapter 4

Codon Optimizing for Increased Membrane Protein Production: A Minimalist Approach

Kiavash Mirzadeh, Stephen Toddo, Morten H.H. Nørholm, and Daniel O. Daley

Abstract

Reengineering a gene with synonymous codons is a popular approach for increasing production levels of recombinant proteins. Here we present a minimalist alternative to this method, which samples synonymous codons only at the second and third positions rather than the entire coding sequence. As demonstrated with two membrane-embedded transporters in *Escherichia coli*, the method was more effective than optimizing the entire coding sequence. The method we present is PCR based and requires three simple steps: (1) the design of two PCR primers, one of which is degenerate; (2) the amplification of a mini-library by PCR; and (3) screening for high-expressing clones.

Key words Membrane protein, Protein expression, Codon optimization, Synonymous codon

1 Introduction

In recent years, codon optimization has emerged as a popular and often necessary approach for boosting production levels of recombinant proteins. In a nutshell it involves reengineering the gene or coding sequence (CDS) of interest with optimal synonymous codons. The codons that are deemed optimal are those whose corresponding tRNA concentration in the host cell is highest, as this is thought to favor fast decoupling by the ribosome and more efficient translation [1]. As the cellular tRNA concentrations have coevolved to mirror the frequency of usage of the corresponding codons in the genome [2, 3], this design principle may seem straightforward. However, when choosing synonymous codons, care must be taken to avoid nucleotide stretches that resemble ribosome-binding sites (RBSs), stretches that form strong mRNA structures, RNase sites, DNA recombination sites, or transcriptional terminators, as they can be detrimental for protein production [1]. The codon optimization approach is conceptually

Isabelle Mus-Veteau (ed.), *Heterologous Expression of Membrane Proteins: Methods and Protocols*, Methods in Molecular Biology, vol. 1432, DOI 10.1007/978-1-4939-3637-3_4, © Springer Science+Business Media New York 2016

attractive and has gained traction, as it is increasingly cheaper to order codon-optimized CDSs from commercial vendors.

Does codon optimization increase production levels of membrane proteins? The short answer is "yes," there are numerous examples of successful optimizations in the literature (*reviewed in* [4]). However unsuccessful optimizations are unlikely to be published, so it is hard to estimate success rates. Our own experiences as well as feedback from colleagues indicate that codon optimization of the entire CDS is often not effective. For example, when we ordered codon-optimized CDSs from two different commercial sources we did not observe significant production of two *E. coli* transporters, called AraH and NarK [5]. As an alternative we explored a minimalist codon optimization approach that samples synonymous codons immediately adjacent to the AUG start codon (i.e. +2, +3). This approach was inspired by the work of Isaksson and co-workers, who systematically sampled all 61 codons in the +2 position of an artificial gene and noted that a single synonymous codon change could affect expression levels by as much as 20-fold (note that a weak Shine Dalgarno sequence was used in these experiments) [6]. When we systematically tested synonymous codons in the +2, +3, and +4 positions of the native *araH* and *narK* CDSs we observed significant increases in production levels [5, 7]. Thus, in these two cases, single synonymous codon substitutions immediately adjacent to the AUG start codon were far more effective at increasing production than codon optimization of the entire coding sequence.

A molecular explanation for why codons adjacent to the AUG start codon influence protein production levels remains elusive. However it most likely relates to translation initiation since it is known that the 16S rRNA of the 30S ribosomal subunit must recognize the nucleotide sequence around the AUG start codon during translation initiation. Minor nucleotide changes in this region, which extends from the Shine Dalgarno sequence to the +4/+5 codons [8, 9], can decrease the efficiency of translation initiation by causing strong mRNA hairpins [10, 11]. In nature, mRNA structure around the AUG start codon has been selected against [12, 13], but in a recombinant protein production experiment this region is a composite of the vector and the 5′ end of the CDS. Thus the optimal synonymous codons for reducing mRNA structure will depend on the sequence context and will differ from CDS to CDS, and vector to vector. This fact underscores the need for sampling different combinations of synonymous codons and testing their expression level.

In this chapter we present our protocol for generating small clone libraries with all possible combinations of synonymous codons in the +2 and +3 positions. Depending on which amino acids are present at these positions of the CDS, the clone libraries can contain up to 36 different variations. Thus, a limited amount of screening is required to identify the clones that express to the highest level. An overview of the method, which we call post-cloning optimization (PCO), is presented in Fig. 1a. It involves three simple steps: (1) the design of two overlapping primers (one of which is degenerate) for amplifying the original expression plasmid, (2) the amplification of a mini-library by PCR and then re-circularization of the plasmids by transformation into *E. coli*, and (3) screening for high-expressing clones. In all of our experiments, the CDS is fused to a region encoding for a -TEV-GFP-His_8 tag (Fig. 1b) so that whole-cell fluorescence can be used as a proxy for protein production [14]. Protocols on how to use GFP in this manner have been presented elsewhere [15]. When we carried out PCO on the *araH* and *narK* CDSs, the mini-libraries contained 6 and 12 clone variations, respectively. And by simply screening 24 colonies we were able to identify clones with considerably increased expression. For example *araH* increased by 9-fold from 1 mg/mL to 9 mg/mL (Fig. 1c), and *narK* by 17-fold from 1.6 mg/mL to 29 mg/mL (Fig. 1d).

The method is a minimalist approach to codon optimization that is inexpensive and simple enough to be carried out in any laboratory that has access to a PCR block. It could be implemented during cloning or as a post-cloning step as we have demonstrated here. Moreover it has a major advantage over codon optimization of the entire CDS; it does not affect the efficiency of elongation, so membrane protein folding should not be perturbed (note that there are reports that suggest elongation rate is linked to the folding of membrane proteins [16, 17]). We foresee that the method could be useful for boosting production of recombinant proteins in *E. coli* (both membrane and soluble). It might also be useful for tuning expression levels of genomically encoded proteins by using (randomized) oligonucleotide-based recombineering [18].

2 Materials

2.1 Components for PCR

1. Primer set for PCR amplification (*see* **Note 1** for design principles).
2. Q5 DNA polymerase and 5x Q5 reaction buffer (New England Biolabs) (*see* **Note 2**).
3. Nucleotide triphosphates: Stock solution containing 100 mM of dATP, dTTP, dCTP, and dGTP.
4. *Dpn*I restriction enzyme (New England Biolabs) (*see* **Note 3**).
5. Sterile H_2O.
6. 0.2 mL Soft-walled PCR tubes.
7. Thermocycler.
8. Agarose powder.

a

Step 1: Primer design
Design primers for amplification of original plasmid. Forward primer should be degenerate to allow synonymous codon changes at triplets +2 and +3. Reverse primer should overlap by approximately 15 base pairs with the forward primer to facilitate re-circularization.

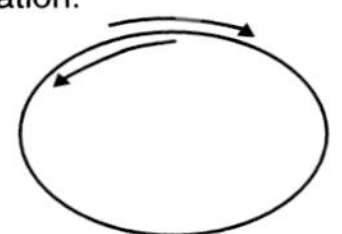

Step 2: PCR with degenerate primers
Digest original plasmid with *DpnI* & transform amplified PCR products into MC1061 to make plasmids circular.

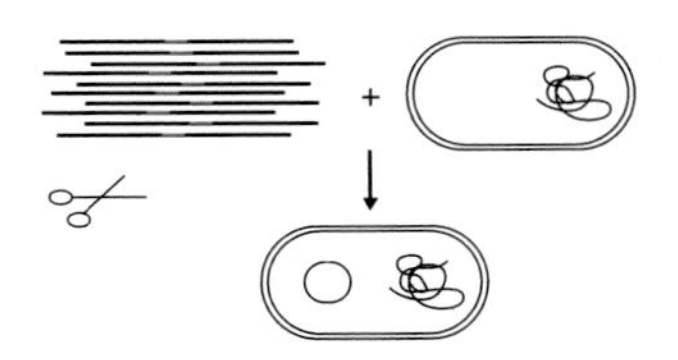

Step 3: Expression level screening
Purify plasmid libraries & screen expression levels in BL21(*DE3*) *pLysS*.

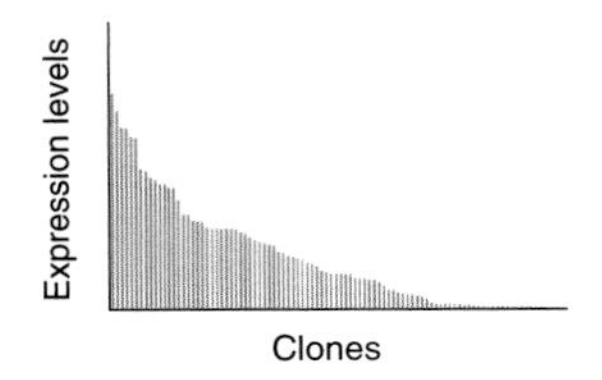

b

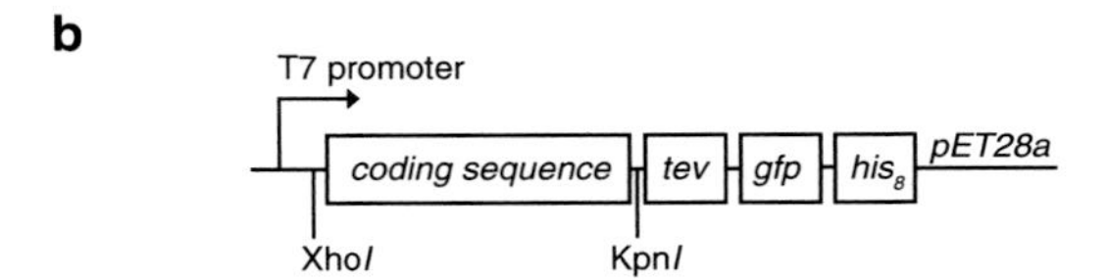

c

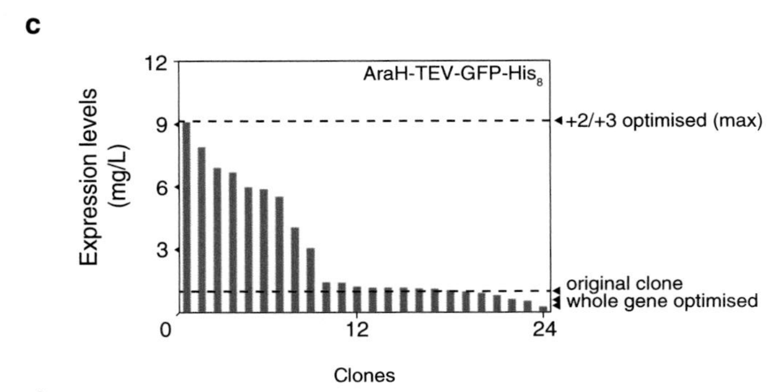

d

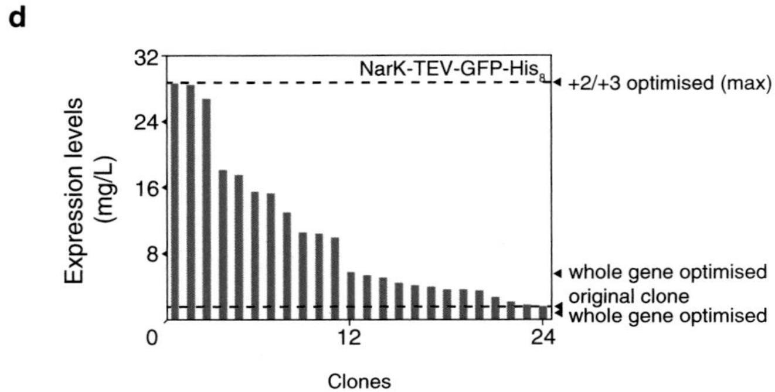

Fig. 1 A minimalistic approach to codon optimization. (**a**) A schematic overview of the method. (**b**) CDSs used in this study were cloned into a modified version of the *pET28a* vector and genetically fused to a region encoding for a -TEV-GFP-His$_8$ tag [14]. This enabled us to use GFP fluorescence as a measure of expression. (**c** and **d**) Screening of expression levels from plasmids harboring *araH* and *narK*, where all possible combinations of synonymous codons in the +2 and +3 positions have been generated. Expression was carried out in BL21(*DE3*) *pLysS* by induction with 1.0 mM IPTG for 5 h at 25 °C. To estimate the amount of protein produced in mg/L, the whole-cell fluorescence was compared to a standard curve obtained with purified GFP. For comparison, the expression levels of the original clone and two whole-gene codon-optimized versions are indicated to the right. Parts a and b adapted from [7]

9. Ethidium bromide.
10. TAE buffer: Prepare 50x stock solution by mixing 242 g Tris base in 600 mL of H_2O. Add 57 mL of glacial acetic acid and 100 mL of 0.5 M EDTA pH 8.0, and bring the final volume to 1 l with H_2O. Store at room temperature.

2.2 Component Transformation and Re-circularization of Libraries

1. Chemically competent *E. coli* cells (*see* **Note 4**).
2. 20 g/L Luria Bertani (LB) broth in H_2O.
3. LB agar plates: LB broth with 15 g/L agar.
4. 1.5 mL Microfuge tubes.
5. Thermomixer for incubating microfuge tubes.
6. 50 mL Reaction tube.
7. 10 mL of LB broth supplemented with appropriate antibiotics.
8. Incubator for 50 mL reaction tube.
9. ENZA DNA mini kit (Omega bio-tek)

3 Methods

3.1 Primer Design

1. The forward primer(s) should straddle the ATG start and should be approximately 40–50 nucleotides long (Fig. 2a). The six nucleotides downstream of the AUG start codon need to be degenerate so that different synonymous codons will be sampled. A table indicating the code used to implement degeneracy is shown in Fig. 2b. Note that in some cases it was not possible to design a single forward primer, so multiple forward primers were used and PCR products were mixed.
2. The reverse primer should match to the region upstream of the ATG start and should also be approximately 40–50 nucleotides in length (*see* Fig. 2a). The 5′ end of the reverse primer should match the 5′ end of the forward primer (overlapping by approximately 15 nucleotides), so that the PCR products can circularize by homologous recombination when transformed into *E. coli.*

3.2 PCR Amplification of Mini-Libraries

1. Mix all reagents for PCR in a 1.5 mL microfuge tube: 71 μL of H_2O, 20 μL of 5× Q5 reaction buffer, 1 μL of Q5 DNA polymerase (2 U/μL), 2 μL of 50 mM dNTP mix, 1 μL of forward primer (50 pmol / μL), 1 μL of reverse primer (50 pmol / μL), 4 μL of original plasmid (4 ng/μL) (*see* **Note 5**).
2. Aliquot 20 μL of the PCR mix into five separate 0.2 mL PCR tubes.

a

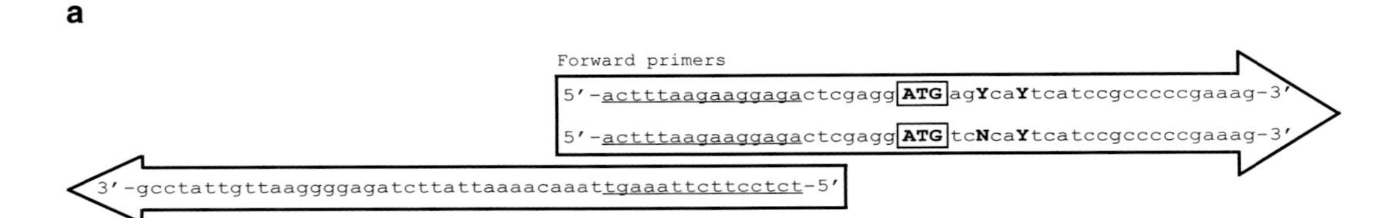

b

Code letter	M	R	W	S	Y	K	V	H	D	B	N
Degenerated bases	AC	AG	AT	GC	CT	GT	AGC	ACT	AGT	GCT	AGCT

Fig. 2 Primer design principles. (**a**) An example of primers used in this study. These primers were used to implement all possible combinations of synonymous codons in the +2 and +3 positions of *narK*. The ATG triplet corresponding to the AUG start codon is boxed. Degenerate bases are marked in capital letters and bold text. Regions of homology between the forward and reverse primers are underlined. (**b**) The code for designing degenerate primers

3. Amplify the mini-library in a thermocycler using a program that consists of 95 °C for 2 min, then 30 cycles of 95 °C for 45 s, a range of temperatures from 48 to 68 °C for 45 s, and 72 °C for 6.5 min. Finish with a final elongation step at 68 °C (*see* **Notes 6** and 7).
4. Add 10 units of *Dpn*I to the reaction mix to digest the original plasmid.
5. To ensure that the mini-library was amplified, analyze 1 μL of the PCR product by agarose gel electrophoresis using standard protocols.

3.3 Transformation of Libraries into E. coli to Facilitate Re-circularization

1. Mix 10 μL of the PCR mix (approximately 500 ng) with 100 μL of competent MC1061 *E. coli* cells in a 1.5 mL reaction tube (*see* **Note 8**).
2. Incubate on ice for 30 min.
3. Heat shock for 1 min at 42 °C.
4. Incubate on ice again for 2 min.
5. To allow cells to recover, add 0.5 mL of LB broth and incubate at 37 °C with shaking for one hour.
6. Transfer cells to a 50 mL reaction tube containing 10 mL of LB broth and appropriate antibiotics, and incubate at 37 °C with shaking for 16 h (*see* **Note 9**).
7. Harvest the cells by centrifugation at 4000 × *g* in a bench-top centrifuge.
8. Purify the plasmids using the ENZA DNA mini kit as per the manufacturer's instructions. This plasmid prep is a mini-library containing variants of your original plasmid, which differ only in

the use of synonymous codons in the +2 and +3 positions, as determined by the design of your degenerate forward primer.

9. The mini-library can then be directly transformed into an expression strain such as BL21(*DE3*) or a derivative that has been selected or engineered for high-level production [19, 20]. To do so, take 1 μL of the mini-library, follow steps 1–5 above, and then plate out 200 μL of the culture on an LB agar plate with appropriate antibiotics.
10. Incubate at 37 °C for 16 h, and then pick colonies for expression testing (*see* **Notes 10** and **11**).

4 Notes

1. The protocol requires at least one forward degenerate primer. However if codons with more than four synonymous variants are being tested, then an additional primer will be required. The protocol always requires one reverse primer.
2. While any high-fidelity polymerase can be used, we choose to use the Q5 polymerase because its error rate is so low that it is difficult to measure in a statistically significant manner (see manufacturer's specifications). This minimizes random errors on the vector backbone and in the CDS.
3. *Dpn*I comes with its own reaction buffer, which we use to dilute it to a concentration of 10 U/μL. This can then be added directly to the PCR when required, since *Dpn*I also works effectively in the reaction buffer supplied for the Q5 polymerase.
4. Any *E. coli* strain that is capable of homologous recombination will suffice. We use MC1061.
5. Typically, we prepare a PCR mix of 100 μL, and then we aliquot 20 μL into five separate 0.2 mL PCR tubes. This allows us to test different annealing temperatures simultaneously by incubating the PCR mix in a thermocycler that has a gradient function.
6. To obtain maximal diversity in the library, we suggest that you test a range of annealing temperatures during the PCR and then choose the reaction that was amplified at the lowest annealing temperature.
7. Note that the extension time will depend on the size of your plasmid and the processivity of the DNA polymerase that you use. We typically allow 6.5 min for a 6.5 kb plasmid (i.e., 1 min/kb).
8. If two forward primers were required for the PCR, then mix 5 μL from each reaction.

9. This step allows propagation of the library.
10. Typically we would expect >500 colonies per plate.
11. Typically we would compare expression between 24 colonies. The type of expression testing done depends on the detection systems available.

Acknowledgments

This work was supported by a grant from the Swedish Research Council to DOD, and by the Novo Nordisk Foundation to MHHN.

Conflict of interest statement: The method considered here has been described in a patent submitted by CloneOpt AB (SE1451553-0). KM, ST, and DOD are founders and shareholders in CloneOpt AB.

References

1. Gustafsson C, Minshull J, Govindarajan S, Ness J, Villalobos A, Welch M (2012) Engineering genes for predictable protein expression. Protein Expr Purif 83:37–46
2. Bulmer M (1987) Coevolution of codon usage and transfer RNA abundance. Nature 325:728–730
3. Ikemura T (1981) Correlation between the abundance of *Escherichia coli* transfer RNAs and the occurrence of the respective codons in its protein genes: a proposal for a synonymous codon choice that is optimal for the *E. coli* translational system. J Mol Biol 151:389–409
4. Norholm MH, Light S, Virkki MT, Elofsson A, von Heijne G, Daley DO (2012) Manipulating the genetic code for membrane protein production: what have we learnt so far? Biochim Biophys Acta 1818:1091–1096
5. Norholm MH, Toddo S, Virkki MT, Light S, von Heijne G, Daley DO (2013) Improved production of membrane proteins in *Escherichia coli* by selective codon substitutions. FEBS Lett 587:2352–2358
6. Stenstrom CM, Jin H, Major LL, Tate WP, Isaksson LA (2001) Codon bias at the 3'-side of the initiation codon is correlated with translation initiation efficiency in *Escherichia coli*. Gene 263:273–284
7. Mirzadeh K, Martinez V, Toddo S, Guntur S, Herrgard MJ, Elofsson A, Norholm MH, Daley DO (2015) Enhanced protein production in *Escherichia coli* by optimization of cloning scars at the vector-coding sequence junction. ACS Synth Biol 4(9):959–965
8. Laursen BS, Sorensen HP, Mortensen KK, Sperling-Petersen HU (2005) Initiation of protein synthesis in bacteria. Microbiol Mol Biol Rev 69:101–123
9. McCarthy JE, Gualerzi C (1990) Translational control of prokaryotic gene expression. Trends Genet 6:78–85
10. Kudla G, Murray AW, Tollervey D, Plotkin JB (2009) Coding-sequence determinants of gene expression in *Escherichia coli*. Science 324:255–258
11. Plotkin JB, Kudla G (2011) Synonymous but not the same: the causes and consequences of codon bias. Nat Rev Genet 12:32–42
12. Mortimer SA, Kidwell MA, Doudna JA (2014) Insights into RNA structure and function from genome-wide studies. Nat Rev Genet 15:469–479
13. Ding Y, Tang Y, Kwok CK, Zhang Y, Bevilacqua PC, Assmann SM (2014) In vivo genome-wide profiling of RNA secondary structure reveals novel regulatory features. Nature 505:696–700
14. Daley DO, Rapp M, Granseth E, Melen K, Drew D, von Heijne G (2005) Global topology analysis of the *Escherichia coli* inner membrane proteome. Science 308:1321–1323
15. Drew D, Lerch M, Kunji E, Slotboom DJ, de Gier JW (2006) Optimization of membrane

protein overexpression and purification using GFP fusions. Nat Methods 3:303–313

16. Kimchi-Sarfaty C, Oh JM, Kim IW, Sauna ZE, Calcagno AM, Ambudkar SV, Gottesman MM (2007) A "silent" polymorphism in the MDR1 gene changes substrate specificity. Science 315:525–528
17. Kim SJ, Yoon JS, Shishido H, Yang Z, Rooney LA, Barral JM, Skach WR (2015) Protein folding. Translational tuning optimizes nascent protein folding in cells. Science 348:444–448
18. Ellis HM, Yu D, DiTizio T, Court DL (2001) High efficiency mutagenesis, repair, and engineering of chromosomal DNA using single-stranded oligonucleotides. PNAS 98:6742–6746
19. Sorensen HP, Mortensen KK (2005) Advanced genetic strategies for recombinant protein expression in *Escherichia coli*. J Biotechnol 115:113–128
20. Rosano GL, Ceccarelli EA (2014) Recombinant protein expression in *Escherichia coli*: advances and challenges. Front Microbiol 5:172

Chapter 5

Generation of Tetracycline-Inducible Mammalian Cell Lines by Flow Cytometry for Improved Overproduction of Membrane Proteins

Juni Andréll, Patricia C. Edwards, Fan Zhang, Maria Daly, and Christopher G. Tate

Abstract

Overexpression of mammalian membrane proteins in mammalian cells is an effective strategy to produce sufficient protein for biophysical analyses and structural studies, because the cells generally express proteins in a correctly folded state. However, obtaining high levels of expression suitable for protein purification on a milligram scale can be challenging. As membrane protein overexpression often has a negative impact on cell viability, it is usual to make stable cell lines where the protein of interest is expressed from an inducible promoter. Here we describe a methodology for optimizing the inducible production of any membrane protein fused to GFP through the isolation of clonal cell lines. Flow cytometry is used to sort uninduced cells and the most fluorescent 5 % of the cell population are used to make clonal cell lines.

Key words Membrane protein, Overexpression, GFP, GPCR, HEK293 cell line, Flow cytometry, Transporter

1 Introduction

Expressing eukaryotic membrane proteins in mammalian cells has the advantage of providing a near-native environment that supports the production of fully functional protein [1]. For many integral membrane protein targets, such as G protein-coupled receptors, ion channels, and transporters, a mammalian expression system is the most effective system for producing properly folded and functional protein [2–4]. We have found that creating stable cell lines in HEK293 cells using an inducible TetR expression system [5], such as the T-Rex™ system, produces the best results [2, 3]. However, crystallization of eukaryotic membrane proteins for structural studies by X-ray crystallography requires milligram quantities of protein. Although some membrane proteins have been crystallized after overexpression in mammalian cells [6–9],

Isabelle Mus-Veteau (ed.), *Heterologous Expression of Membrane Proteins: Methods and Protocols*, Methods in Molecular Biology, vol. 1432, DOI 10.1007/978-1-4939-3637-3_5, © Springer Science+Business Media New York 2016

the purification of milligram amounts of protein can be challenging if expression levels are below one million copies/cell. In this protocol we describe a method to obtain cell lines producing milligram quantities of target protein by expressing membrane protein-GFP fusion proteins from a tetracycline-inducible promoter and using flow cytometry to create overexpressing clonal cell lines. These cell lines can be passaged many times with no reduction in expression levels of the target protein, which is not the case when flow cytometry has been used to overexpress membrane proteins using constitutive expression from a strong promoter [10].

To make a stable cell line using the T-Rex™ system, a plasmid expressing the protein-GFP fusion of interest from a tetracycline-inducible CMV promoter is transfected into HEK293(TetR) cells where it integrates randomly into the host cell genome. The expression level of the transgene depends to a large extent on where it has integrated. This positional effect on expression levels is then exploited by selecting for those cells where the random integration results in high levels of expression. Normally in a polyclonal cell line there is a wide range of expression levels both before and after induction. We have found that sorting the top 5 % of fluorescent cells of the uninduced population of a polyclonal cell line will generate clonal cell lines with the highest likelihood of high levels of induced expression (*Andréll et al. unpublished*). Usually we start the process by performing a transfection of the plasmid DNA expressing the cDNA of interest and setting up six polyclonal cell lines from this transfection. Sorting by flow cytometry will then be performed only on the polyclonal cell line that shows the widest distribution of fluorescence in the uninduced state. Sorting at low cell numbers in a 96-well plate will give rise to 40–60 different cell lines, which ensures that an overexpressing cell line will be found. Initial analysis of these cell lines is accomplished by flow cytometry and fluorescence microscopy, which together assess both the quantity and quality of the expressed fusion protein. As the cell lines are passaged, their viability and growth potential are also selected for, which allows easy identification of cell lines suitable for large-scale culture and downstream purification of milligram quantities of target protein.

The methodology we have developed is suitable for any integral membrane protein that is usually found at the plasma membrane and can be fused to GFP without affecting its functionality. Although the methodology will also work for membrane proteins localized to intracellular organelles, fluorescence microscopy may not be able to differentiate between correctly folded protein and misfolded protein localized in the endoplasmic reticulum, so it may be necessary to use a different assay to assess if misfolded protein is present, e.g., a differential detergent solubility assay [3].

2 Materials

Prepare all stocks and buffers at room temperature using ultrapure water, MilliQ grade (18 ΩM cm at 25 °C), and filter using a 0.22 μm filter. All tissue culture work must be performed inside a microbiological safety cabinet. Use sterile equipment and buffers if in contact with cells. Wear nitrile gloves and a lab coat when working with mammalian cells.

2.1 Equipment (See Note 1)

1. Microbiological safety cabinet.
2. Temperature and CO_2 regulated incubator set at 37 °C and 5 % CO_2.
3. Personal protection equipment: lab coat, nitrile gloves, safety glasses.
4. Pipette filler, such as S1 Pipette filler (Thermo Scientific).
5. Pipettes: 5, 10, 25, and 50 ml sterile.
6. Benchtop centrifuge suitable for 50 ml conical-bottom tubes.
7. Aspirator, such as VACUSAFE (Integra) used with sterile aspirator pastettes (Alpha laboratories LW4811).
8. Cell counting equipment, such as Countess™ automated cell counter (Life technologies) with Countess™ cell counting chamber slides (Life technologies C10228).
9. Fluorescent microscope, such as a Leica DMI LED Fluo (Leica Microsystems) coupled to a Lumen 200 fluorescence illumination system (Prior Scientific) and a QI click camera (QI imaging) and computer.
10. Flow cytometry analyzer, such as FACSCalibur™ (BD Biosciences).
11. Flow cytometry cell sorter, such as MoFlo™ Legacy (Beckman Coulter).
12. Flow cytometry data analysis program, such as FlowJo vX.0.7.
13. Liquid nitrogen and liquid nitrogen cell line storage dewar.

2.2 Tissue Culture Consumables (See Note 2)

1. T175 flask, such as 175 cm^2 tissue culture flask with vented cap. 10 cm plate, such as 100 × 20 mm tissue culture dish.
2. 96-well, 24-well, and 6-well tissue culture treated plates.
3. Polystyrene round bottom tube, Falcon® 5 ml, 12 × 75 mm, sterile (*see* **Note 3**).
4. 1.5 ml Eppendorf microcentrifuge tubes, sterile.
5. Cell strainer, such as CellTrics® 50 μm filters, sterile or nonsterile (*see* **Note 4**).
6. Cryogenic vial.

7. 15 ml and 50 ml conical-bottom tubes, sterile.
8. Steriflip-GP, 50 ml disposable vacuum filter system, 0.22 μm, sterile.

2.3 Mammalian Expression System (See Note 5)

1. T-REx™-293 cell line (Life technologies) (*see* **Note 6**).
2. Construct DNA: pcDNA™4/TO mammalian expression vector (Life technologies) containing the gene of interest C-terminally tagged with eGFP.

2.4 Reagents, Stock Solutions, and Buffers

1. Blasticidin stock solution (5 mg/ml): Prepare inside microbiological safety cabinet. Dissolve 50 mg Blasticidin S HCl powder in 10 ml water. Filter sterilize using a Steriflip-GP 50 ml. Aliquot into 20 × 0.5 ml and store at −20 °C.
2. Doxycycline stock solution (1 mg/ml): Per 1 ml water, add 1 mg of doxycycline hyclate for a 1 mg/ml stock. Filter sterilize using Steriflip-GFP 50 ml. Make up fresh for each use.
3. Phosphate buffered saline (PBS) solution.
4. Cell buffer: 20 mM Tris pH 7.4, 100 mM NaCl with added Complete EDTA-free protease inhibitors according to manufacturer's instruction.
5. GeneJuice™ transfection reagent.

2.5 Media

1. Complete media: Prepare inside microbiological safety cabinet. Take one 500 ml bottle of Dulbecco's Modified Eagle Medium high glucose GLUTAMAX™ supplement pyruvate (DMEM), and add one aliquot of blasticidin stock solution for a final 5 μg/ml concentration and add 50 ml of certified Fetal Bovine Serum (FBS) (*see* **Notes 7** and **8**).
2. Serum-free media: DMEM only.
3. Antibiotic selection media: Prepare inside microbiological safety cabinet. Add 1 ml Zeocin™ (100 mg/ml) to one 500 ml bottle of complete media for a final 200 μg/ml concentration.
4. Cell freezing media: Add 47.5 ml complete media to 2.5 ml sterile Dimethyl sulfoxide Hybri-Max™ (DMSO) in a 50 ml conical-bottom tube for a final concentration of 5 % DMSO.
5. Induction media: Per 1 ml selection media add 1 μl doxycycline stock solution for a 1 μg/ml final concentration of doxycycline. Make up fresh for each use.

3 Methods

A protocol overview is presented in Fig. 1.

3.1 Polyclonal Cell Line Generation

1. Grow T-REx™-293 cells observing proper mammalian tissue cell culture technique. Use good tissue cell culture practice

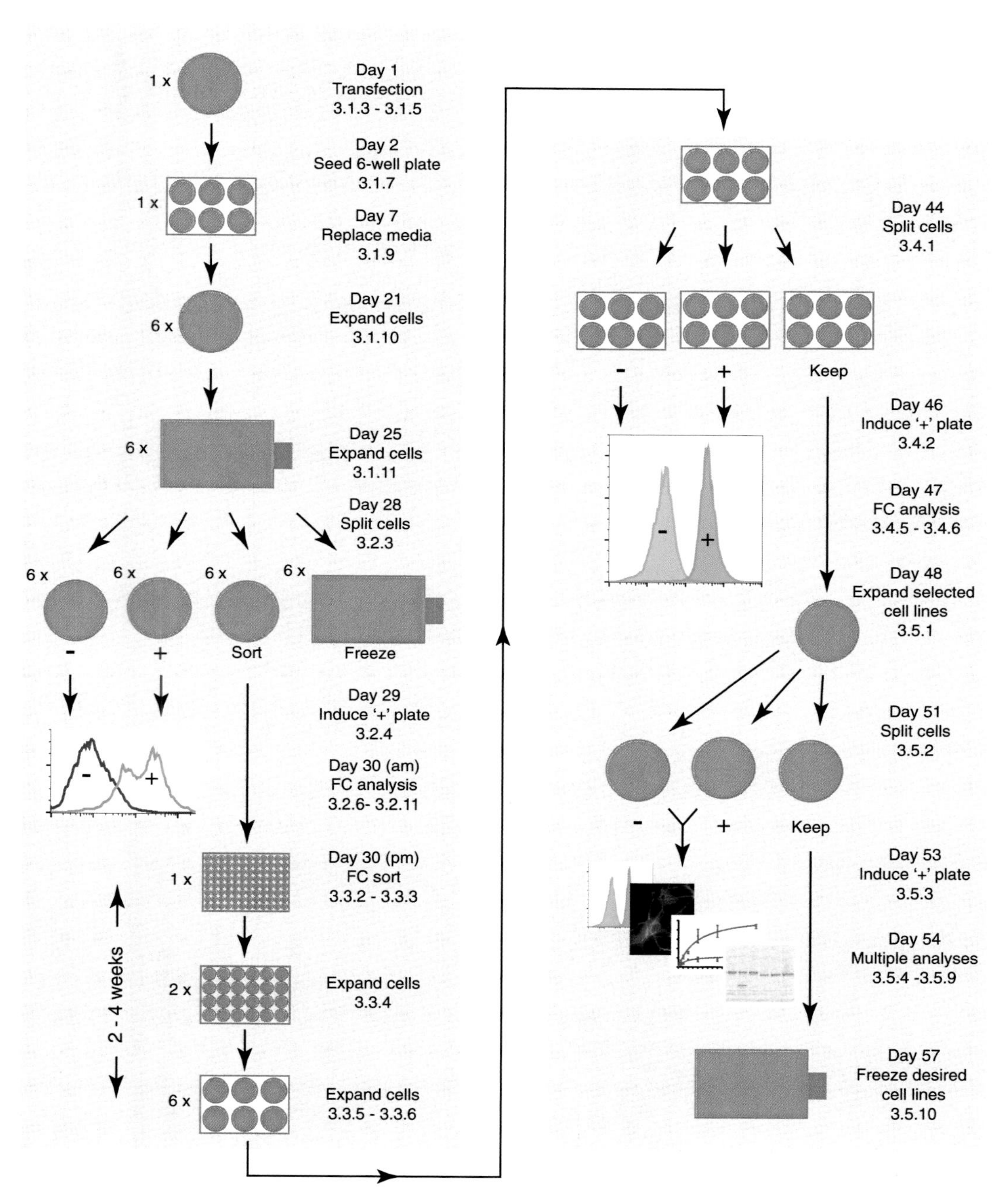

Fig. 1 Protocol overview. For each step we have given an approximate time and a reference to the section in the main text where it is discussed. The timings will vary depending on the rate of growth of individual cell lines and the protein/construct being expressed. The number of plates typically used for a single target is indicated above the cartoon of the tissue culture plate used (either a 10 cm plate, 6-well plate, 24-well plate, 96-well plate or a T175 flask). Abbreviations: *FC* flow cytometry; '–', no added tetracycline; '+', induced with tetracycline

throughout to keep cells healthy at high viability for good results. All tissue culture work must be performed inside a microbiological safety cabinet. Use sterile solutions if in contact with cells. All media must be warmed to 37 °C prior to use. All cells are grown by incubating them in an incubator set at 37 °C and 5% CO_2. Scale up cells into T175 tissue culture flasks.

2. To set up plates for transfection, harvest a confluent T175 tissue culture flask in 12 ml complete media. Unless otherwise stated cells are harvested by removing the media with an aspirator, washing the cells gently without dislodging them by swirling PBS over them (*see* **Note 9**), removing the PBS by aspiration, and finally dislodging the cells by pipetting the media over them repeatedly and pipetting the cell solution up and down to homogenize it. When harvesting flasks, the flask can be tapped by hand to dislodge the cells prior to pipetting them. Seed 1/6th of cells (2 ml) into a 10 cm tissue culture plate containing 8 ml complete media. Set up one 10 cm tissue culture plate per construct and one additional 10 cm plate as a negative control. After 24 h the 10 cm plates should be 70–80% confluent prior to transfection (*see* **Note 10**).
3. For each plate, pipette 800 μl serum-free media into a sterile microcentrifuge tube and add 18 μl GeneJuice™ transfection reagent. Vortex to mix and incubate at room temperature for 5 min.
4. Add 6 μg of construct DNA. Mix gently by pipetting and incubate at room temperature for 15 min. During this incubation replace the media of the 70–80% confluent 10 cm plate with 10 ml complete DMEM media.
5. Add the DNA/GeneJuice™ mixture dropwise to the plate. Tilt plate side-to-side and backwards and forwards for a gentle mix (*see* **Note 11**), put it in an incubator and leave for 24 h. For the negative control plate, repeat all the same steps but do not add any construct DNA to the transfection reagent.
6. For each transfected plate and the negative control: prepare a 6-well tissue culture plate by pipetting 4 ml antibiotic selection media into each well. Harvest the transfected plate in 5 ml of selection media. Count the cells using, for instance, a Countess™ automated cell counter.
7. Seed each well in the 6-well plate with ~400,000 cells to allow six polyclonal cell line selections to be done in parallel (*see* **Note 12**).
8. Set up a transfection efficiency control plate by adding 4 ml of the harvested cells to a 10 cm tissue culture plate. Add 6 ml complete media containing 10 μg doxycycline for a final concentration of 1 μg/ml and induce for 24 h.

9. Estimate the transfection efficiency by looking at the positive control plate in a fluorescence microscope. Compare the white light image with the green fluorescence image to estimate percentage transfection efficiency (*see* **Note 13**). After 5–7 days no live cells should be visible under the microscope in negative control plates (*see* **Note 14**). Depending on the transfection efficiency the transfected cells will either (a) die massively in the 6-well plates or (b) become confluent rapidly without any cell death. In the case of (a) the media should be replaced once colonies have started to grow (*see* **Note 15**). In the case of (b) proceed directly to **step 10**.
10. Once the media in each 6-well has started to become more yellow in color, harvest the cells in fresh 4 ml selection media and add them to a 10 cm tissue culture plate containing 6 ml selection media. If the cells did not die during the 6-well plate stage (9b), then this first passage will trigger cell death. The surviving cells will grow as colonies. Once grown to when they are visible to the eye, resuspend the colonies in 10 ml fresh selection media in order to get uniform confluency.
11. When confluent harvest the cells from each 10 cm plate in 10 ml selection media and add to a T175 flask containing 15 ml selection media.
12. Defrost a vial of T-REx™-293 cells to be used as a nonfluorescent control during flow cytometry analysis and cell sorting.

3.2 Choosing a Polyclonal Cell Line for Sorting

1. To set up plates for flow cytometry analysis, when the T175 flasks are confluent harvest the cells in 9 ml selection media. For each T175 flask containing a polyclonal cell line label three 10 cm plates as "–" (uninduced), "+" (induced), and "sort" and a T175 flask as "freeze."
2. Add 9 ml selection media to the "–", "+", and "sort" plates and 20 ml selection media to the T175 "freeze" flask.
3. Harvest each T175 flask in 9 ml selection media. Add 1.5 ml harvested cells to the "–" plate, add 2 ml of the harvested cells to the "+" plate, add 1.5 ml of harvested cells to the "sort" plate, and add the remaining cells to the "freeze" flask.
4. When the "+" plates are 70–90% confluent, induce them by replacing the media with 10 ml induction media.
5. After 24 h induction image the "+" and "–" plates in a fluorescence microscope (Fig. 2a–c) (*see* **Note 16**).
6. Harvest the cells from the "+" plate in 10 ml PBS. Make sure the cells are well dispersed and have a cell density of $\sim 1 \times 10^6$ cells/ml (*see* **Note 17**). Harvest parental T-REx™-293 cells to use as a negative control in a similar manner.
7. Transfer 0.5 ml of each sample into a polystyrene round bottom tube. These are the samples for the flow cytometry analyzer.

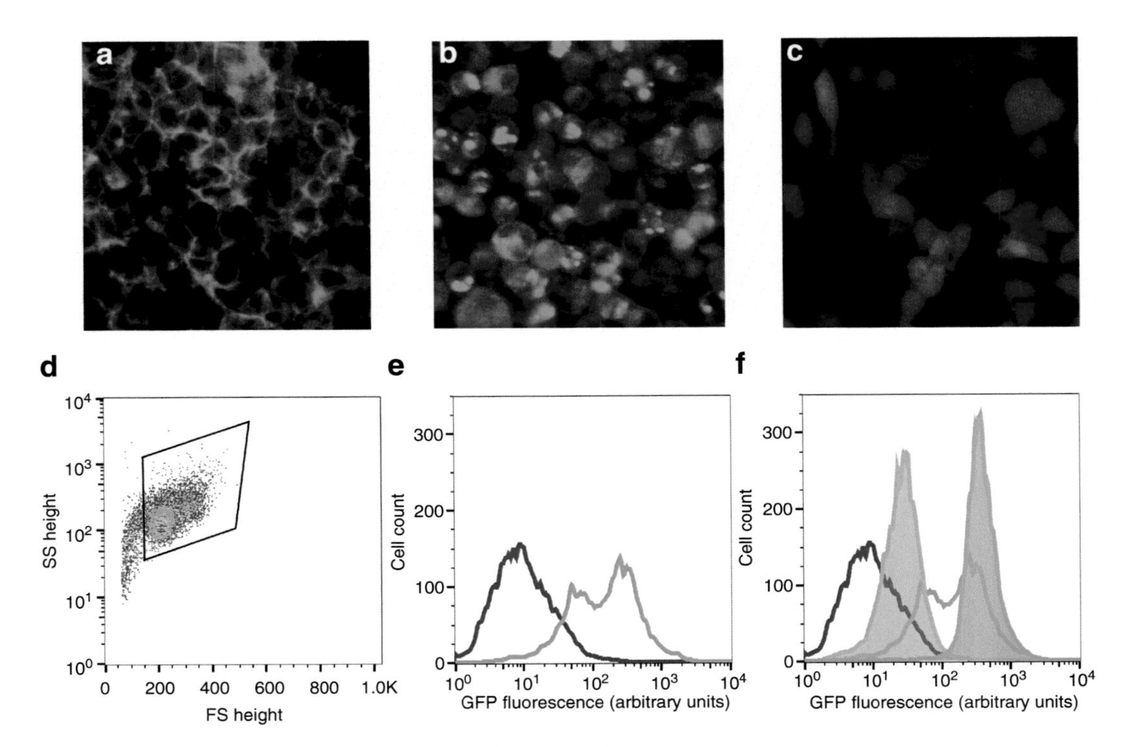

Fig. 2 (**a–c**) Fluorescence microscope images of cell lines expressing target membrane proteins showing different cellular localizations; (**a**) correctly localized at the plasma membrane with no obvious intracellular aggregates, (**b**) undesirable aggregates of the membrane protein-GFP fusion are clearly visible intracellularly as intense areas of fluorescence, (**c**) GFP fluorescence is decoupled from target membrane protein expression and low levels of soluble GFP are observed throughout the cell. Fluorescence brightness has been adjusted separately for each image for clarity. (**d**) During flow cytometry, intact cells are selected using a gate defined by the rhomboid in the Forward Scatter (FS) versus Side Scatter (SC) plot. (**e**) Histograms of an uninduced polyclonal cell line (*blue line*) and an induced polyclonal cell line (*green line*). (**f**) Histograms of the final clonal cell line, uninduced (*filled blue area*) and induced (*filled green area*), overlaid on the original polyclonal cell line histogram

8. Put the remaining cells in a 50 ml conical-bottom tube and keep on ice during the analysis **steps 9–11,** Subheading 3.2.
9. Set up the flow cytometry analyzer according to manufacturer's instruction with the parental T-REx™-293 cell sample as an untransfected control. Adjust the voltage on the photomultiplier tube (PMT) detecting the GFP fluorescence so that the fluorescence value falls between 0 and 10 on the fluorescence log scale (arbitrary units).
10. Collect flow cytometry data from each test sample using the same PMT voltage set for the untransfected control sample. Use a Forward Scatter (FS) versus Side Scatter (SS) plot to select a region to gate the intact cells, avoiding the smaller cell debris and larger cell aggregates (Fig. 2d). Collect 10,000 events.

11. Analyze the data with a suitable analyzer program, *e.g.*, FlowJo. Use the FS versus SS plot to select intact cells (Fig. 2d) and represent the selected cells as a histogram with GFP fluorescence on the x-axis (Fig. 2e). Select polyclonal cell lines to sort from based on broadness of fluorescent peak when induced, inducibility, and presence of a highly fluorescent subpopulation (*see* **Note 18**). Discard all cell lines not meeting the selection criteria.
12. Count the "+" and "−" cell samples in the 50 ml conical-bottom tubes on ice corresponding to the selected cell lines.
13. Spin down the cells for 5 min at 4000 rpm, aspirate or decant the supernatant.
14. Resuspend the cell pellet in 50 ml PBS, spin down the cells for 5 min at 4000 rpm, aspirate or decant the supernatant.
15. Resuspend the cell pellet in cell buffer at 10 million cells/ml and transfer to a microcentrifuge tube. Label and freeze samples in liquid nitrogen and store at −80 °C (*see* **Note 19**).
16. Once the T175 "freeze" flask is confluent, harvest the cells in 10 ml PBS, spin them down for 5 min at 1500 rpm, aspirate or decant the supernatant.
17. Resuspend the cell pellet gently in 4 ml of cell freezing media. Gently aliquot 4 × 1 ml into cryogenic vials, freeze cells in −80 °C freezer, and transfer the next day to liquid nitrogen cell line storage dewar (*see* **Note 20**).

3.3 Clonal Cell Line Generation by Flow Cytometry Sorting

1. Prepare 1 × 96-well plate per selected polyclonal cell line with 200 μl selection media/well. Keep plates in an incubator until ready for use.
2. Harvest the uninduced cells from one "sort" plate at a time in 10 ml selection media. Filter cells through a sterile cell filter and transfer 1 ml of ~1×10^6 cells/ml to a 12 × 75 mm polystyrene round bottom tube.
3. Set up the sorter following the manufacturer's instructions. Use a FS versus SS plot to select a region to gate the intact cells avoiding the smaller cell debris and larger cell aggregates. Use a second plot of Pulse Width versus Forward Scatter to set a gate to exclude cells of a greater width, *i.e.*, to exclude doublets. Single intact cells from the top 5 % of the uninduced cell population were selected using the sort logic. 10 cells/well were sorted into the prepared 96-well plate. Repeat the procedure for each selected polyclonal cell line (*see* **Note 21**) and place plates into an incubator for growth.
4. The first confluent wells will appear 10–14 days post-sorting (*see* **Note 22**).

5. As the wells become confluent, harvest the cells from each well in the media they have grown in and transfer them to a 24-well plate containing 1 ml selection media per well.
6. As these become confluent harvest the cells from each well in the media they have grown in and transfer them to a 6-well plate containing 3 ml selection media per well (*see* **Note 23**).

3.4 Selection of Sorted Cell Lines by Flow Cytometry Analysis

1. To set up plates for flow cytometry analysis of sorted cell lines, when each 6-well becomes confluent, harvest the cells from each well in 6 ml selection media. Label three new 6-well plates "+", "−" and "keep." Add 2 ml selection media into each well. Transfer 2.5 ml/2 ml/1.5 ml of the harvested cells into the "+", "−", and "keep" plates respectively (*see* **Note 24**).
2. Induce the "+" wells when 70–90% confluent (~48 h after passaging) by replacing the media with 4 ml induction media.
3. After 24 h induction image the "+" and "−" plates in a fluorescence microscope (*see* **Note 25**).
4. After a 24 h induction harvest each well in the "+" and "−" plates in 2 ml PBS. Transfer 0.5 ml into a polystyrene round bottom tube for flow cytometry analysis.
5. Use the set up from Subheading 3.2, **step 9**. Check the settings by running a test sample of parental T-REx™-293 cells. Collect the flow cytometry data as in Subheading 3.2, **step 10** on all induced ("+") and uninduced ("−") samples.
6. Analyze the flow cytometry data similar to Subheading 3.2, **step 11**. Select which cell lines to keep based on the inducibility and median value (*see* **Note 26**). Discard the other cell lines.

3.5 Further Analysis of Selected Sorted Cell Lines

1. To expand selected sorted cell lines, go back to the "keep" plates and, when confluent, harvest each selected cell line in 3 ml selection media and add them to a 10 cm plate containing 7 ml selection media.
2. To set up plates for further analysis, when confluent, harvest each 10 cm plate in 10 ml selection media, split the cells into three 10 cm plates containing 8 ml selection media each by adding 4 ml of cells to the plate labeled "+" and 3 ml each into the plates labeled "−" and "keep."
3. Induce the "+" plate when 70–90% confluent by replacing the media with 10 ml of induction media.
4. After 24 h induction, image the "+" and "−" plates in a fluorescence microscope to verify the correct localization of the fusion protein (Fig. 2a–c) (*see* **Note 27**).
5. Harvest the "+" and "−" plates in 10 ml PBS. Make sure the cells are well dispersed and of ~1×10^6 cells/ml density. Harvest parental T-REx™-293 cells to use as a negative control

in a similar manner. Transfer 0.5 ml for flow cytometry analysis in a polystyrene round bottom tube. Put the remaining cells in a 50 ml tube and keep on ice during analysis **step 6**, Subheading 3.5.

6. For flow cytometry analysis, use the set up in Subheading 3.2, **step 9**. Check the calibration by running a sample of parental T-REx™-293 cells. Collect flow cytometry data as in Subheading 3.2, **step 10** on all induced ("+") and uninduced ("–") samples. Analyze the flow cytometry data similar to Subheading 3.2, **step 11** (*see* **Note 28**).
7. Count the "+" and "–" cell samples in the 50 ml conical-bottom tubes on ice corresponding to the selected cell lines. Spin down the cells for 5 min at 4000 rpm, aspirate or decant the supernatant.
8. Resuspend the cells in 50 ml PBS, spin down the cells for 5 min at 4000 rpm, aspirate or decant the supernatant.
9. Resuspend the cells in cell buffer at 10 million cells/ml and transfer to a microcentrifuge tube. Label and freeze samples in liquid nitrogen and store at –80 °C (*see* **Note 29**).
10. To freeze selected cell lines, expand the 10 cm "keep" plate by harvesting the cells when confluent in 10 ml selection media and adding them to a T175 flask containing 15 ml selection media. When confluent, harvest the cells in 10 ml PBS, spin down the cells at 1500 rpm 5 min, aspirate away the supernatant carefully, and resuspend the cells in 4 ml freezing media. Aliquot the cells in the freezing media into 4 × 1 ml cryogenic vials, freeze at –80 °C and transfer the next day to a liquid nitrogen cell storage dewar (*see* **Note 30**).

4 Notes

1. The brands we use are suggested, but a different brand with the same function may be substituted.
2. It is important that plate and flasks are tissue culture treated and sterile. We recommend vented caps on the tissue culture flasks to reduce the risk of contamination.
3. These tubes are compatible with the BD FACSCalibur and MoFlow machines. If other flow cytometry machines are used, use the sample holders recommended for them.
4. Use sterile cell filters if preparing samples for flow cytometry sorting. Nonsterile filters are fine for preparing samples for flow cytometry analysis.
5. It is important to use an inducible mammalian expression system that allows for the random integration of the transgene, such as the T-REx™ system used here.

6. Alternatively, an equivalent cell line expressing the TetR repressor protein can be used, such as the HEK293S-TetR [4] or HEK293S-GnTI⁻ [11].
7. The certified FBS is guaranteed not to contain tetracycline, which is essential to retain the cells in an uninduced state.
8. Blasticidin is used to select for the presence of the TetR gene in the T-REx™-293 parental cell line.
9. Washing the cells with PBS removes debris and provides cleaner flow cytometry data.
10. The amount of cells split into the 10 cm tissue culture plate may have to be adjusted to reach 70–80 % confluency the next day.
11. Do not swirl the plate as this will concentrate the transfection mix at the center of the plate and prevent uniform spread.
12. Each of the six polyclonal cell lines will be slightly different since they originate from different batches of cells.
13. If there is no fluorescence upon induction of the transfected control plates then the transfection has failed. Make sure the fluorescent microscope is working. When repeating the transfection make sure that the cells are healthy with no contamination, that the 10 cm tissue culture plate is 70–80 % confluent prior to transfection, that the GeneJuice™ batch is fresh, and that the doxycycline stock is fresh and used at the correct concentration. Importantly, double-check that the construct is compatible with expression in the T-REx™ system and that the purified plasmid DNA is of sufficient quality.
14. In general dying cells will be round and floating in the media and live cells will be growing adherently to the bottom of the plate in an extended morphology. If there are live cells present in the negative control the antibiotic selection has failed. Repeat the transfection and cell seeding steps with a freshly prepared antibiotic selection media using fresh certified FBS and Zeocin™. If the negative control cells fail to die in the antibiotic selection media again, defrost a fresh vial of T-REx™-293 cells and repeat all steps.
15. When changing the media be careful to not disturb the cells, particularly if the colonies are small, as this may lead them to be washed away.
16. If the GFP fluorescence localizes to the plasma membrane it indicates that the fusion protein is intact, that the membrane protein has folded properly, and that it is likely to be functional (Fig. 2a). In this instance, the fluorescence intensity is proportional to the fusion protein expression levels. However, misleading fluorescence values could come from high levels of free GFP in the cell or from intensely fluorescent retained material of aggregated fusion protein. When looking at the cells in a

fluorescence microscope these two false positives can easily be discriminated between. Any retained aggregated fusion protein will be seen as a clump of very bright material in the cytosol (Fig. 2b). If there are high levels of free GFP, the entire cytoplasm will be uniformly fluorescent (Fig. 2c). Discard any polyclonal cell line containing these false positives.

17. If the harvested cells contain clumps or are very dense they may clog up the flow cell of the flow cytometry analyzer. Avoid this by filtering cells through a cell filter to remove any clumps and/or by diluting the cell sample with PBS to ~0.5 million cells/ml density.
18. Polyclonal cell line selection criteria: (1) A suitable polyclonal cell line to sort from should have a broad fluorescent peak once it is induced, indicating expression variability. (2) It is preferable if the cell line is fully inducible. The inducibility of a cell line is measured by how many cells are positive for GFP fluorescence upon induction, *i.e.*, they have a fluorescence value higher than the highest fluorescence measured from negative control cells. The percentage of positive cells in a cell line is indicative of how many cells carry the transgene, which ideally should be more than 90 %. (3) The presence of a highly fluorescent subpopulation compared to the other polyclonal cell lines. Important: the decision of which polyclonal cell line to sort from is based on the flow cytometry data from the induced cells, but the flow cytometry sorting should be done using the uninduced cells.
19. These cell samples are useful to compare with the final sorted cell lines when verifying expression levels.
20. Polyclonal cell lines are particularly prone to detrimental changes to expression levels with increasing number of passaging. Hence it is important to freeze several aliquots of all cell lines that meet the selection criteria.
21. During sorting, live cells are chosen according to FS and SC profile (not propidium iodide).
22. As a rule ~50 % of the wells will grow and become confluent. Even though the plate was seeded with 10 cells/well most of the cell lines that grow will originate from only 1–2 cells and occasionally from 3 to 4 cells (determined under the microscope by checking the colony formation in each well). If only one cell per well is used, over 95 % of the cells will die. While the cell lines are not necessarily monoclonal, the resulting clonal cell lines behave as such, with relatively narrow fluorescent peaks that keep constant for 20+ passages (Fig. 2f).
23. The expansion process is an important indirect cell line selection process. The speed of growth will vary depending on how many cells per well are growing, but also on the individual

cells in the well and how well they cope with the uninduced constitutive expression of the transgene. Those cell lines that have constitutive expression levels compatible with healthy growth will expand in a timely manner. Cell lines unable to expand at this stage can automatically be culled, since they would not be useful for extended culture or large-scale growth. Because the expansion time will vary, this part of the protocol can be very labor intensive and requires careful planning. The steps from here until the end of the protocol will happen on different days or on overlapping days depending on the batch of cell lines. At this stage it is therefore very important to keep on top of the process and passage cells when confluent but not overgrown. High viability of the cells is extremely important at all times in order to ensure good expression levels and comparable cell lines during the cell line selection steps.

24. The varying amounts of cells in the three different plates will ensure their confluency at staggered time points. From this point it is important to label each cell line individually.
25. The imaging of potentially hundreds of cell lines in a day is very time-consuming. If there is a time constraint this step can be skipped. If doing so, be aware that, when making the cell line selection based on fluorescence in Subheading 3.4, **step 6**, the top fluorescent cell lines may contain free GFP and/or retained aggregated fusion protein giving rise to misleadingly high fluorescent values. Therefore select a surplus of cell lines to carry forward for further analysis.
26. Evaluating the clonal cell lines for GFP fluorescence using flow cytometry can be extremely quick and data from hundreds of cell lines can easily be collected in a couple of hours. Analyzing the data in a suitable program such as FlowJo allows for quick and easy compilation of statistics and histograms from which a first cell line selection can be made. There are three selection criteria for a suitable clonal cell line to be carried forward from this point. (1) The cell line should be inducible with 99 % of cells positive for GFP fluorescence after induction. If the cell line is not fully inducible, it suggests the presence of a small subpopulation not carrying the transgene. During extended time in culture such a small subpopulation can take over the cell line since they grow better as they experience less metabolic demand, toxicity, and stress than those cells carrying the transgene. These cells could be re-sorted if other better cell lines are not available. (2) Populations of clonal cells before and after induction should exhibit narrow fluorescence profiles compared to the polyclonal cell line (Fig. 2f). The narrowness of the peaks suggests that the cell line is clonal. Broad peaks or multiple peaks are indicative of a polyclonal cell line and should

be re-sorted or discarded at this stage. (3) The cell line should have a high median fluorescence value, which is the value where half of the cells have higher fluorescence values and half of the cells have lower fluorescence values. There is no absolute median value that can be used as a cut-off, since fluorescence is in arbitrary units and depends on the flow cytometry analyzer used, the calibration of the machine, and the protein expressed. Once a high expressing cell line has been verified its fluorescence median (on a particular flow cytometry analyzer) can become the fluorescence selection criteria. The first time this protocol is carried out, keep the 10–20 cell lines with the highest median values that also fulfill the other two selection criteria. Note that the induced cells with the highest fluorescent median are also likely to have a high fluorescent median when uninduced (Fig. 2f) *(Andréll et al. unpublished)*.

27. If the GFP fluorescence localizes to the plasma membrane, it suggests that the fusion protein is intact and that the membrane protein has folded properly and is likely to be functional (*see* **Note 16** and Fig. 2a). The selected sorted cell lines are likely to have a relatively high basal expression level when uninduced and these will show very low fluorescence under a fluorescent microscope.
28. It is not unusual for the median of each cell line to differ at this stage compared to the initial data from Subheading 3.4, **step 5**. The data from the scaled up 10 cm plate is more reliable since the cell line has stabilized at this point. Commonly, the cell lines with a very high basal expression level struggle to expand in a timely manner, which will be apparent at this point. These can be discarded as they are unsuitable for extended culturing and large-scale growth.
29. Use these cell samples for further verification of functional expression levels: such as (1) In-gel fluorescence and/or western blot. The in-gel fluorescence will verify the presence of the full-length fluorescent fusion protein, the absence/presence of free GFP, and the level of degradation of the fusion protein (containing the GFP moiety). (2) Radioligand binding assay. Provided there is a radiolabeled ligand of sufficient affinity to the target protein, this can provide quantification of functional copies of target membrane protein per cell. (3) Fluorescence size exclusion chromatography (FSEC). This will allow the evaluation of the quality of fusion protein when solubilized with different detergents. This is useful to evaluate which cell line to carry forward into large-scale production for protein purification purposes.
30. It is very important to have frozen aliquots stored of each selected sorted cell line.

Acknowledgment

The authors would like to thank Annette Strege for providing us with the image in Fig. 2c. This research was funded by the Medical Research Council (MRC U105197215) and by an EC FP7 grant for the EDICT consortium (HEALTH-201924).

References

1. Andrell J, Tate CG (2013) Overexpression of membrane proteins in mammalian cells for structural studies. Mol Membr Biol 30:52–63
2. Tate CG, Haase J, Baker C, Boorsma M, Magnani F, Vallis Y, Williams DC (2003) Comparison of seven different heterologous protein expression systems for the production of the serotonin transporter. Biochim Biophys Acta 1610:141–153
3. Thomas J, Tate CG (2014) Quality control in eukaryotic membrane protein overproduction. J Mol Biol 426:4139–4154
4. Reeves PJ, Kim JM, Khorana HG (2002) Structure and function in rhodopsin: a tetracycline-inducible system in stable mammalian cell lines for high-level expression of opsin mutants. Proc Natl Acad Sci U S A 99: 13413–13418
5. Yao F, Svensjo T, Winkler T, Lu M, Eriksson C, Eriksson E (1998) Tetracycline repressor, tetR, rather than the tetR-mammalian cell transcription factor fusion derivatives, regulates inducible gene expression in mammalian cells. Hum Gene Ther 9:1939–1950
6. Standfuss J, Edwards PC, D'Antona A, Fransen M, Xie G, Oprian DD, Schertler GF (2011) The structural basis of agonist-induced activation in constitutively active rhodopsin. Nature 471:656–660
7. Standfuss J, Xie G, Edwards PC, Burghammer M, Oprian DD, Schertler GF (2007) Crystal structure of a thermally stable rhodopsin mutant. J Mol Biol 372:1179–1188
8. Gruswitz F, Chaudhary S, Ho JD, Schlessinger A, Pezeshki B, Ho CM, Sali A, Westhoff CM, Stroud RM (2010) Function of human Rh based on structure of RhCG at 2.1 A. Proc Natl Acad Sci U S A 107:9638–9643
9. Deupi X, Edwards P, Singhal A, Nickle B, Oprian D, Schertler G, Standfuss J (2012) Stabilized G protein binding site in the structure of constitutively active metarhodopsin-II. Proc Natl Acad Sci U S A 109: 119–124
10. Mancia F, Patel SD, Rajala MW, Scherer PE, Nemes A, Schieren I, Hendrickson WA, Shapiro L (2004) Optimization of protein production in mammalian cells with a coexpressed fluorescent marker. Structure 12: 1355–1360
11. Reeves PJ, Callewaert N, Contreras R, Khorana HG (2002) Structure and function in rhodopsin: high-level expression of rhodopsin with restricted and homogeneous N-glycosylation by a tetracycline-inducible N-acetylglucosaminyltransferase I-negative HEK293S stable mammalian cell line. Proc Natl Acad Sci U S A 99:13419–13424

Chapter 6

Membrane Protein Production in *Lactococcus lactis* for Functional Studies

Daphne Seigneurin-Berny, Martin S. King, Emiline Sautron, Lucas Moyet, Patrice Catty, François André, Norbert Rolland, Edmund R.S. Kunji, and Annie Frelet-Barrand

Abstract

Due to their unique properties, expression and study of membrane proteins in heterologous systems remains difficult. Among the bacterial systems available, the Gram-positive lactic bacterium, *Lactococcus lactis*, traditionally used in food fermentations, is nowadays widely used for large-scale production and functional characterization of bacterial and eukaryotic membrane proteins. The aim of this chapter is to describe the different possibilities for the functional characterization of peripheral or intrinsic membrane proteins expressed in *Lactococcus lactis*.

Key words *Lactococcus lactis*, Membrane proteins, Expression, Transport assays

1 Introduction

In the past decades, *Lactococcus lactis*, a Gram-positive bacterium traditionally used in food fermentations, has emerged as a useful system for functional expression of prokaryotic and eukaryotic membrane proteins (MPs) [1]. *L. lactis* is an attractive alternative host for *Escherichia coli*, especially for eukaryotic MPs, because of (1) its moderate proteolytic activity, (2) the absence of inclusion body formation and endotoxin production [2, 3], (3) the efficient targeting of MPs into a single glycolipid cytoplasmic membrane [2, 4, 5], and (4) its ability to express MPs in their oligomeric state [2, 6]. This facultative anaerobe-aerobe lactic acid bacterium (LAB) grows at 30 °C with a doubling time of 35–60 min [7]; it is easy and inexpensive to grow and genetic methods and vector systems are well developed [8]. In addition to the classical cloning techniques, different strategies have been developed to obtain a larger number of recombinant clones [9, 10].

Isabelle Mus-Veteau (ed.), *Heterologous Expression of Membrane Proteins: Methods and Protocols*, Methods in Molecular Biology, vol. 1432, DOI 10.1007/978-1-4939-3637-3_6, © Springer Science+Business Media New York 2016

The expression of heterologous proteins in *L. lactis* has been facilitated both by advances in genetic methods and by new developments in molecular biology techniques. Using these tools, various vectors containing either constitutive or inducible promoters have been developed to obtain increased levels of proteins and to control their production. They currently constitute the basis of all expression systems in *L. lactis* [11]. The tightly regulated nisin-controlled gene expression (NICE) system is the most commonly used [12]. This promising and powerful expression system is based on genes involved in the biosynthesis and regulation of the antimicrobial peptide, nisin. When a gene of interest is placed upstream of the inducible promoter PnisA on a plasmid, its expression can be induced by the addition of subinhibitory amounts of nisin (0.1–5 ng/mL) to the culture medium [13]. The NICE system has proved to be highly versatile and is widely used in pharmaceutical, medical, and bio- and food-technology applications [14]. This well-characterized system is nowadays widely used for functional studies of homologously and heterologously expressed soluble and membrane proteins from diverse origins (prokaryotic or eukaryotic), topologies, and sizes (for reviews, see [1] and [14]). Moreover, in the last years, three structures have been obtained after expression of MPs using the NICE system [15–17] as well as several domain structures of human membrane proteins [18, 19].

In this chapter, we will give some examples of eukaryotic MPs for which functional analysis has been carried out after their expression in *L. lactis* using the NICE system. These functional characterizations can be performed on: (1) whole-cell bacteria, (2) membrane extracts, (3) fused membrane vesicles, (4) proteoliposomes after reconstitution of the MPs in phospholipids, using radioactive substrates, or (5) directly solubilized and purified membrane proteins [2, 3, 10, 12, 18, 19].

2 Materials

2.1 Growth of Recombinant Bacteria and Expression of the Proteins of Interest

1. *Lactococcus lactis* NZ9000 and nisin-producing NZ9700 strains (NIZO; *see* **Notes 1** and **2**).
2. M171GChl medium: M17 broth, 1% [w/v] glucose, 10 μg mL^{-1} chloramphenicol (*see* **Note 3**).
3. Laboratory glassware bottles (Schott bottles).
4. Incubator for cell growth.
5. Appropriate buffers for bacterial resuspension (*see* **Note 4**).

2.2 Isolation of Lactococcal Membranes with the Cell Disruptor

1. One Shot (Constant Cell Disruption Systems, Northants, UK) (*see* **Note 5**).
2. Appropriate buffers for protein resuspension (*see* **Note 4**).

2.3 SDS-PAGE and Detection of Recombinant Protein

1. Acrylamide-bis ready-to-use solution, 30% [w/v] (37.5:1).
2. 8× Laemmli resolving gel buffer: 3 M Tris–HCl pH 8.8 (60.6 g Tris–HCl resuspended in Milli-Q water; adjust to pH 8.8 at 25 °C with 12 N HCl. Store at room temperature).
3. 4× Laemmli stacking gel buffer: 0.5 M Tris–HCl pH 6.8 (363 g Tris–HCl resuspended in Milli-Q water; adjust to pH 6.8 at 25 °C with 12 N HCl. Store at room temperature).
4. Aqueous solution 20% [w/v] sodium dodecyl sulfate (SDS).
5. Ammonium persulfate: Prepare 10% [w/v] solution in water and immediately freeze in single-use (200 μL) aliquots at −20 °C.
6. Tetramethylethylenediamine (TEMED).
7. Resolving gels (10% acrylamide): 3.3 mL of 30% [w/v] acrylamide solution, 1.25 mL of 8× Laemmli resolving gel buffer, 50 μL of 20% [w/v] SDS, 5.3 mL of water, 10 μL of TEMED, and 100 μL of 10% [w/v] ammonium persulfate (*see* **Note 6**).
8. Stacking gels (5% [w/v] acrylamide): 2.8 mL of 30% [w/v] acrylamide solution, 1.25 mL of 4× Laemmli stacking gel buffer, 25 μL of SDS 20% [w/v], 2.8 mL of water, 5 μL of TEMED, and 50 μL of 10% [w/v] ammonium persulfate.
9. Laemmli running buffer (10×): For 1 L, 144.2 g of glycine (192 mM), 30.3 g of Tris–HCl (25 mM); add 50 mL of 20% [w/v] SDS (0.1% final concentration) and Milli-Q water. Store at room temperature.
10. Molecular weight marker.
11. Reducing sample buffer (4×): 0.08 M Tris–HCl, pH 6.8, 40% [v/v] glycerol, 1% [w/v] SDS, 0.1 mM bromophenol blue, 10 mM dithiothreitol. Store at −20 °C.
12. Sample buffer: 100 mL reducing buffer and 20 mL of 20% [w/v] SDS.
13. Control protein: Recombinant *Strep*-tag II fusion protein, MW about 28 kDa (0.1 mg mL^{-1}) (IBA, Goettingen, Germany).
14. System for protein transfer to nitrocellulose membranes (central core assembly, holder cassette, nitrocellulose filter paper, fibber pads, and cooling unit).
15. Protein transfer buffer: dilute running buffer 1× with ethanol to a final concentration of 20% [v/v]. Store at 4 °C.
16. Nitrocellulose or polyvinylidene difluoride (PVDF) membranes.
17. 3 MM paper from Whatman.
18. Protein-specific antibody or conjugate specific to the affinity tag (*see* **Note 7**).
19. Bio-Safe Coomassie (Biorad).
20. Electrochemiluminescence (ECL) detection kit.

2.4 Functional Characterization of Membrane Proteins Expressed in *L. lactis*

2.4.1 Dehydrogenase Assay on Purified Protein (ceQORH)

1. Solubilization buffer: 50 mM MOPS pH 7.8 containing 0.5 or 1 M NaCl.
2. 10 mM Tris–HCl pH 8.0.
3. Ni-NTA resin (Qiagen).
4. Binding buffer: 5 mM imidazole, 0.5 M NaCl, 20 mM Tris–HCl pH 7.9.
5. Wash buffer: 60 mM imidazole, 0.5 M NaCl, 20 mM Tris–HCl pH 7.9.
6. Elution buffer: 0.5 M imidazole, 0.25 M NaCl, 10 mM Tris–HCl pH 7.9.
7. Dehydrogenase reaction buffer: 100 μM NADPH, 100 μM nitroblue tetrazolium (NBT), 10 mM Tris–HCl pH 8.0.
8. PD10 column (GE Healthcare).
9. Eppendorf centrifuge.
10. Lipids (P3644, Sigma).
11. Spectrophotometer.

2.4.2 Phosphorylation Assays with AtHMA6 and AtHMA8 Using *L. lactis* Membranes

1. ATP phosphorylation buffer: 20 mM Hepes pH 7.0, 100 mM KCl, 5 mM $MgCl_2$, 300 mM sucrose (*see* **Note 8**).
2. Pi phosphorylation buffer: 20 mM Hepes pH 6.0, 10 mM $MgCl_2$, 20% [v/v] DMSO (*see* **Note 9**).
3. Metal solutions: prepare solutions at concentration ranging from 1 to 100 μM in phosphorylation buffer or water (*see* **Note 10**).
4. 1 mM ^{32}Pi (10–100 μCi nmol^{-1}, Perkin Elmer, 7 μL): Add 1 mL of 1 mM H_3PO_4 (prepared in 100 mM Hepes pH 5.6) directly in the tube containing the isotope ^{32}Pi. Filter the solution through a 0.2 μm membrane. To avoid loss of the solution in the filter, push the volume of the solution stayed in the filter using an empty syringe. Use 10 μL/reaction. Store at 4 °C.
5. 10 μM [γ-^{32}P]ATP (50–500 μCi nmol^{-1}, Perkin Elmer, 7 μL): for 10 reactions, add 1 μL of [γ-^{32}P]ATP to 110 μL of a solution of 10 μM ATP (*see* **Note 11**). Prepare a stock solution of 400 mM ATP in H_2O, aliquot in small volumes, and store at −20 °C. 10 μM ATP is prepared freshly from one stored aliquot of 400 mM ATP.
6. Stop buffer: 1 mM KH_2PO_4 in 7% [v/v] trichloroacetic acid (TCA). Store at 4 °C.
7. Denaturing buffer: 5 mM Tris-PO_4 pH 5.0, 6.7 mM urea, 400 mM DTT, 5% [w/v] SDS, 0.004% [w/v] orange G (*see* **Note 12**).
8. Chelator mix: 1 mM bicinchoninic acid (BCA) and 100 μM bathocuproine disulfonate (BCS) (*see* **Note 13**).

9. Resolving gel: 6 mL of Acrylamide/Bis 24/0.8% [w/v] solution (*see* **Note 14**), 4.5 mL of 4× resolving gel buffer (260 mM Tris-H_3PO_4 pH 6.5, 0.4% SDS), 7.4 mL of H_2O, 18 μL of TEMED, and 100 μL of 10% [w/v] ammonium persulfate.
10. Stacking gel: 1.2 mL of Acrylamide/Bis 24/0.8% [v/v] solution, 2 mL of 4× stacking gel buffer (260 mM Tris-H_3PO_4 pH 5.5, 0.4% SDS), 4.64 mL of H_2O, 7.5 μL of TEMED, and 160 μL of 10% [w/v] ammonium persulfate.
11. Running buffer: 0.17 mM MOPS (pH 6.0 adjusted with 2 M Tris), 0.1% [w/v] SDS. Store at 4 °C before use (do not store for more than 1 week) (*see* **Note 15**).
12. Acetic acid 15% [v/v].
13. Gel staining medium: acetic acid/isopropanol/water, 10/25/65 [v/v/v], supplemented with 2.5 $g.L^{-1}$ of Coomassie Brilliant Blue R250.
14. Gel destaining medium: 30% [v/v] ethanol.
15. Eppendorf centrifuge.
16. Gel electrophoresis apparatus (Bio-Rad Protean 3 or equivalent), with the various accessories needed for protein separation by electrophoresis (combs, plates, and casting apparatus).
17. Phosphorimaging device (*see* **Note 16**).

2.4.3 Transport Assays with Whole Cells Expressing AtAATP1/NTT1

1. 50 mM potassium phosphate buffer pH 7.0.
2. 3.33 nM [α-^{32}P]ATP (3000 mCi $mmol^{-1}$, Perkin Elmer). Add 1 μL to 50 mM potassium phosphate buffer pH 7.0 containing 1.5 mM cold ATP (A9062; Sigma) to obtain a 50 μM solution. Store on ice until use.
3. Filter membranes 0.45 μm (HAWP02500; Millipore).
4. Scintillation vials.
5. d.d. water.
6. Polymeric Vacuum Filter Holder (1225 Sampling Manifold; Millipore).
7. Multi-Purpose Scintillation Counter.

2.4.4 Transport Assays with Whole Cells and Fused Membrane Vesicles Expressing Mitochondrial Carriers

1. *E. coli* polar lipid extract (Avanti Polar Lipids).
2. Egg yolk phosphatidylcholine (Avanti Polar Lipids).
3. Nitrogen.
4. Diethyl ether.
5. Substrate/inhibitor, 10× stock (*see* **Note 17**).
   - ADP to give a final concentration of 5 mM
   - Carboxyatractyloside (CATR; Sigma) to give a final concentration of 2 μM

– Bongkrekic acid (BKA; Sigma) to give a final concentration of 2 μM

6. Extruder Set (Avanti Polar Lipids).
7. 1 μm polycarbonate filter (Whatman); filter supports (Whatman).
8. PD-10 column (GE Healthcare).
9. 2 mL Eppendorf tubes.
10. Hamilton robot (with vacuum manifold).
11. 96-well MultiScreenHTS-Hi Flow-FB opaque, Barex plastic plates (pore size = 1 μm; Millipore).
12. 96-well MultiScreenHTS-HA opaque, Barex plastic plates (pore size = 0.45 μm; Millipore).
13. 1.5 μM [^{14}C]-ADP (60 mCi mmol^{-1} = 2.22 GBq mmol^{-1}; Perkin Elmer) prepared in PIPES buffer to start the transport assays with mitochondrial carriers.
14. BackSeal black backing paper (Perkin Elmer).
15. MicroScint-20 (Perkin Elmer).
16. MultiScreen sealing tape, clear backing paper (Perkin Elmer).
17. TopCount (Perkin Elmer).

3 Methods

3.1 Growth of Recombinant Bacteria and Expression of the Proteins of Interest

The gene(s) of protein(s) of interest have to be cloned first into an expression vector containing nisin-inducible promoter either through classical cloning methods [2] or other strategies developed in the last years to facilitate cloning [9, 10]. First trials could be performed following protocols already described [2, 10, 20–22]. Here we present protocols that can be optimized for each protein of interest.

1. Inoculate M17G1Chl precultures with concentrated glycerol stocks or frozen cell stocks of recombinant bacteria carrying the gene of protein of interest and bacteria carrying the empty vector as negative control.
2. Incubate overnight at 30 °C without shaking (*see* **Notes 18** and **19**).
3. Inoculate M17G1Chl with 1/40^{e} precultures (*see* **Note 20**).
4. Incubate cultures at 30 °C; measure OD_{600nm} every 45 min to construct growth curve (doubling roughly every 45 min).
5. Induce protein expression by addition of homemade nisin (*see* **Note 21**) at OD_{600nm} from 0.5 to 0.8 depending on proteins of interest (*see* **Note 22**).

6. Swirl immediately to prevent cell lysis; return the flasks to the 30 °C incubator for further 2.5–4 h (*see* **Note 23**).
7. Depending on functional tests, bacteria are either centrifuged, resuspended with an appropriate buffer (*see* **Note 4**) and centrifuged again before storage at −20 °C (*see* Subheading 3.2), or directly used (*see* Subheading 3.4.1) or snap-frozen and stored in liquid nitrogen (*see* Subheading 3.4.4).

3.2 Isolation of Lactococcal Membranes with the Cell Disruptor

1. Resuspend the bacteria into the appropriate buffer (*see* **Note 4**).
2. Disrupt the bacteria by twofold passages through a One Shot at 35,000 p.s.i. (2.3 kbars).
3. Centrifuge 100,000 × *g*, 15 min, 4 °C and transfer the supernatant into ultracentrifuge tubes.
4. Centrifuge 100,000 × *g*, 1 h, 4 °C; resuspend the pellet and homogenize into the appropriate buffer (*see* **Note 4**).
5. Snap-freeze and store at −80 °C or in liquid nitrogen until use.

3.3 Detection of the Recombinant Protein Produced

1. Prior to the experiment, prepare acrylamide gels for protein electrophoresis, the gel apparatus according to the manufacturer's specifications, and the different gel solutions (stacking gel, acrylamide separating gel; *see* **Note 6**).
2. Heat the protein samples at 95 °C for 5 min to solubilize the proteins (*see* **Note 24**). Load protein samples, one molecular weight marker, and positive controls in defined quantities.
3. Run gels for 1 h at room temperature at 150 V with constant voltage (*see* **Note 25**).
4. After electrophoresis, perform the transfer for 1 h 30 min at 100 V in protein transfer medium prior to Western blotting analysis.
5. Recover the nitrocellulose membrane and rinse the membrane with water. The following incubation and washing steps require agitation on a rocking plate at room temperature.
6. Perform Western blotting analysis and/or Coomassie blue staining using protocols already established or given by manufacturers.
7. Perform ECL detection (Figs. 1 and 2)

3.4 Functional Characterization of Membrane Proteins

3.4.1 Dehydrogenase Assay on Purified Protein (ceQORH)

The chloroplast envelope Quinone OxidoReductase Homologue (ceQORH) protein from *Arabidopsis thaliana* is a peripheral protein associated with the chloroplast envelope through electrostatic interactions [24]. This nuclear-encoded protein is devoid of a classical and cleavable transit peptide and uses an alternative targeting pathway for its import into the chloroplast [25, 26]. The ceQORH protein is structurally related to bacterial, fungal, and animal proteins with known quinone oxidoreductase function. In an earlier

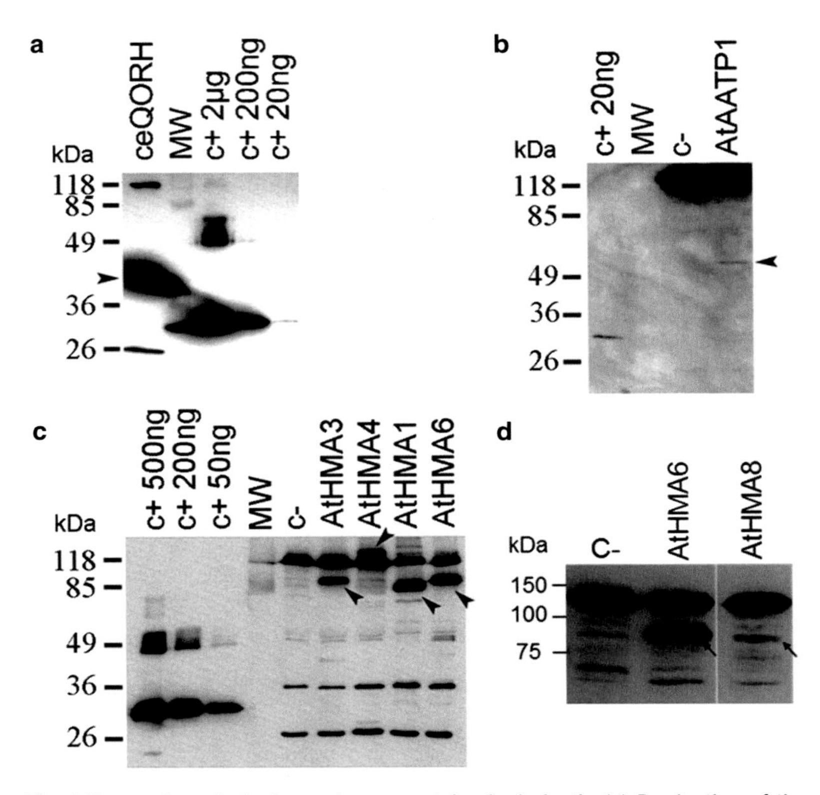

Fig. 1 Expression of plant membrane proteins in *L. lactis*. (**a**) Production of the peripheral protein ceQORH. (**b**–**d**) Production of Arabidopsis transmembrane proteins (the nucleotide transporter, AtAATP1 in panel **b** and the P_{1B}-ATPases in panels **c** and **d**). Total membrane proteins (10 µg, 20 µg, 15 µg, and 50 µg in panels **a**, **b**, **c**, and **d** respectively) were separated in a 10 % SDS-PAGE and analyzed by western blot performed using a horseradish peroxidase (HRP) conjugate specific to *Strep*-tag II. *Arrows* indicate the positions of the expressed proteins. In panels **a** and **b**, expressed proteins contain an additional N-terminal sequence resulting from the translation of the attB sites. In panels **b**, **c**, and **d**, c- meanscrude membrane proteins derived from bacteria containing the empty pNZ8148 vector. In panels **a**, **b**, **c**, defined amounts of a positive control protein (c+, *Strep*-tag II protein) were loaded to estimate the expression levels of the recombinant proteins. Adapted from [10] and [23]

work, an NADPH quinone oxidoreductase activity was detected in the chloroplast envelope [27], and this activity was assumed to be associated with the presence of the ceQORH protein. To investigate its enzymatic properties, we first expressed it in *Escherichia coli* but the protein was mostly recovered in inclusion bodies. By contrast, in *L. lactis*, the ceQORH protein was well produced and recovered in the membrane fraction, representing 20–30 % of total membrane proteins. This relatively high percentage is closed to levels of expression obtained for prokaryotic membrane proteins produced in *L. lactis* [1]. Performing dehydrogenase activity assays

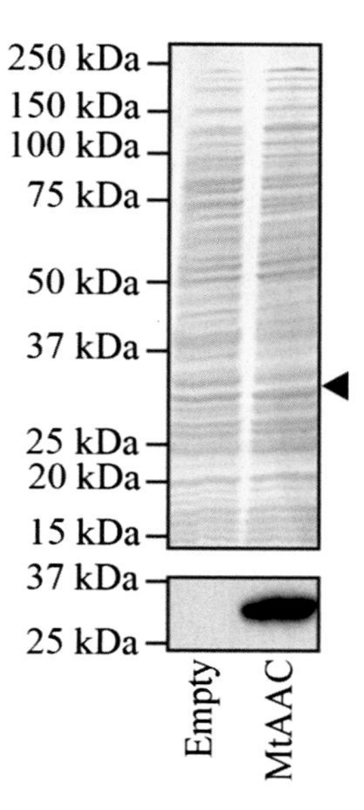

Fig. 2 Expression of the mitochondrial ADP/ATP carrier from the thermophilic fungus *Myceliophthora thermophile* (MtAAC). Instant blue-stained 12 % SDS-PAGE gel (*top*) and western blot (*bottom*) of lactococcal membranes of the control strain (without protein) and strain expressing MtAAC. The *arrowhead* indicates the position of the expressed protein. Approximately 15 μg of total protein was loaded per lane. For Western blotting, the proteins were transferred to PVDF membranes and probed with a 1:20,000 dilution of a chicken IgY antibody for 1 h, followed by an anti-chicken-HRP conjugate at 1:20,000 dilution for another hour. The blotted membrane was developed using Amersham ECL Western blotting detection system for 30 min

on the purified protein using artificial substrates (NBT), we were able to establish that the ceQORH protein indeed exhibits NADPH-dependent dehydrogenase activity, and that this activity requires a lipid environment [10].

1. Incubate *L. lactis* membrane proteins (1 mg mL^{-1}) in 50 mM MOPS pH 7.8 containing 0.5 or 1 M NaCl for 45 min at 4 °C (*see* **Note 26**). Mix gently the sample every 15 min.
2. Following treatment, centrifuge membranes at 160,000 × *g*, for 1 h, at 4 °C to separate solubilized proteins in the supernatant from insoluble endogenous membrane proteins in the pellet. Keep the supernatant for further purification steps (Fig. 3a).
3. Desalt the supernatant on a PD-10 column against Binding buffer.
4. Pre-equilibrate the Ni-NTA resin in Binding buffer (*see* **Note 27**).
5. Incubate the solubilized proteins (*see* **Note 28**) with the pre-equilibrated Ni-NTA resin and mix gently by shaking (200 rpm on a rotary shaker) at 4 °C for 60 min.

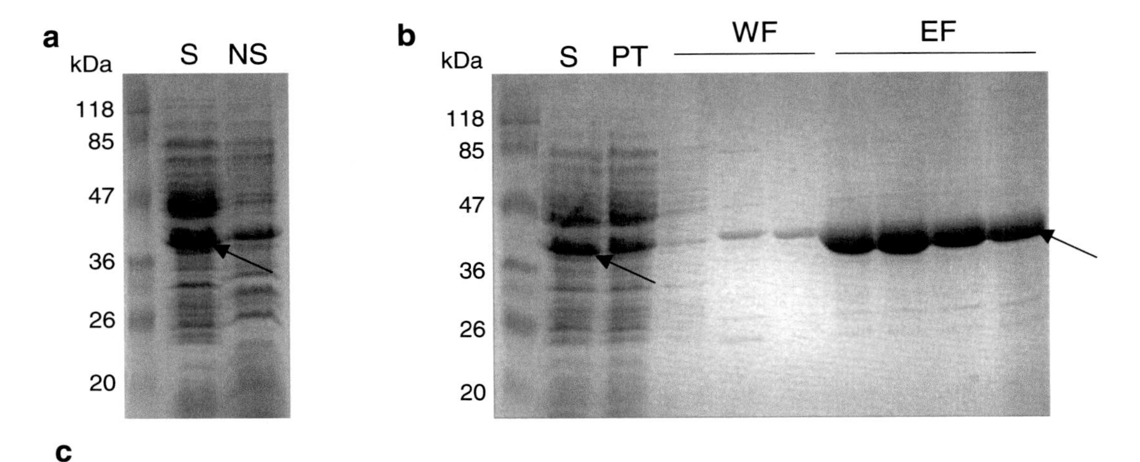

c

Dehydrogenase activity of the ceQORH protein produced in L. lactis	
System	Specific activity of the ceQORH protein (nmol substrate/mg protein/min)
Complete	19.28 (100%)
Without ceQORH	0 (0%)
Without PC	1.97 (10,2%)
Without NADPH	0 (0%)
Without NBT	1.14 (5,9%)

Fig. 3 Functional characterization of the purified plant peripheral protein ceQORH. (**a**) *Lactococcus* membranes expressing the 6× His-tag ceQORH are solubilized in the presence of 1 M NaCl. After centrifugation, the solubilized (S) and nonsolubilized (NS) proteins are analyzed by SDS-PAGE and Coomassie blue staining of the gel. The 6× His-tag ceQORH protein is indicated by the arrow. (**b**) Purification of the 6× His-tag ceQORH protein using a Ni-NTA resin. Each fraction obtained during the purification process is analyzed by SDS-PAGE. PT: solubilized proteins not bound to the resin, WF: wash fractions, EF: elution fractions. (**c**). Dehydrogenase activity of the purified recombinant 6× His-tag ceQORH protein using the artificial substrate NBT. The impact of the lack of every reaction components (protein, phosphatidylcholine (PC), NAPDH, NBT) was tested. The specific activity determined for the 6× His-tag ceQORH protein (100 %) was 19.3 nmol of formazan/mg 6× His-tag ceQORH protein/min. Adapted from [10]

6. Load the solubilized proteins/Ni-NTA mixture onto a column.
7. Wash twice with Washing buffer.
8. Elute the His-tagged ceQORH protein with Elution buffer (*see* Fig. 3b for a typical purification).
9. Desalt the eluted His-tag ceQORH protein against 10 mM Tris–HCl pH 8.0.

For dehydrogenase activity assay (see **Note 29**):

10. Incubate 1 μg of purified ceQORH with 1 μg of lipids at 30 °C for 30 min (final volume of 20 μL).
11. Start the reaction by addition of 150 μL of 10 mM Tris–HCl containing 100 μM NADPH and 100 μM NBT.

12. Follow the formation of Formazan at 560 nm using a Spectrophotometer.
13. Deduce the enzymatic activities from the OD measurement using the molar absorption coefficient (*see* **Note 30**, Fig. 3c).

3.4.2 Phosphorylation Assays with AtHMA6 and AtHMA8 Using *L. lactis* Membranes

P_{IB}-ATPases (reviewed in [28]) belong to the large family of P-type ATPases that are transmembrane proteins responsible for the transport of ions and phospholipids across plasma and organelle membranes using the energy provided by ATP hydrolysis. Like all P-type ATPases, P_{IB}-ATPases (or HMA for Heavy Metal ATPase) are composed of a transmembrane domain M containing the transport site and determining ion selectivity, and of three cytosolic loops constituting the catalytic domain. P_{IB}-ATPases have six to eight predicted transmembrane helices and have been classified into several subgroups according to their ionic specificity [29]. Recently, the crystal structure of a prokaryotic P_{IB}-ATPase, LpCopA, has been solved providing new topological information on these enzymes [30]. In *Arabidopsis* species, AtHMA6 and AtHMA8 are two chloroplastic ATPases of the PIB-1 subgroup involved in Cu transport across the chloroplast envelope and the thylakoid respectively [31, 32]. The enzymatic properties of these two transmembrane proteins could be assessed using in vitro biochemical assays [23, 33] after their successful and efficient expression in *Lactococcus lactis*. Phosphorylation assays performed on lactococcal membranes expressing these exogenous P_{IB}-ATPases (*see* **Note 31**) could provide information about (1) the kinetic parameters of the enzyme using phosphorylation from ATP (*see* Fig. 4a–c), and (2) the apparent affinity for the translocated metal using phosphorylation from Pi (Fig. 4b). These phosphorylation assays can be performed on all P_{IB}-ATPases whatever their ionic specificity to assess their enzymatic properties.

Phosphorylation from ATP:

1. Prepare a mix containing 0.5 mg mL^{-1} of *L. lactis* membranes, with metals or chelators at the desired concentration (*see* **Note 32**) and complete with ATP phosphorylation buffer to a final volume of 90 μL.
2. Start the reaction by addition of 10 μL of 10 μM [γ-^{32}P]ATP (1 μM final). Vortex the suspension.
3. Stop the reaction 30 s later (*see* **Note 33**) by addition of 1 mL ice-cold Stop buffer. Vortex the suspension and incubate 30 min on ice.
4. Centrifuge for 15 min, at 15,000 × *g*, 4 °C and keep the pellet (*see* **Note 34**).
5. Wash the pellet with 1 mL of ice-cold Stop buffer and centrifuge for 15 min, at 10,000 × *g*, 4 °C.

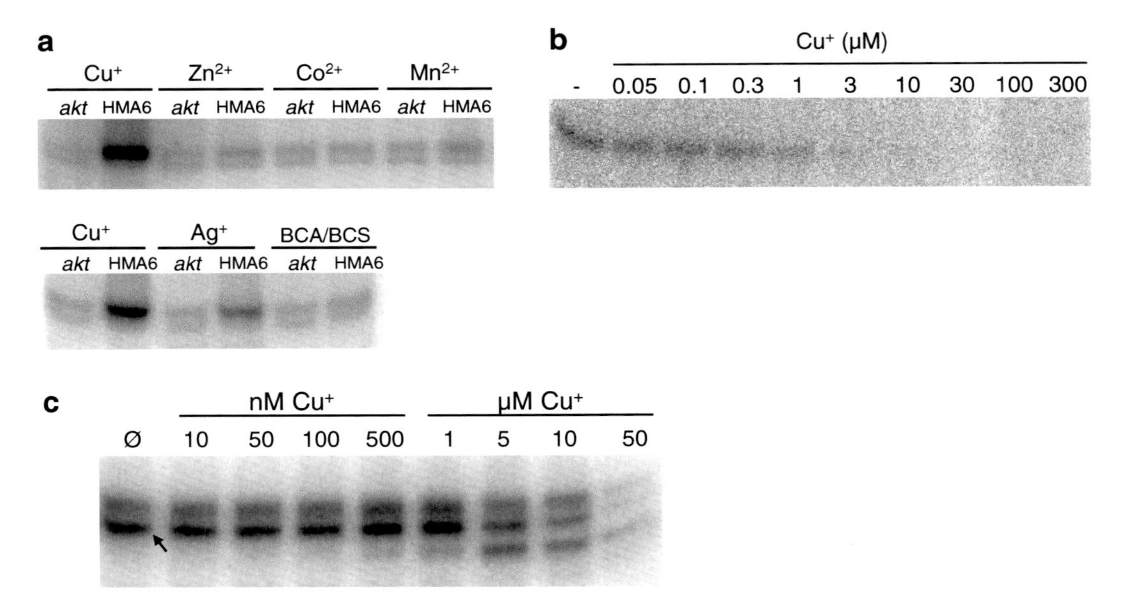

Fig. 4 Functional characterization of plant PIB-ATPases, AtHMA6 and AtHMA8 using phosphorylation assays. (**a**) Phosphorylation assays from ATP on lactococcal membranes expressing HMA6 and an inactive form of HMA6 ("akt") in the presence of 5 μM of various metals. Phosphorylation signal was observed only in the presence of Cu^+ and to a lesser extend with Ag^+ (*see* [33] for more information). (**b**) Phosphorylation assays from Pi on lactococcal membranes expressing HMA6 and in presence of various concentration of Cu^+. In that case, phosphorylation was inhibited by the transported metal, here Cu^+ [33]. (**c**) Phosphorylation assays from ATP on lactococcal membranes expressing HMA8. Phosphorylation signal of HMA8 is indicated by the *arrow*. One can notice that HMA8 was phosphorylated in the absence of added Cu^+ in the media (Ø) and that its phosphorylation was inhibited by high concentration of Cu^+ (*see* [23] for more details)

6. Repeat the wash step once.
7. Resuspend the pellet in 35 μL of Denaturing buffer. Vortex vigorously during 1–2 min.
8. Load the samples onto an acidic SDS-acrylamide gel (*see* **Note 35**).
9. After electrophoresis (about 1 h at 40 mA per gel), incubate the gel in the acetic acid solution for 10 min at room temperature.
10. Place the gel between "saran film" and expose it against a phosphorimager screen (*see* **Note 16**) in common film cassette overnight at room temperature.
11. Analyze the phosphorylation signal using a phosphorimaging device (*see* Fig. 4a–c).
12. Stain the gel (to check the amount of loaded proteins) by incubation for 30 min in the Gel staining medium and then in the Gel destaining medium.

Phosphorylation assays from Pi:

1. Prepare a mix containing 1 mg mL^{-1} of *L. lactis* membranes, metals, or chelators (*see* **Note 32**) in Pi phosphorylation buffer to a final volume of 90 μL.
2. Incubate the mix 5 min at 30 °C.
3. Start the reaction by addition of 10 μL of 1 mM ^{32}Pi (100 μM final). Vortex the suspension and incubate the reaction mix at 30 °C for 10 min.
4. Stop the reaction by addition of 1 mL ice-cold stop buffer. Vortex the suspension and incubate 30 min on ice.
5. Then proceed as described in **steps 4–12** for phosphorylation assays from ATP (Fig.4b).

3.4.3 Transport Assays with Whole Cells Expressing AtAATP1/NTT1

The nucleotide transporter 1, AtAATP1/NTT1, a highly hydrophobic membrane protein with 12 predicted transmembrane domains [34], is localized within the inner membrane of the chloroplast envelope [35]. This translocator imports ATP in exchange of ADP and Pi [36] in contrast to mitochondrial ATP/ADP translocators [37]. It supplies energy to chloroplasts used by storage plastids required for starch synthesis and to allow nocturnal anabolic reactions in chloroplasts [38].

1. Centrifuge 4350 × *g*, 10 min twice and resuspend recombinant bacteria into 30 mL to a final concentration of 100 mg mL^{-1} (3 mg/30 μL) in ice-cold 50 mM potassium phosphate buffer pH 7.0.
2. Add 50 μM [α-^{32}P]ATP (3000 mCi $mmol^{-1}$; Perkin Elmer) diluted in ice-cold 50 mM potassium phosphate buffer pH 7.0 to each sample (*see* **Note 36**).
3. Incubate at 25 °C for planned time periods and stop nucleotide uptake by addition of 1 mL of ice-cold potassium phosphate buffer.
4. Filtrate the cells through a 0.45 μm filter under vacuum.
5. Wash three times with 1 mL of ice-cold potassium phosphate buffer.
6. Transfer filter to a scintillation vial and add 3.5 mL of d.d. water
7. Measure the radioactivity retained on the filters in a scintillation counter.
8. Generate graphs using the KaleidaGraph version 4.02 (Synergy Software) and fit experimental data with the appropriate curve (i.e., single exponential) (Fig. 5).

3.4.4 High-Throughput Transport Assays of Mitochondrial Carriers

Mitochondrial carriers link the biochemical pathways of the cytosol and the mitochondrion matrix by transporting metabolites, nucleotides, inorganic ions, and cofactors across the mitochondrial inner

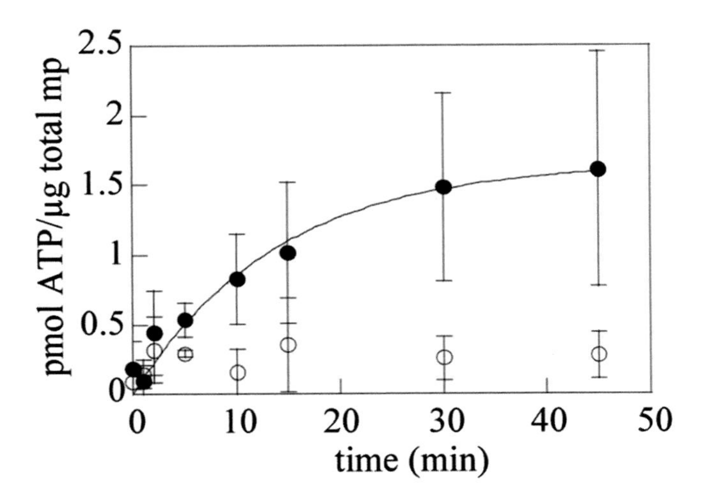

Fig. 5 Functional characterization of the nucleotide translocator AtAATP1/NTT1. *L. lactis* cells expressing AtAATP1/NTT1 (*closed circles*) or *L. lactis* cells containing the empty vector (as a negative control, *open circles*) were incubated with 50 mM [α-32P] ATP for the indicated time periods. Data are represented by the mean and standard deviation of four independent experiments. Adapted from [10]

membrane [39, 40]. The topology is pseudo-symmetric threefold, consisting of a barrel of six transmembrane α-helices with three short α-helices in the matrix loops [41]. Mitochondrial carriers function as monomers [42], but the aspartate/glutamate carrier was found to be dimeric in its resolved structure [18]. A single substrate-binding site has been identified in the central part of the cavity [43], which is located between two salt bridge networks on the matrix and cytoplasmic side of the carriers [44]. The opening and closing of the carrier could be coupled to the disruption and formation of the two salt bridge networks, changing the accessibility of the central substrate to either side of the membrane in an alternative way [44, 45]. Many mitochondrial carriers have not yet been identified and many aspects of their transport mechanism are unsolved. Therefore, it is important to develop transport assays that allow their characterization in high-throughput experiments.

Transport assays can be performed either with washed whole lactococcal cells when *L. lactis* has an endogenous pool of substrates, e.g., adenosine di- and tri-phosphate nucleotides or with fused membrane vesicles. For the latter, membranes are isolated, fused with liposomes by freeze thawing, and extruded in the presence of internal substrate to create single membrane vesicles, after which the external substrate is removed by gel filtration. In both cases transport assays are initiated by the addition of radiolabeled substrate. The automated procedures are illustrated with the characterization of the mitochondrial ADP/ATP carrier from *Myceliophthora thermophila*, a thermophilic fungus.

For transport assays on whole cells, snap-thaw the cell pellets (*see* Subheading 3.1) by gentle pipetting whilst the falcon tube is immersed in warm water (*see* **Note 37**) and adjust the OD_{600nm} to approximately 25.

For transport assay on ***fused membrane vesicles***, liposomes have to be prepared prior to the fusions with *L. lactis* membranes isolated (*see* Subheading 3.2) to form fused membrane vesicles.

1. Allow the frozen lipids to defrost at room temperature.
2. Prepare liposomes in a 3:1 weight ratio of *E. coli* polar lipid extract to egg yolk phosphatidylcholine; dry chloroform under a stream of nitrogen, ensuring all the chloroform to be evaporated.
3. Add 5 mL diethyl ether, vortex, and evaporate under a stream of nitrogen.
4. Vortex the lipids for 3 h in PIPES buffer to get a 20 mg mL^{-1} stock and homogenize.
5. Aliquot, flash-freeze, and store in liquid nitrogen.
6. Use 5 mg liposomes (250 μL of 20 mg mL^{-1} stock) and 1 mg of isolated *L. lactis* membranes (*see* Subheading 3.2); dilute to 1 mL with PIPES buffer.
7. Fuse the membranes by 6 cycles of snap-freezing in liquid nitrogen, and thawing to room temperature for 45 min; the fused membranes can be stored in liquid nitrogen.
8. Thaw the membranes, and add 50 μL of a 10× stock of substrate/inhibitor to 450 μL fused membranes for a final volume of 500 μL.
9. Equilibrate a 1 μm polycarbonate filter in PIPES.
10. Sandwich two supports (four needed in total) on either side of filter, assemble the extruders and push through some buffer to remove any air pockets and extrude each sample 11 times at room temperature (*see* **Note 38**); keep samples on ice before and after extrusion.
11. Prepare the PD-10 column; equilibrate with buffer as per instructions.
12. Gently add 400 μL post-extrusion vesicles onto each column; let the vesicles absorb into the column, add a further 2.1 mL buffer, and discard the flow-through.
13. Let the column run dry, add 1.6 mL of buffer, and collect the sample in a 2 mL Eppendorf tube.
14. Dilute each sample fourfold before use (1.5 mL into a final volume of 6 mL PIPES, which will give approximately 5 μg protein per well if using 100 μL vesicles, and enough protein for quadruple trials with 11 time points per trials); keep remaining undiluted sample for BCA protein determination.

High-throughput transport assays can be carried out using a Hamilton MicroLab Star robot. The first five steps are programmed to be carried out by the robot in 96-well format, allowing eight different uptake experiments to be performed simultaneously.

1. Pipet 100 μL bacteria (OD_{600nm} = 25) into the wells of a 96-well $MultiScreen_{HTS}$-Hi Flow-FB plate (pore size = 1 μm) or fused membrane vesicles (5 μg) into a 96-well $MultiScreen_{HTS}$-HA plate (pore size = 0.45 μm), while the plate is placed on a vacuum manifold.
2. Initiate transport with the addition of 100 μL PIPES buffer containing 1.5 μM of ^{14}C-labeled nucleotide.
3. Incubate at room temperature for planned time periods, the longer time points being added first.
4. Stop transport by filtration followed immediately by the addition of 200 μL of ice-cold PIPES buffer to all wells.
5. Wash wells two times with 200 μL ice-cold PIPES buffer.
6. Leave the plates to dry overnight.
7. Stick black backing paper on the underside of the filter plate.
8. Add 200 μL of MicroScint-20 to each well using the robot.
9. Stick clear backing paper on the topside of the filter plate; stand for at least 4 h to allow filter dissolution.
10. Load the plates into the TopCount Scintillation Counter. Initial rates are determined from the linear part of the uptake curves (first 60 s) (*see* Fig. 6).

4 Notes

1. Nisin can be either commercial or produced by the nisin-producing strain NZ9700. We have noticed that homemade nisin gave rise to higher amounts of proteins compared to commercial one [21].
2. Recombinant bacteria are generated through transformation with an expression vector containing the gene of interest; in our studies, we use pNZ8148.
3. The concentration of antibiotic (chloramphenicol) could vary from 5 to 10 μg mL^{-1}.
4. The buffers for bacterial and protein resuspension have to be adjusted to the protein of interest (50 mM Tris–HCl pH 8.0 for ceQORH [10] and 20 mM HEPES pH 6.0 for AtHMA6 and AtHMA8 [33] containing 6 or 20% glycerol for protein resuspension respectively; 50 mM potassium phosphate buffer pH 7.0 for AtNTT1 [10]; 10 mM PIPES pH 7.0, 50 mM NaCl for mitochondrial carriers [3]).

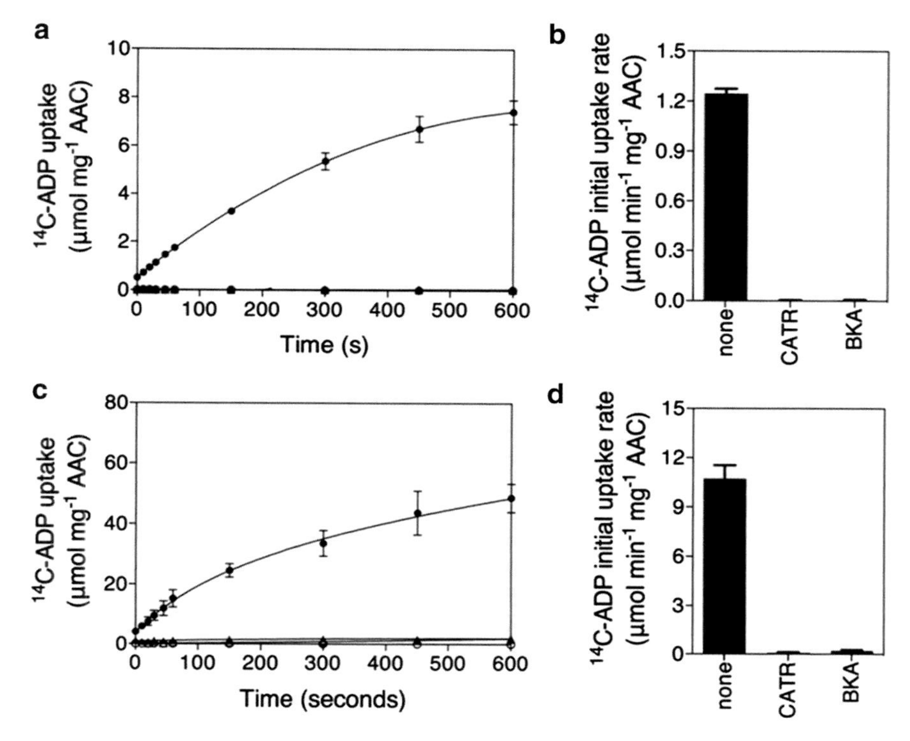

Fig. 6 Functional characterization of the mitochondrial ADP/ATP carrier from *Myceliophthora thermophile* (MtAAC) expressed in *L. lactis* by high-throughput transport assays. (**a**) Whole-cell uptake assays using *L. lactis* expressing MtAAC in the absence (*closed circles*) or presence of the specific inhibitors carboxyatractyloside (CATR; *closed triangles*) or bongkrekic acid (BKA; *open triangles*), both at 20 µM. The control shown is vector without the insert encoding MtAAC (*empty circles*). (**b**) Specific initial rates of ^{14}C-ADP uptake on whole cell over the first 60 s in the absence (none) or the presence of inhibitors (CATR or BKA), background subtracted. (**c**) Transport assays with fused membrane vesicles of *L. lactis* expressing MtAAC preloaded with 5 mM ADP in the absence (*open circles*) or presence of CATR (*closed triangles*) or BKA (*open triangles*). As control, transport was carried out with vesicles expressing MtAAC, but not loaded with internal substrate (*open circles*). (**d**) Specific initial of ^{14}C-ADP uptake on fused membrane vesicles over 60 s in the absence (none) or in the presence of inhibitors (CATR or BKA), background subtracted. Transport was initiated with the external addition of 1.5 µM ^{14}C-ADP. The *error bars* represent the standard deviation of four assays

5. One Shot and other disruption systems from Constant Cell Disruption System are the most suitable to disrupt the lactococcal cell wall with a yield in crude membranes improved by more than fivefold when compared to that obtained by lysozyme and French Press treatment [10].
6. The percentage of acrylamide in the resolving gel has to be adapted to the molecular weight of the protein of interest.
7. Western blot analysis can be performed using either protein-specific antibody or conjugate specific to the affinity tag.
8. HEPES buffer is preferentially used since it has low affinity for metals. $MgCl_2$ is required to favor Mg-ATP as substrate.

9. KCl is prohibited in the Pi phosphorylation buffer, since it accelerates the dephosphorylation rate.
10. According to the ionic specificity of the P_{IB}-ATPases, various metal solutions can be used (i.e., $CuSO_4$/$CuCl_2$, $AgNO_3$/AgCl, $ZnSO_4$, $MnSO_4$, or $NiCl_2$). Usually, different dilutions are prepared to perform the phosphorylation tests in the presence of metal concentrations ranging from 0.05 to 100 μM. To test Cu in its monovalent form Cu^+, a reducing agent has to be added in the reaction mix. Preferentially use Na_2SO_3 at 500 μM. Note that the pH of the solution can change the speciation of the metal (*see* Pourbaix diagrams) and thus its chemical properties (i.e., its solubility and then its interaction with the ATPase).
11. For phosphorylation assays on P_{IB}-ATPase well expressed in *Lactococcus* membranes, half amount of radiolabeled ATP can be used (i.e., 0.5 μL in 110 μL of unlabeled 10 μM ATP).
12. DTT is sensitive to oxidation and can then lose its reducing power within prolonged exposure to the air. This may result in an insufficient denaturation of the sample.
13. Metal chelators are used as controls. BCA/BCS are usually used as monovalent metal chelators and Cu chelators, whereas EGTA is used as a broad range divalent metal chelator. These chelators are therefore used to inhibit the catalytic cycle of ATPases. BCA and BCS are prepared in H_2O at a concentration of 1 mM and 100 μM respectively, stored in small aliquots at −20 °C.
14. The mix Acrylamide/Bis 24/0.8 % is prepared from two independent solutions: 60 mL of 40% [w/v] Acrylamide and 40 mL of 2 % [w/v] Bis-acrylamide. Premixed solutions of acrylamide/Bis-acrylamide can also be used.
15. The pH value of the gel and the running buffer must be lower than 7 to avoid the hydrolysis of the aspartyl-phosphate bound.
16. Instead of using phosphorimaging device, it is possible to use autoradiography films, exposed in common film cassettes and then revealed using developer and fixer solutions.
17. The mitochondrial ADP/ATP carriers exchange ADP against ATP and they are inhibited by carboxyatractyloside (CATR) and bongkrekic acid (BKA) [46].
18. Shaking with a gentle rotation (90 rpm) avoids the sedimentation of bacteria at the bottom of the bottle.
19. The M17 medium could be prewarmed at 30 °C overnight. This step allows gaining time to reach the correct OD. Nevertheless, it might be better to add glucose and antibiotics on the day of culture because of the instability of antibiotic at 30 °C, and to check if the medium is not contaminated.

20. Cultures can be inoculated with a larger volume of precultures (1/20[e] to get an OD600 around 0.1). This can help to shorten the time to get the appropriate OD for induction.
21. The dilution of nisin to add for induction (0.0001–0.005) has to be optimized in order to get higher amounts of proteins produced and must be determined for every new preparation of nisin [47].
22. Several values for OD_{600nm} have been reported for induction of protein expression: from 0.5 for mitochondrial carriers [2] to 0.8 for plant membrane proteins [10].
23. The induction time needs to be optimized for each protein of interest. For the proteins analyzed here, an induction time of 2.5 and 4 h allowed production of sufficient amounts of protein for analyses of mitochondrial carriers and plant membranes proteins respectively.
24. Heating of the samples for 5 min at 95 °C is not necessary. Solubilization also works with an incubation of the samples in the sample buffer for 5 min at room temperature.
25. The migration parameters (voltage/time) have to be adapted to the protein of interest.
26. Membrane proteins either peripherally or intrinsically associated with membranes have to be solubilized to become soluble in aqueous solution before purification steps. The ceQORH protein is a peripheral protein which interacts with the inner envelope membrane of chloroplast through electrostatic interactions. This kind of protein can be solubilized with mild treatments using high salt-containing solution [24].
27. For more information concerning the purification of 6× Histidine-tag proteins, refer to the handbook of IBA (IBA, Goettingen, Germany).
28. Here we used a 6× Histidine-tag for the detection and purification of the recombinant proteins (this tag was fused to the *N*-terminal part of the protein). However, other affinity tags can be used like the *Strep*-tag II [10].
29. Dehydrogenase assays are performed with NBT chloride as an artificial substrate, and the reaction (NBT reduction to formazan) is monitored using a spectrophotometer. An absorption coefficient of 15,000 M^{-1} cm^{-1} for the formazan product at 560 nm is used to calculate the initial velocity [48]. Alternatively, NADPH oxidation can be recorded in the initial linear phase of decay of absorbance at 340 nm using a molar extinction coefficient of 6250 M^{-1} cm^{-1}.
30. The specific activity determined for the purified ceQORH protein was 19.3 nmoles of formazan per mg of ceQORH protein. min^{-1}. Note that the specific activity of the ceQORH protein

produced in *E. coli* was found 2.5 times lower than that of the ceQORH protein produced in *L. lactis* [10].

31. AtHMA6 and AtHMA8 belong to the family of P-type ATPases, that are multispanning membrane proteins that translocate ions across plasma or organelle membranes at the expense of ATP consumption [49]. The catalytic cycle of P-ATPases can be schematically reduced to a four-step process accounting for the coupling of ion motion to ATP hydrolysis, involving the formation of transient phosphorylated states initiated by ATP γ-phosphate transfer to the conserved aspartic residue of the DKTGT motif. In its free state (E), the P-ATPase binds cytoplasmic ions at its high affinity membrane site (**step 1**). ATP, bound to the large cytoplasmic domain of the transporter, is then hydrolyzed, leading to the formation of a phosphorylated enzyme (**step 2**). The ion bound phosphorylated form of the enzyme (Me.E ~ P) undergoes important conformation changes leading to metal release at the extracytoplasmic side of the membrane (**step 3**). In the metal-free phosphorylated enzyme E-P, the aspartyl-phosphate bound is then hydrolyzed to bring the enzyme back to its free state E (**step 4**). In many cases, ATPases can be phosphorylated from ATP in the forward direction (physiological condition; **steps 1–4** of the cycle) and from Pi in the backward direction (**step 4** only). Both reactions rely on the presence of the ion to be transported. Whereas phosphorylation from ATP only occurs on the ion(s) bound form of the transporter, transported ion(s) do competitively inhibit phosphorylation from Pi. Hence, the latter also gives access to the ionic specificity of the enzyme. In addition, involving simple equilibrium reaction (**step 4**), phosphorylation from Pi also allows to determine more precisely the apparent affinity of the transporter to the transported ion.
32. Usually it is better to test a wide range of metal concentration from 0.05 to 100 μM (concentration of the metal in the assay, i.e., in 100 μL). If needed in the assay, chelators are added to a final concentration of 1 mM (EGTA) and 1 mM/100 μM (BCA/BCS mix).
33. Depending on the activity of the studied P_{IB}-ATPase, the reaction time can vary. The incubation can also be performed at 4 °C to slow down the reaction.
34. The supernatant highly radioactive has to be removed carefully.
35. The use of acidic condition for the SDS-PAGE analysis is essential to avoid the hydrolysis of the aspartyl-phosphate bound and to detect the phosphorylated intermediates.
36. Each condition has to be tested in triplicate experiments at minimum.

37. Quickly thawing the pellets is vital for transport activity.
38. The syringes are not airtight at 4 °C for prolonged periods of time and will leak.

Acknowledgments

This study received financial support from the Commissariat à l'Energie Atomique et aux Energies Alternatives (CEA), the Centre National de la Recherche Scientifique (CNRS), the French National Institute for Agricultural Research (INRA), the University Joseph Fourier (Grenoble), and the Medical Research Council UK. E.S. was funded by a joint grant from the GRAL Labex (Grenoble Alliance for Integrated Structural Cell Biology: ANR-10-LABEX-04) and the CEA.

References

1. Bakari S, André F, Seigneurin-Berny D, Delaforge M, Rolland N, Frelet-Barrand A (2014) Lactococcus lactis, recent developments in functional expression of membrane proteins. In: Mus-Veteau I (ed) Membrane proteins production for structural analysis. Springer, New-York, USA, pp 107–132, Chapter 5
2. Kunji ERS, Slotboom DJ, Poolman B (2003) Lactococcus lactis as host for overproduction of functional membrane proteins. Biochim Biophys Acta 1610:97–108
3. Kunji ERS, Chan KW, Slotboom DJ, Floyd S, O'Connor R, Monné M (2005) Eukaryotic membrane protein overproduction in Lactococcus lactis. Curr Opin Biotechnol 16:546–551
4. Monné M, Chan KW, Slotboom DJ, Kunji ERS (2005) Functional expression of eukaryotic membrane proteins in Lactococcus lactis. Protein Sci 14:3048–3056
5. Bernaudat F, Frelet-Barrand A, Pochon N, Dementin S, Hivin P, Boutigny S, Rioux JB, Salvi D, Seigneurin-Berny D, Richaud P, Joyard J, Pignol D, Sabaty M, Desnos T, Pebay-Peyroula E, Darrouzet E, Vernet T, Rolland N (2011) Heterologous expression of membrane proteins: choosing the appropriate host. PLoS One 6, e29191
6. Herzig S, Raemy E, Montessuit S, Veuthey JL, Zamboni N, Westermann B, Kunji ERS, Martinou JC (2012) Identification and functional expression of the mitochondrial pyruvate carrier. Science 337:93–96
7. Gasson MJ, de Vos WM (eds) (1994) Genetics and biotechnology of lactic acid bacteria. Blackie Academic and Professional, London, United Kingdom
8. Morello E, Bermúdez-Humarán LG, Llull D, Solé V, Miraglio N, Langella P, Poquet I (2008) Lactococcus lactis, an efficient cell factory for recombinant protein production and secretion. J Mol Microbiol Biotechnol 14:48–58
9. Geertsma ER, Poolman B (2007) High-throughput cloning and expression in recalcitrant bacteria. Nat Methods 4:705–707
10. Frelet-Barrand A, Boutigny S, Moyet L, Deniaud A, Seigneurin-Berny D, Salvi D, Bernaudat F, Richaud P, Pebay-Peyroula E, Joyard J, Rolland N (2010) Lactococcus lactis, an alternative system for functional expression of peripheral and intrinsic Arabidopsis membrane proteins. PLoS One 5, e8746
11. Pontes DS, de Azevedo MS, Chatel JM, Langella P, Azevedo V, Miyoshi A (2011) Lactococcus lactis as a live vector: heterologous protein production and DNA delivery systems. Protein Expr Purif 79:165–175
12. Mierau I, Kleerebezem M (2005) 10 years of the nisin-controlled gene expression system (NICE) in Lactococcus lactis. Appl Microbiol Biotechnol 68:705–717
13. de Ruyter PG, Kuipers OP, de Vos WM (1996) Controlled gene expression systems for Lactococcus lactis with the food-grade inducer nisin. Appl Environ Microbiol 62: 3662–3667

14. Zhou XX, Li WF, Ma GX, Pan YJ (2006) The nisin-controlled gene expression system: construction, application and improvements. Biotechnol Adv 24:285–295
15. Erkens GB, Berntsson RP, Fulyani F, Majsnerowska M, Vujičić-Žagar A, Ter Beek J, Poolman B, Slotboom DJ (2011) The structural basis of modularity in ECF-type ABC transporters. Nat Struct Mol Biol 18:755–760
16. Berntsson RP, ter Beek J, Majsnerowska M, Duurkens RH, Puri P, Poolman B, Slotboom DJ (2012) Structural divergence of paralogous S components from ECF-type ABC transporters. Proc Natl Acad Sci U S A 109:13990–13995
17. Malinauskaite L, Quick M, Reinhard L, Lyons JA, Yano H, Javitch JA, Nissen P (2014) A mechanism for intracellular release of Na+ by neurotransmitter/sodium symporters. Nat Struct Mol Biol 21:1006–1012
18. Thangaratnarajah C, Ruprecht JJ, Kunji ERS (2014) Calcium-induced conformational changes of the regulatory domain of human mitochondrial aspartate/glutamate carriers. Nat Commun 5:5491
19. Harborne SP, Ruprecht JJ, Kunji ERS (2015) Calcium-induced conformational changes in the regulatory domain of the human mitochondrial ATP-Mg/Pi carrier. Biochim Biophys Acta 1847:1245–1253
20. Frelet-Barrand A, Boutigny S, Kunji ER, Rolland N (2010) Membrane protein expression in Lactococcus lactis. Methods Mol Biol 601:67–85
21. Boutigny S, Sautron E, Frelet-Barrand A, Moyet L, Salvi D, Rolland N, Seigneurin-Berny D (2015) Functional expression of plant membrane proteins in Lactococcus lactis. Methods Mol Biol 1258:147–165
22. King MS, Boes C, Kunji ERS (2015) Membrane protein expression in Lactococcus lactis. Methods Enzymol 556:77–97
23. Sautron E, Mayerhofer H, Giustini C, Pro D, Crouzy S, Ravaud S, Pebay-Peyroula E, Rolland N, Catty P, Seigneurin-Berny D (2015) HMA6 and HMA8 are two chloroplast Cu+-ATPases with different enzymatic properties. Biosci Rep 35(3): pii:e00201
24. Miras S, Salvi D, Ferro M, Grunwald D, Garin J, Joyard J, Rolland N (2002) Non-canonical transit peptide for import into the chloroplast. J Biol Chem 277:47770–47778
25. Miras S, Salvi D, Piette L, Seigneurin-Berny D, Grunwald D, Reinbothe C, Joyard J, Reinbothe S, Rolland N (2007) Toc159- and Toc75-independent import of a transit sequence-less precursor into the inner envelope of chloroplasts. J Biol Chem 282:29482–29492
26. Chang W, Soll J, Bolter B (2014) A new member of the psToc159 family contributes to distinct protein targeting pathways in pea chloroplasts. Front Plant Sci 5:239
27. Jager-Vottero P, Dorne AJ, Jordanov J, Douce R, Joyard J (1997) Redox chains in chloroplast envelope membranes: spectroscopic evidence for the presence of electron carriers, including iron-sulfur centers. Proc Natl Acad Sci U S A 94:1597–1602
28. Rosenzweig AC, Arguello JM (2012) Toward a molecular understanding of metal transport by P(1B)-type ATPases. Curr Top Membr 69: 113–136
29. Arguello JM (2003) Identification of ion-selectivity determinants in heavy-metal transport P1B-type ATPases. J Membr Biol 195: 93–108
30. Gourdon P, Liu XY, Skjorringe T, Morth JP, Moller LB, Pedersen BP, Nissen P (2011) Crystal structure of a copper-transporting PIB-type ATPase. Nature 475:59–64
31. Shikanai T, Muller-Moule P, Munekage Y, Niyogi KK, Pilon M (2003) PAA1, a P-type ATPase of Arabidopsis, functions in copper transport in chloroplasts. Plant Cell 15: 1333–1346
32. Abdel-Ghany SE, Muller-Moule P, Niyogi KK, Pilon M, Shikanai T (2005) Two P-type ATPases are required for copper delivery in Arabidopsis thaliana chloroplasts. Plant Cell 17:1233–1251
33. Catty P, Boutigny S, Miras R, Joyard J, Rolland N, Seigneurin-Berny D (2011) Biochemical characterization of AtHMA6/PAA1, a chloroplast envelope Cu(I)-ATPase. J Biol Chem 286:36188–36197
34. Kampfenkel K, Möhlmann T, Batz O, Van Montagu M, Inze D, Neuhaus HE (1995) Molecular characterization of an Arabidopsis thaliana cDNA encoding a novel putative adenylate translocator of higher plants. FEBS Lett 374:351–355
35. Neuhaus HE, Thom E, Möhlmann T, Steup M, Kampfenkel K (1997) Characterization of a novel eukaryotic ATP/ADP translocator located in the plastid envelope of Arabidopsis thaliana L. Plant J 11:73–82
36. Trentmann O, Jung B, Neuhaus HE, Haferkamp I (2008) Nonmitochondrial ATP/ADP transporters accept phosphate as third substrate. J Biol Chem 283:36486–36493
37. Haferkamp I, Fernie AR, Neuhaus HE (2011) Adenine nucleotide transport in plants: much more than a mitochondrial issue. Trends Plant Sci 16:507–515
38. Tjaden J, Mohlmann T, Kampfenkel K, Henrichs G, Neuhaus HE (1998) Altered plas-

tidic ATP/ADP-transporter activity influences potato (Solanum tuberosum L.) tuber morphology, yield and composition of tuber starch. Plant J 16:531–540

39. Kunji ERS (2012) Structural and mechanistic aspects of mitochondrial transport proteins. In: Ferguson S (eds), Comprehensive Biophysics, Elsevier. pp 174–205
40. Palmieri F (2014) Mitochondrial transporters of the SLC25 family and associated diseases: a review. J Inherit Metab Dis 37:565–575
41. Pebay-Peyroula E, Dahout-Gonzalez C, Kahn R, Trezeguet V, Lauquin GJ, Brandolin G (2003) Structure of mitochondrial ADP/ATP carrier in complex with carboxyatractyloside. Nature 426:39–44
42. Kunji ERS, Crichton PG (2010) Mitochondrial carriers function as monomers. Biochim Biophys Acta 1797:817–831
43. Robinson AJ, Kunji ERS (2006) Mitochondrial carriers in the cytoplasmic state have a common substrate binding site. Proc Natl Acad Sci U S A 103:2617–2622
44. Robinson AJ, Overy C, Kunji ERS (2008) The mechanism of transport by mitochondrial carriers based on analysis of symmetry. Proc Natl Acad Sci U S A 105:17766–17771
45. Ruprecht JJ, Hellawell AM, Harding M, Crichton PG, Mccoy AJ, Kunji ERS (2014) Structures of yeast mitochondrial ADP/ATP carriers support a domain-based alternating-access transport mechanism. Proc Natl Acad Sci U S A 111:E426–E434
46. Klingenberg M (2008) The ADP and ATP transport in mitochondria and its carrier. Biochim Biophys Acta 1778:1978–2021
47. Mierau I, Leij P, van Swam I, Blommestein B, Floris E, Mond J, Smid EJ (2005) Industrial-scale production and purification of a heterologous protein in Lactococcus lactis using the nisin controlled gene expression system NICE: the case of lysostaphin. Microb Cell Fact 4:15
48. Hillar A, Loewen PC (1995) Comparison of isoniazid oxidation catalyzed by bacterial catalase-peroxidases and horseradish peroxidase. Arch Biochem Biophys 323:438–446
49. Kuhlbrandt W (2004) Biology, structure and mechanism of P-type ATPases. Nat Rev Mol Cell Biol 5:282–295

Chapter 7

Expression of Viral Envelope Glycoproteins in *Drosophila melanogaster* S2 Cells

Renato Mancini Astray, Sandra Fernanda Suárez-Patiño, and Soraia Attie Calil Jorge

Abstract

The expression of recombinant viral envelope glycoproteins in S2 (*Drosophila melanogaster*) has been performed with good results. This chapter contains protocols for the utilization of this system for the expression and analysis of proteins presented in cell plasma membrane.

Key words S2 cells, *Drosophila melanogaster*, Viral envelope glycoprotein, S2 protocols, Insect cells

1 Introduction

Viral glycoproteins are complex molecules which show a straight correlation between structure and biological activity or antigenicity [1]. Therefore, to obtain a recombinant viral glycoprotein with preserved biological characteristics, the choice of an appropriate expression system is of crucial importance. The *Drosophila melanogaster* expression system has been successfully utilized for the expression of viral glycoproteins, such as the HIV gp120 [2], the Japanese Encephalitis virus (JEV) glycoprotein [3], Rift Valley fever virus surface glycoproteins [4], and Dengue virus envelope glycoprotein [5], which were expressed at good levels with appropriate biological activity. Maybe the most studied recombinant viral glycoprotein expressed in S2 cells is the one from the Rabies virus (RVGP) [6–13]. Several scientific works performed on the expression and characterization of the RVGP comprised detailed studies on the molecular biology (plasmids and promoters), S2 cell metabolism, alternative culture media, kinetic studies of cell growth and recombinant expression and culture conditions (reviewed in 14). This chapter presents most protocols established during the studies of RVGP expression in S2 cells, which can be used as a basis for similar studies. The native Rabies glycoprotein is a trimer (3 × 65 kDa)

Isabelle Mus-Veteau (ed.), *Heterologous Expression of Membrane Proteins: Methods and Protocols,* Methods in Molecular Biology, vol. 1432, DOI 10.1007/978-1-4939-3637-3_7, © Springer Science+Business Media New York 2016

located in the viral envelope and cell membranes before virus budding. In S2 cells, the Rabies glycoprotein gene, including its signal peptide, was cloned under the control of strong promoters and after expression it was analyzed by several methods in cell membranes or after extraction and solubilization.

1.1 S2 Cells and the Design of an Expression System

Drosophila melanogaster (S2) cell lines have been used recently as an alternative expression system, since they can be stably transfected, perform complex post-translational modifications, grow in suspension cultures and need easy culture conditions. The recombinant gene in plasmids can be under the control of the inductive metallothionein (pMt) or the constitutive actin (pAc) drosophila promoters. Alternatively, heat shock promoters can be used in this system. The drosophila expression vectors were designed in order to bear or not the BiP (i) external secretion signal, allowing the protein secretion in some cases [15].

These cells can be grown in culture media supplemented with heat-inactivated fetal calf serum as well as Schneider's Drosophila [5, 10], TC-100 [16], TNM-FH [17], and M3 [18]. Defined commercial serum-free media can also be used to grow this cell line, such as HyQ SFX [19], EX-CELL 400 [20], and InsectXpress [10, 18]. Other approaches for S2 cell culture have been based on the use of basal culture media as IPL-41 or TC-100 with other supplements than FCS [7, 8]. S2 cells grown in Sf-900™ III FCS-free medium (Life Technologies) have shown quite satisfactory growth parameters with high values of μ_{max} in several culture conditions [11, 17].

Regarding culture conditions, S2 cells are easy to grow and maintain in the laboratory. All incubations of S2 cells are performed in a 25–28 °C incubator, at pH values ranging from 6.2 to 6.5 without CO_2 buffering requirement. These cells can be grown and scaled up in a variety of containers. Static culture is performed in regular tissue culture plasticware. For higher scale growth and recombinant expression, S2 cells can be grown as suspension cultures in Duran® laboratory glass bottles, Spinner flasks, or bioreactors [7, 11, 17].

2 Materials

2.1 Cell Culture

1. Laminar flow hood.
2. Hemocytometer or Neubauer chamber (preferentially with mirrored base).
3. Plasticware: T-25 and T-75 cm² flasks, 15 and 50 mL conical tubes, microtubes, serologic pipettes, pipette tips, culture plates, and cryovials.
4. Incubator at 26–28 °C, no CO_2 required.

5. Trypan blue solution (typically 0.4%. For use dilute 10× in sample and PBS).
6. Sf-900™ III medium (Life Technologies).
7. Freezing medium: 50% heat-inactivated fetal calf serum (FCS), 40% Sf-900™ III conditioned medium, and 10% Dimethyl sulfoxide (DMSO).
8. Polystyrene box or freezing device (e.g., Freezing Container Nalgene®).
9. Cell culture tested copper sulfate ($CuSO_4$, 200 mM).
10. Freezing buffer (for samples): 25 mM Tris–HCl, 25 mM NaCl, 5 mM $MgCl_2$, 1 mM PMSF (phenylmethylsulfonyl fluoride); 20% glycerol; pH 7.2.
11. Phosphate buffered saline (PBS): 137 mM NaCl, 10 mM Na_2HPO_4, 2.7 mM KCl, 1.8 mM KH_2PO_4, pH 7.0.
12. Duran® laboratory glass bottles (100 mL).
13. Incubator orbital shaker (100–140 rpm, 26–28 °C).
14. BioFlo® 110 bioreactor (New Brunswick, Enfield, Connecticut, EUA).
15. Sodium butyrate solution in water (NaBu, 100 mM). Store up to one month as aliquots in tightly sealed vials at −20 °C. Before use, let equilibrate to room temperature (1 h).
16. Cycloheximide solution in PBS (25 mM).
17. Expression and selection plasmids such as pMt/V5-His and pCoHygro (Life Technologies).
18. Cellfectin® reagent (Life Technologies).
19. Selection reagent such as hygromycin B (50 mg/mL).
20. Sterile PEB: PBS containing 0.5% BSA and 2 mM EDTA.
21. Centrifuge with rotors and adaptors for 15 mL, 50 mL, 1.5 mL tubes.
22. Mouse antibody targeting the recombinant protein or incorporated tag sequence.
23. Anti-mouse IgG magnetic microbeads (e.g., MACS®, Miltenyi Biotec).
24. Magnetic column.
25. Magnetic support.
26. Gentamicin sulfate (10 mg/mL).

2.2 Analysis of Expression

1. Round high quality glass cell culture tested coverslips (12–35 mm diameter).
2. Crystal clear glass slides with no fluorescent background.
3. PBS.

4. 50–50 % methanol:acetone solution (−20 °C).
5. 2 % formaldehyde in PBS (4 °C).
6. Anti-fading.
7. PBS + FCS (PBS with 1 % Fetal Calf Serum).
8. Antibody targeting the recombinant protein or incorporated tag sequence.
9. Fluorophore-conjugated secondary antibody.
10. Fluorescence microscope.
11. Flow cytometer.
12. PMSF (phenylmethylsulfonyl fluoride), frozen stock solution 100 mM in DMSO.
13. Membrane buffer: 50 mM Tris–HCl, 0.5 M NaCl, 1 mM PMSF, pH 8.
14. Membrane storage buffer: 50 mM Tris–HCl, 320 mM sucrose, pH 7.4
15. Protease inhibitor cocktail.
16. Ultrapure water.
17. Tissue homogenizer.
18. Potter homogenizer.
19. 26-gauge needle.
20. 0.45 μm membrane filter.
21. Ultracentrifuge and ultracentrifuge tubes.
22. Solubilization buffer: 25 mM Tris–HCl, 25 mM NaCl, 5 mM $MgCl_2$, 1 mM PMSF, Igepal® 0.2 % (Sigma-Aldrich), pH 7.2.
23. Lysis buffer R: 10 mM Tris-Cl pH 8.0, 1 mM EDTA, 1 % Igepal®, 0.1 % Sodium Deoxycholate, 0.1 % SDS, 140 mM NaCl, 1 mM PMSF.
24. Wash buffer: 50 mM Tris–HCl pH 7.4, 0.5 M NaCl, Igepal® 0.2 %.
25. Elution Buffer: 50 mM Tris–HCl pH 7.4, 0.5 M NaCl, Igepal® 0.2 %, Imidazole (100–500 mM).
26. IMAC column or nickel resin.
27. Target gene forward and reverse primers designed for RT-qPCR.
28. α-tubulin forward primer (5′-TGTCGCGTGTGAAACACTTC) and reverse primer (5′-AGCAGGCGTTTCCAATCTG).
29. Trizol®, RNAZol®, or other phenol–guanidine isothiocyanate solution.
30. DNase.
31. RNA inhibitor.

32. Reverse polymerase.
33. Quantitative PCR kit.
34. Real time thermocycler.

3 Methods

3.1 Cell Line Unfreezing

1. Remove the cryovial from liquid nitrogen and thaw quickly in a water bath at 30 °C.
2. When just a small ice block is still visible, clean the vial with 70% ethanol and transfer the content into a 15 mL tube with 10 mL of culture medium.
3. Centrifuge at $120 \times g$ for 10 min. Remove the supernatant containing DMSO (*see* **Note 1**).
4. Resuspend the cells in 5 mL of fresh medium and transfer to a T-25 cm^2 flask with loosen caps (vent position) or caps with filter to allow oxygenation.
5. Incubate at 28 °C, until cells reach a density of $0.5–2 \times 10^7$ cells/mL before splitting. This may take 3–4 days.

3.2 Freezing S2 Cells

1. Grow cells to a density between 0.8 and 1×10^7 cells/mL (log phase). Detach cells from the flask surface by gently tapping it. Softly pipet cell suspension up and down a few times to break up cell clumps.
2. Determine viable and total cell counts in a hemocytometer using a standard trypan blue exclusion assay (*see* **Note 2**). The viability must be between 95 and 99%.
3. Prepare the amount of freezing medium required to resuspend the cells at a density $\geq 1 \times 10^7$ viable cells/mL.
4. Centrifuge cells at $1000 \times g$ for 3 min at 4 °C. Remove the supernatant from the cell pellet.
5. Resuspend the cells at a density of 1×10^7 cells/mL in freezing medium.
6. Immediately place 1 mL sterile aliquots of the cell suspension in previously identified cryovials. Place the cryovials into a polystyrene box or freezing device and allow cells to freeze in a −80 °C ultrafreezer. After 24 h transfer cryovials to liquid nitrogen for long-term storage.

3.3 Static Cultivation

1. To maintain S2 cells in static cultivation use 5 mL of fresh medium into T-25 cm^2 flasks. Alternatively, use 15 mL of fresh medium into T-75 cm^2 flasks.
2. Incubate at 26–28 °C until cells reach a density of 2×10^7 cells/mL. This may take 3–4 days.

3. Split cells into fresh medium at a 1:2 or 1:5 dilution into new flask each 5–6 days. S2 cell lines adhere weakly to the substrate. Release the cells by gently pipetting up and down or tapping the flask (*see* **Note 3**).
4. For recombinant expression studies determine viable and total cell counts in a hemocytometer using a standard trypan blue exclusion assay (cells must be 95–99 % viable for good reproducibility). Inoculate $0.5–1 \times 10^6$ cells/mL.
5. Two days after inoculum, if necessary induce the recombinant protein expression adding sterile cell culture tested copper sulfate ($CuSO_4$) to the medium to a final concentration of 700 μM.
6. After 48 h of induction resuspend/homogenize cell culture and determine cell concentration. Samples containing at least 1×10^6 cells can be collected for recombinant protein analysis (*see* **Note 4**).
7. Take the desired sample volume in a microtube, centrifuge at $2000 \times g$ for 5 min, and discard supernatant. Wash cells with 500 μL of PBS and centrifuge again. Discard buffer and freeze cell samples as dry pellet or after adding 500 μL of freezing buffer.

3.4 Cultivation in Shaker Flasks

1. Inoculate as many as T-75 cm^2 flasks needed with S2 cells. Let cells grow to mid-exponential phase. Transfer cells from the T flasks to a centrifuge tube making a pool. Centrifuge at $800 \times g$ for 5–10 min (depending on total volume) to pellet the cells. Resuspend with fresh medium and determine cell concentration and viability.
2. Inoculate a 100 mL Duran® bottle with 20 mL of working volume containing $0.5–1 \times 10^6$ cells/mL (*see* **Note 5**).
3. Incubate the bottle in an orbital shaker at 110–140 rpm and 26–28 °C. Determine cell concentration at least each 24 h.
4. To induce recombinant protein expression, add sterile cell culture tested $CuSO_4$ to the final concentration of 700 μM when cells reach a concentration between 3×10^6 and 5×10^6 cells/mL.
5. Collect samples as described in Subheading 3.3 (*see* **Note 6**).

3.5 Cultivation in Bioreactor

1. Grow cells in Duran® bottles to reach log phase (48–72 h) as described in Subheading 3.1, **step 4**. Determine cell concentration and viability.
2. Inoculate the BioFlo 110 bioreactor to final cell concentration of 0.5×10^6 cells/mL in 1 L culture medium. Good culture conditions can change depending on the growth

characteristics of the established recombinant cell, the target protein, and culture medium used. As a reference using Sf-900™ III medium (Ventini et al. 2010): pH 6.2, 28 °C, 90 rpm agitation, pitched blade impellers, DO controlled at 10% of air saturation and four gas sparging aeration (0.1 L/min) (*see* **Note 7**).

3. Induce expression with 700 μM of $CuSO_4$ when cell concentration reaches 4–5 × 10^6 cells/mL.
4. During the procedure, samples can be sterile collected to determine cell concentration and expression kinetics. Treat samples as described in Subheading 3.3. Culture supernatant can be stored for nutrient consumption/metabolite production measurements.

3.6 Enhancement of Recombinant Expression

A transitory and often lethal enhancement of recombinant expression in S2 cells can be achieved by the addition of sodium butyrate to cultures.

1. Grow cells in Duran® bottles to reach log phase (48–72 h) as described in Subheading 3.4. Determine cell concentration and viability.
2. To induce recombinant protein expression, add sterile cell culture tested $CuSO_4$ to the final concentration of 700 μM when cells reach a concentration between 3 × 10^6 and 5 × 10^6 cells/mL. Simultaneously add sodium butyrate to the final concentration of 5 mM.
3. Collect samples as described in Subheading 3.3.
4. Enhanced recombinant expression can be detected from 12 to 72 h.

3.7 Inhibition of Recombinant Expression

A transitory and proapoptotic inhibition of RNA translation can be achieved by adding cycloheximide to cultures. This can be interesting for studies of recombinant RNA abundance and protein turnover.

1. Grow cells in Duran® bottles to reach log phase (48–72 h) as described in Subheading 3.4. Determine cell concentration and viability.
2. To induce recombinant protein expression, add sterile cell culture tested $CuSO_4$ to the final concentration of 700 μM when cells reach a concentration between 3 × 10^6 and 5 × 10^6 cells/mL. After 24 h add cycloheximide to the final concentration of 10 μM.
3. Collect samples as described in Subheading 3.3.
4. The inhibition of expression can be detected after 3 h.

3.8 Cell Transfection and Selection

1. The day before transfection, plate 0.5×10^6 cells in each one of two T-25 cm^2 flasks with culture medium.
2. Dilute 2–5 μg of the expression and selection plasmids (normally in a ratio of 30:1) to a final volume of 500 μL with serum-free medium (*see* **Note 8**).
3. Dilute 10 μL of Cellfectin® to a final volume of 500 μL with serum-free medium.
4. Add the Cellfectin® solution to the DNA solution all at once. Mix gently and incubate for 20 min at room temperature. After incubation add 1 mL more of serum-free medium.
5. Remove the culture medium from one of the T-25 flasks and dropwise add 2 mL of Cellfectin®/DNA mixture. Incubate 4–6 h at 26–28 °C. Further add 3 mL of fresh medium and incubate again.
6. After 48 h of transfection, remove the medium from the two T-25 flasks and start selection by adding selection agent, such as hygromycin B (600 μg/mL).
7. Change medium containing the selection agent every 3 days.
8. Two or three weeks after starting the selection, progressively decrease the antibiotic concentration after noticing that untransfected cells in negative control T-25 flask have already died (*see* **Note 9**).

3.9 Enrichment of S2 Cells Expressing the Recombinant Protein

1. Inoculate S2 cells in T flasks and induce expression if necessary. Cultivate during 24 h after expression induction or inoculum.
2. Remove supernatant and softly wash cell layer with sterile PBS to remove dead cells. Resuspend cells in PEB and determine cell concentration. Centrifuge at 400 × *g* for 5 min the volume equivalent to 3.4×10^7. Suspend cell pellet with anti-target protein antibodies diluted in 340 μL PEB (1:10) and incubate for 10 min at 4 °C.
3. Wash cells twice to remove antibody excess: add 3 mL of PEB and centrifuge at 300 × *g* for 10 min at 4 °C. Suspend cell pellet with 240 μL PEB and 60 μL anti-mouse IgG magnetic microbeads suspension in PEB. Incubate for 15 min at 4 °C.
4. Wash cells as described in **step 2**. Suspend cell pellet with 500 μL PEB and apply the cell suspension into a magnetic column placed in the magnetic support and previously conditioned with PEB. Let the cell suspension slowly flow down the column. This suspension can be applied twice for better yield.
5. Wash the column three times with cold sterile 500 μL PEB.
6. Recover the bound cells by removing the column from the support and eluting it with 1 mL PEB directly into a recovery tube.

7. Centrifuge at 400 × g for 5 min at 4 °C. Suspend cell pellet with 1.5 mL of culture medium with gentamicin sulfate (5.0 μg/mL). Incubate cells at the appropriate temperature.

3.10 Analysis of Expression

The analysis of a recombinant viral membrane glycoprotein expressed in S2 cells can be performed with the protein into cell membranes or after its extraction and solubilization. The main advantages of the analysis directly on the S2 cells are the possibility of accessing important epitopes with specific antibodies, which can be indicative of protein quality, and the determination of the number of cells expressing the recombinant glycoprotein in an S2 population for semi-quantitative analysis of expression. However, more precise quantitative analysis is usually performed with extracted and solubilized glycoprotein, which can be used for immuno-based analytical methods or purified and quantified by other methodologies.

The semi-adherent nature of S2 cell growth is a valuable characteristic for the analysis of recombinant expression of membrane or membrane-associated proteins, as no damage is caused to membranes while cells are being suspended. In our experience no Fc receptor blocking agents were needed when analyzing live cells with specific antibodies, but performing the usual negative controls with wild S2 cells is recommended.

3.10.1 Immunofluorescence

1. Place sterile coverslips on the bottom of 6- or 12-well culture plates.
2. Inoculate 3×10^6 (6-well plate) or 1×10^6 (12-well plate) in each well and incubate for 24 h. Induce expression if necessary and incubate for more than 24 h.
3. Remove culture medium and wash two to three times with PBS, taking care to not displace attached cells (*see* **Note 10**).
4. Fix the cells with ice-cold methanol:acetone (1:1) or 2 % paraformaldehyde in PBS (4 °C) for 20 min in ice bath. Remove the fixation solution and wash three times with PBS.
5. Dilute the first antibody targeting the recombinant protein or incorporated tag sequence (1:400–1:1000) and add 200 μL to each sample, covering the coverslip. Incubate 1 h at room temperature. Wash carefully three times with PBS.
6. Dilute the fluorophore-conjugated secondary antibody in PBS (1:1000–1:2000) containing or not a counterstain (e.g., Evans Blue 0.03–1 %) and add 200 μL to samples. Incubate in the dark for 1 h at room temperature. Wash three times with PBS.
7. Remove coverslips from wells and pipet anti-fading reagent over a fluorescence grade glass slide. Place coverslips with cells faced down in contact with anti-fading reagent. Seal coverslips borders and observe in a fluorescence microscope.

3.10.2 Flow Cytometry

1. Inoculate S2 cells and induce expression if necessary. Cultivate cells at least 24 h after expression induction or inoculum.
2. Take at least 10^6 cells and wash once with sterile PEB (*see* **Note 11**). Centrifuge at 400 × *g* for 5 min. Suspend the pellet with antibody targeting the recombinant protein or incorporated tag sequence diluted in 50–100 μL PEB (1:400–1:1000) and incubate for 30 min at 4 °C.
3. Centrifuge at 400 × *g* for 5 min at 4 °C. Suspend cell pellet with fluorophore-conjugated secondary antibody diluted in 50–100 μL PEB (1:1000–1:3000) and incubate in the dark for 30 min at 4 °C.
4. Centrifuge at 400 × *g* for 5 min at 4 °C. Suspend cell pellet with 500 μL PBS + FCS and directly analyze in a Flow Cytometer equipment. Alternatively, suspend cells in 1 mL of 2 % formaldehyde in PBS, incubate at 4 °C for 1 h, centrifuge, resuspend with PBS + FCS, and store at 4 °C until analysis.

3.10.3 Membrane Preparation and Solubilization for Protein Expression Analysis

The analysis of a viral membrane glycoprotein can be made by several techniques as ELISA, Western blotting, Spectrometry, and SPR. Most of the analytical tools or purification procedures require the solubilization of the protein. While the purification of such glycoproteins is notably facilitated by membrane preparation protocols, which can be considered a first purification step and a procedure which confers increased protection against proteases, the direct lysis of cell pellets is a more straightforward procedure that allows the rapid analysis. The utilization of detergents for the solubilization of this type of glycoproteins is often mandatory, as these proteins share in common the presence of hydrophobic domains, generally unstable in aqueous solutions and prone to aggregation in the absence of detergent micelles [19].

1. Take 7–8 × 10^8 cells and wash once with sterile PBS. Centrifuge at 1000 × *g* for 5 min. Suspend the pellet with 1 mL Membrane Buffer containing protease inhibitor cocktail or 1 mM PMSF (*see* **Note 12**). Proceed to cell lysis.
2. Lyse cells with 20 strokes in a Potter homogenizer or by 5 min homogenization using a tissue homogenizer. Always check by microscopy if cells were effectively broken. Alternatively cells can be lysed by hypotonic shock suspending cell pellet in 25 mL of ultrapure water + PMSF 1 mM, instead of Membrane buffer in first step.
3. Pellet nuclei and unbroken cells by centrifugation at 2000 × *g* for 10 min at 4 °C. Keep the supernatant and repeat the lysis procedure with pellet fraction adding 1 mL membrane buffer.

4. Combine both lysis supernatants and ultracentrifuge at 120,000 × *g* for 30 min at 4 °C.
5. Resuspend the membrane pellet in 500 μL of Membrane Storage buffer. Homogenize through a 26-gauge needle and take 10 μL for the determination of total protein concentration (use any method of your choice, but BCA is recommended). Store as aliquots at −80 °C. Alternatively the membrane pellet can be resuspended in the required analytical buffer for subsequent analysis.
6. Dilute the membrane sample to the total protein concentration of 2 mg/mL in solubilization buffer. Incubate for 30 min at room temperature under soft agitation (*see* **Notes 13** and **14**).
7. Ultracentrifuge samples at 125,000 × *g* for 45 min at 4 °C.
8. Take samples from the supernatant for further analysis. Discard the pellet.
9. For SDS-PAGE and Western-blot analysis, mix supernatant with loading buffer (reducing or nonreducing), boil or not the samples for 5 min at 95 °C (*see* **Note 16**).

3.10.4 Protein Preparations from Direct Solubilization of Cell Pellets

1. Separate 10^6–3×10^6 cells, centrifuge at 2000 × *g* for 5 min, and discard supernatant. Wash cells once with PBS.
2. Resuspend cell pellet with 0.5 mL of lysis buffer without detergent and vortex briefly.
3. Add 0.5 mL of lysis buffer with detergent twice the final concentration and mix well. Incubate samples at 4 °C for 1 h or at room temperature for 30 min. Mix samples continuously or at least each 15 min by inversion. Do not vortex samples at this step.
4. Centrifuge samples at 10,000 × *g* for 10 min at 4 °C. Use supernatant for analysis (*see* **Note 15**).
5. For SDS-PAGE and Western-blot analysis, mix supernatant with loading buffer (reducing or nonreducing), boil or not the samples for 5 min at 95 °C (*see* **Note 16**).

3.10.5 Purification of Histidine-Tagged Glycoprotein from S2 Cells

1. Prepare and solubilize a membrane sample as described in previous sections (*see* **Note 17**). Filter samples through a 0.45 μm membrane.
2. Equilibrate IMAC column or nickel resin with the same buffer used for sample solubilization. Follow manufacturer recommendations.
3. Apply samples in slow flow rate (1 mL/min) or let it interact with resin for 1 h at 4 °C. Collect flow through for analysis.
4. Wash at least with five column volumes of washing buffer.

5. Elute column with increasing imidazole concentrations from 100 to 500 mM in a gradient run or by applying different elution buffers containing specified imidazole quantities.
6. Analyze sample fractions for the presence of the protein of interest.

3.10.6 Analysis of Relative Recombinant RNA

The quantitative analysis of recombinant RNA by RT-qPCR is a useful method to indirectly evaluate protein expression.

1. Grow cells in Duran® bottles to reach log phase (48–72 h) as described in Subheading 3.4. Determine cell concentration and viability.
2. To induce recombinant protein expression, add sterile cell culture tested $CuSO_4$ to the final concentration of 700 μM when cells reach a concentration between 3×10^6 and 5×10^6 cells/mL.
3. Collect samples of 3×10^6 cells as described in Subheading 3.3.
4. Immediately extract RNA by the method of choice or freeze samples preferentially at −80 °C in a phenol–guanidine isothiocyanate solution, such as Trizol®, RNAZol® or other, until analysis.
5. For removal of residual DNA treat 3 μg of total RNA with RNase-free DNase I in the presence of RNase inhibitor.
6. For reverse transcription use 600 ng of total DNA-free RNA and reverse specific primers for the recombinant target RNA and for the *D. melanogaster* β-tubulin RNA as housekeeping gene (*see* **Note 18**).
7. Perform qPCR with the reagents of choice using 3 μL of sample to a 15 μL reaction volume. Use previously validated forward and reverse primers for recombinant gene amplification (usually, 0.2–0.3 μM). Always place target and housekeeping reactions in the same run.
8. Analyze results based on cycle threshold considering previously determined efficiency amplification values for target and housekeeping genes (*see* **Note 19**).

4 Notes

1. Alternatively, cells can be placed directly on T flasks without prior centrifugation if they are not growing well through the described procedure. However, as freezing media usually contains harmful DMSO in the formulation, the medium has to be changed the day after unfreezing when cells have already attached to the flask.
2. Trypan blue exclusion method can be performed as follows: after completely resuspending the cells, take a 20 μL sample

and dilute 1:10 in PBS. Take an aliquot of 10 μL of the diluted sample and dilute 1:10 with a 0.4% Trypan blue solution in PBS. Place 10 μL of the solution in one side of a hemocytometer chamber and count at least three squares and one hundred cells. Divide the cell number by the squares counted and multiply by the dilution (10×) and by hemocytometer reference value ($\times 10^4$ cells/mL). Consider bright cells as live and blue cells as dead. Do not let cells stand too much time in Trypan solution, always mix cells and stain solution immediately before counting.

3. S2 cell lines adhere loosely to the substrate, what permits easy cell dislodging by gently pipetting up and down or, by tapping the bottom of the flask. S2 cells are known by their limited growth when inoculated at low densities. Studies comparing different inoculum showed that initial concentration of $5–6 \times 10^5$ cell/mL decreased the length of lag phase and increased X_{max} comparatively to lower ones [14, 19].
4. Although it is not recommended, kinetic studies can be performed in T flasks. In this case, one flask must be used for each sample time, avoiding repetitive resuspension of the cells.
5. The working volume of 1:5 in shake flasks or bottles is important to provide suitable oxygen transfer through the medium surface to the cell culture during agitation. S2 cells have shown low oxygen requirements, with good growth and recombinant production achieved with 30% dissolved oxygen. More details about S2 respiration features can be found elsewhere [22].
6. S2 cells are normally cultivated at 26 °C with good results. The cultivation at higher temperatures (not above 28 °C) tends to increase growth rates. In our experience S2 cells can be cultivated at temperatures as low as 22 °C. Sampling S2 cells is a very easy process and cultures usually don't present significant modifications if the total culture volume is kept above 80% initial working volume.
7. The culture conditions suggested for Bioflo 110 are based on the results of many studies performed with S2 cells expressing Rabies recombinant glycoprotein. Other good studies using these same cells in bioreactors showed that changing culture temperature may be used as a strategy to improve recombinant expression. In our experience cells did not undergo cycle arrest after temperature shock but slowing down growth rates showed to increase productivity [9, 23].
8. To obtain a stable recombinant cell line it is necessary first to transfect the cells with the expression gene and the selection gene. For S2 cells, three antibiotic resistance genes have already been used: hygromycin, puromycin, and blasticidin. The hygromycin is the most used and acts as an aminocyclitol to

inhibit protein synthesis by disrupting translocation and promoting mistranslation. The pCoHygro (Life Technologies) selection vector contains the *E. coli* hygromycin resistance gene (*HPH*) which codes for a phosphotransferase that inactivates the hygromycin B. The hygromycin resistance gene can also be inserted together with p*Copia* promoter in the expression vector, avoiding co-transfection protocols [10, 21].

9. Despite the fact that the transfection of S2 cells produces stable cell lines, the use of a basal 150–300 μg/mL of hygromycin B, during the cultivation in T flasks, helps to maintain high expression levels and increases reproducibility.
10. S2 cells are naturally loosely attached cells. This characteristic and the presence of heterologous membrane proteins in the cell surface make it difficult to handle for immunofluorescence purpose until it is fixed on coverslips. It is important to carefully add wash buffer and other solutions to not detach the cells. Use just cell culture tested glass, as the quality of coverslips has an impact on the cell adhesion.
11. The addition of 1 % sodium azide to this buffer helps to prevent the internalization of antigens, which can be important for some recombinant proteins.
12. The presence of proteases in S2 cells is especially a concern for the recombinant protein purification. Always work with protease inhibitors.
13. Choose the appropriate lysis buffer, considering pH, amount and strength of detergents, salt concentration, etc... The prior evaluation of the best lysis buffer is strongly recommended. Determine empirically the best detergent for your target protein, as the hydrophobicity of the protein may require a specific detergent strength. The existence of tag peptides and the epitopes may be masked if the wrong detergent is chosen.
14. The temperature and time for solubilization are critical. We recommend trying room and 4 °C temperatures. Generally the solubilization is best performed at room or higher temperatures but the degradation may be increased. The time for solubilization is also dependent on the agitation. As a general rule, let samples solubilize under gentle agitation in a Kline agitator for 30 min, as more vigorous agitation may cause foam that is harmful for proteins.
15. Samples obtained directly from cell lysis, even containing protease inhibitors, are unstable. Freezing samples in detergent containing buffer is usually worse than keeping them at 4 °C.
16. Membrane proteins can have their epitopes completely mischaracterized when boiled. Consider also using LDS loading buffer instead of reducing loading buffer when performing

Western blotting. Some membrane proteins have high molecular weight (>100 kDa) and are not very well transferred to synthetic membranes by semi-dry technology.

17. Membrane preparation is highly recommended as a first step for membrane protein purification. Although it is possible to use cell lysate, membrane samples are free of many protease removed during the ultracentrifugation step. In our experience the overall increase in RVGP recovery using membrane preparations is 60% higher than using cell lysate. Sample buffers intended to be used in IMAC purification protocols should not contain more than 1 mM EDTA.
18. The choice of an appropriate housekeeping gene is usually empirically determined. As a general rule, the ideal gene has to show stable expression in the control and experimental situations. Good examples of *D. melanogaster* housekeeping genes are α-tubulin, actin42A, and 18S ribosomal RNA [24]. Consider using 18S ribosomal RNA as housekeeping gene especially when performing experiments with cycloheximide, as the expression levels of all messenger RNAs will be changed in this situation, causing misinterpretation of results.
19. Reliable and quantitative data are generated by RT-qPCR analysis since essential technical procedures and standardizations are performed [25]. We recommend the use of a mathematical model instead of common comparative $\Delta\Delta C_t$ analysis [26].

Acknowledgments

This work was supported by FAPESP (2012/24647-0; 2013/18610-0; 2012/00978-8).

References

1. Doms RW, Lamb RA, Rose JK, Helenius A (1993) Folding and assembly of viral membrane proteins. Virology 193:545–562. doi:10.1006/viro.1993.1164
2. Ivey-Hoyle M, Clark RK, Rosenberg M (1991) The N-terminal 31 amino acids of human immunodeficiency virus type 1 envelope protein gp120 contain a potential gp41 contact site. J Virol 65:2682–2685
3. Zhang F, Ma W, Zhang L et al (2007) Expression of particulate-form of Japanese encephalitis virus envelope protein in a stably transfected Drosophila cell line. Virol J 4:17. doi:10.1186/1743-422X-4-17
4. De Boer SM, Kortekaas J, Antonis AF et al (2010) Rift Valley fever virus subunit vaccines confer complete protection against a lethal virus challenge. Vaccine 28:2330–2339. doi:10.1016/j.vaccine.2009.12.062
5. Clements DE, Coller B-AG, Lieberman MM et al (2010) Development of a recombinant tetravalent dengue virus vaccine: immunogenicity and efficacy studies in mice and monkeys. Vaccine 28:2705–2715. doi:10.1016/j.vaccine.2010.01.022
6. Yokomizo AY, Jorge SAC, Astray RM et al (2007) Rabies virus glycoprotein expression in Drosophila S2 cells. I. Functional recombinant protein in stable co-transfected cell line. Biotechnol J 2:102–109. doi:10.1002/biot.200600211
7. Galesi ALL, Aguiar MA, Astray RM et al (2008) Growth of recombinant Drosophila melanogaster Schneider 2 cells producing

rabies virus glycoprotein in bioreactor employing serum-free medium. Cytotechnology 57: 73–81. doi:10.1007/s10616-008-9139-y

8. Batista FRX, Moraes AM, Büntemeyer H, Noll T (2009) Influence of culture conditions on recombinant Drosophila melanogaster S2 cells producing rabies virus glycoprotein cultivated in serum-free medium. Biologicals 37:108–118. doi:10.1016/j.biologicals.2008.11.001
9. Swiech K, Rossi N, Silva BG et al (2008) Bioreactor culture of recombinant Drosophila melanogaster S2 cells: characterization of metabolic features related to cell growth and production of the rabies virus glycoprotein. Cytotechnology 57:61–66. doi:10.1007/s10616-008-9130-7
10. Lemos MAN, Santos ASD, Astray RM et al (2009) Rabies virus glycoprotein expression in Drosophila S2 cells. I: design of expression/selection vectors, subpopulations selection and influence of sodium butyrate and culture medium on protein expression. J Biotechnol 143:103–110. doi:10.1016/j.jbiotec.2009.07.003
11. Ventini DC, Astray RM, Lemos MAN et al (2010) Recombinant rabies virus glycoprotein synthesis in bioreactor by transfected Drosophila melanogaster S2 cells carrying a constitutive or an inducible promoter. J Biotechnol 146:169–172. doi:10.1016/j.jbiotec.2010.02.011
12. Astray RM, Jorge SA, Lemos MA et al (2013) Kinetic studies of recombinant rabies virus glycoprotein (RVGP) cDNA transcription and mRNA translation in Drosophila melanogaster S2 cell populations. Cytotechnology 65:829–838. doi:10.1007/s10616-012-9522-6
13. Suárez-Patiño SF, Mancini RA, Pereira CA et al (2014) Transient expression of rabies virus glycoprotein (RVGP) in Drosophila melanogaster Schneider 2 (S2) cells. J Biotechnol 192 Pt A: 255–262. doi:
14. Moraes AM, Jorge SAC, Astray RM et al (2012) Drosophila melanogaster S2 cells for expression of heterologous genes: from gene cloning to bioprocess development. Biotechnol Adv 30:613–628. doi:10.1016/j.biotechadv.2011.10.009
15. Kirkpatrick RB, Matico RE, McNulty DE et al (1995) An abundantly secreted glycoprotein from Drosophila melanogaster is related to mammalian secretory proteins produced in rheumatoid tissues and by activated macrophages. Gene 153:147–154
16. Bovo R, Galesi ALL, Jorge SAC et al (2008) Kinetic response of a Drosophila melanogaster cell line to different medium formulations and culture conditions. Cytotechnology 57:23–35. doi:10.1007/s10616-008-9146-z
17. Swiech K, da Silva CS, Arantes MK et al (2008) Characterization of growth and metabolism of Drosophila melanogaster cells transfected with the rabies-virus glycoprotein gene. Biotechnol Appl Biochem 49:41–49. doi:10.1042/BA20060148
18. Kim YK, Kim KR, Kang DG et al (2011) Expression of β-1,4-galactosyltransferase and suppression of β-N-acetylglucosaminidase to aid synthesis of complex N-glycans in insect Drosophila S2 cells. J Biotechnol 153:145–152. doi:10.1016/j.jbiotec.2011.03.021
19. Jorge SAC, Santos AS, Spina A, Pereira CA (2008) Expression of the hepatitis B virus surface antigen in Drosophila S2 cells. Cytotechnology 57:51–59. doi:10.1007/s10616-008-9154-z
20. Li B, Tsing S, Kosaka AH et al (1996) Expression of human dopamine beta-hydroxylase in Drosophila Schneider 2 cells. Biochem J 313(Pt 1):57–64
21. Smith SM (2011) Strategies for the purification of membrane proteins. Methods Mol Biol 681:485–496. doi:10.1007/978-1-60761-913-0_29
22. Pamboukian MM, Jorge SAC, Santos MG et al (2008) Insect cells respiratory activity in bioreactor. Cytotechnology 57:37–44. doi:10.1007/s10616-007-9118-8
23. Rossi N, Silva BG, Astray R et al (2012) Effect of hypothermic temperatures on production of rabies virus glycoprotein by recombinant Drosophila melanogaster S2 cells cultured in suspension. J Biotechnol 161:328–335. doi:10.1016/j.jbiotec.2012.05.016
24. Ponton F, Chapuis M-P, Pernice M et al (2011) Evaluation of potential reference genes for reverse transcription-qPCR studies of physiological responses in Drosophila melanogaster. J Insect Physiol 57:840–850. doi:10.1016/j.jinsphys.2011.03.014
25. Nolan T, Hands RE, Bustin SA (2006) Quantification of mRNA using real-time RT-PCR. Nat Protoc 1:1559–1582. doi:10.1038/nprot.2006.236
26. Pfaffl MW (2001) A new mathematical model for relative quantification in real-time RT-PCR. Nucleic Acids Res 29, e45

Chapter 8

Leishmania tarentolae as a Promising Tool for Expressing Polytopic and Multi-Transmembrane Spans Eukaryotic Membrane Proteins: The Case of the ABC Pump ABCG6

Lucia Gonzalez-Lobato, Vincent Chaptal, Jennifer Molle, and Pierre Falson

Abstract

This chapter includes a practical method of membrane protein production in *Leishmania tarentolae* cells. We routinely use it to express membrane proteins of the ABC (adenosine triphosphate-binding cassette) family, here exemplified with ABCG6 from *L. braziliensis*, implicated in phospholipid trafficking and drug efflux. The pLEXSY system used here allows membrane protein production with a mammalian-like *N*-glycosylation pattern, at high levels and at low costs. Also the effects of an N-terminal truncation of the protein are described. The method is described to allow any kind of membrane protein production.

Key words ABC transporters, *Leishmania tarentolae*, Membrane protein expression, Drug efflux

1 Introduction

Structural studies of membrane proteins are limited by the amount of fully functional protein that can be produced. This is especially vivid for mammalian membrane proteins, for which production in the milligram range ensuring correct folding and functionality remains challenging. In general, large-scale production and purification require a low-cost and effective expression system. Several approaches have been reported in mammalian cells [1], *Xenopus laevis* oocytes [2], bacteria [3, 4], *Spodoptera frugiperda Sf9* insect cells [5], and yeast [6]. Up to now, mammalian cells are most often used in functional studies. ABCG2 is an ABC transporter rather difficult to produce for which several systems were evaluated. Oocyte system represents an easy system to study ABC transporter function; however, it produces a high background in efflux experiments due to nonspecific binding of

Lucia Gonzalez-Lobato, Vincent Chaptal, and Jennifer Molle contributed equally with all other contributors.

Isabelle Mus-Veteau (ed.), *Heterologous Expression of Membrane Proteins: Methods and Protocols*, Methods in Molecular Biology, vol. 1432, DOI 10.1007/978-1-4939-3637-3_8, © Springer Science+Business Media New York 2016

hydrophobic ABCG2 substrates to intracellular structures such as yolk granules, which represent around 50 % of cellular volume. *Lactococcus lactis* expression system allows quantifying sterol transport mediated by ABCG2, not possible in mammalian or insect cells where the membrane sterol content can reach up to 25 %. However, ABCG2 expression level remains too low for structural studies [7]. Overexpression in *Escherichia coli* provides a high yield of recombinant protein but devoid of drug efflux or ATPase activity [8]. The baculovirus-*Sf9* expression system allows membrane protein expression in a quite high level in intact cells and membranes, being a good tool to measure ATPase activity and transport of fluorescence substrates [9]. Nevertheless, cholesterol content, crucial for ABCG2 function, is very low in insect cell membranes [10]. BTI-TN-5B1-4 High Five insect cells produce even higher levels of protein but in a heterogeneous manner. Membrane protein overexpression in *Pichia pastoris* yeast has been used to successfully express and purify large quantities of P-gp [11] and MRP1 [12]. However, this is not the case for ABCG2, which has been produced in comparable levels to the ones achieved in HEK cells but not yet purified. Previously cited expression methods constitute a useful tool for membrane protein expression leading to functional and structural studies but they are not suitable for all membrane proteins; that is the case of ABCG2 whose structure is still unsolved.

A fundamental problem in the production of heterologous proteins in prokaryotic systems is downregulation of protein expression via activation of transcriptional control mechanisms in the host. One alternative is using *Trypanosomatidae* protozoa such as *L. tarentolae* with a mammalian-type posttranslational modification of target proteins [13] and successfully used for the expression of other proteins [14]. *L. tarentolae* is a parasite of the gecko *Tarentola annularis* and has been developed as new eukaryotic system for expression of recombinant proteins with a mammalian-like *N*-glycosylation pattern [15]. This system has already been described to successfully express GFP protein in the parasite using pLEXSY vectors [16].

This chapter describes the experimental procedure to produce a membrane protein, ABCG6, from *L. braziliensis*, by using the pLEXSY system in *L. tarentolae*. *lb*ABCG6 is expressed in the plasma membrane of the parasite and mediates phospholipid trafficking and drug resistance [17]. It shares the highest similarity (28 %) with human ABCG2 among all the ABC transporters in *Leishmania* species. The latter protein confers resistance to anticancer drugs [18] and has been analyzed in multiple functional and comparative studies [19]. The expression system described below presents several advantages such as low cost, nonspecial biosafety requirements, and no cross-contamination with other cultures.

2 Materials

2.1 Proteins

The gene coding for *lb*ABCG6 (UniProtKB #A4HPF5) was synthesized by GENEART (Life Technologies SAS). Two *Bam*HI restriction sites were added, at the beginning and right before the nucleotide-binding domain, allowing N-terminal truncation by molecular biology methods (*lb*ABCG6ΔN). Also an N-terminal 6xHis-tag was added to allow protein purification. The material used for *lb*ABCG6 expression in *L. tarentolae* described here is provided from Jena Bioscience and Lonza.

2.2 Lab Equipment

1. All the material needed for molecular biology experiments.
2. Sodium dodecyl sulfate polyacrylamide gel electrophoresis (SDS-PAGE) and Western blots are carried out using the Mini-Protean 3 apparatus and related devices from Bio-Rad.
3. Microfluidizer M-110P (Microfluidics IDEX CORPORATION).
4. Transfection and culture of the cells are achieved in a class I-type room (*see* **Note 1**) equipped with a Steril Bio Ban 48, an incubator Heraeus BK6160 with a H + P Biomag Biomodule 40B, a microscope Olympus CKX31, and a low-speed centrifuge handling 15/30 mL Falcon-type tubes.
5. Tissue culture T-25 and T-75 flasks are used for static cell cultures. Bigger cultures from 50 mL to 1 L with 75–140 rpm agitation are carried out in Erlenmeyer and baffled Fernbach flasks, respectively.

2.3 Cells

1. Eukaryotic protozoan parasite *L. tarentolae* (Jena BioScience).
2. XL1-Blue chemically competent *E. coli* or equivalent to generate the recombinant plasmid of interest.

2.4 Media

1. BHI medium: The powder is dissolved in deionized water (37 g/L), sterilized by filtration, and stored at 4 °C. Right before use, the medium is supplemented with 0.5% penicillin and streptomycin, 50 μg/mL G418-sulfate, 5 μg/mL hemin (stock solution at 0.25% in 50% triethanolamine, tube wrapped with foil to avoid light), 100 μg/mL nourseothricin, and 100 μg/mL hygromycin. Medium is then stored at 4 °C up to 15 days. Sterilization is achieved by filtration as autoclaving leads to partial degradation of nutriments varying from batch to batch, to which *L. tarentolae* cells are sensitive.
2. Tetracycline 10 mg/mL.
3. Yeast extract medium: 24 g/L of yeast extract, 3 g/L glucose, 12.5 g/L K_2HPO_4, and 2.3 g/L KH_2PO_4, sterilized by filtration and stored at 4 °C until use. Right before use, the medium

is supplemented with 1% fetal bovine serum (FBS), and the same supplements as for BHI medium. Medium is then stored at 4 °C for a maximum of 15 days. Sterilization is achieved by filtration.

4. Luria-Bertani medium: 10 g/L Bactotryptone, 5 g/L bacto-yeast extract, 10 g/L NaCl, supplemented with 15 g/L agar for plates. Autoclave for 20 min at 120 °C. Before use add 100 mg/L ampicillin (*see* **Note 2**). Store without antibiotics at room temperature.
5. Hepes buffer: 50 mM Hepes-NaOH pH 7.5, 200 mM NaCl.

2.5 Transfection

1. Reagent: Supplemented Nucleofector® solution (Basic Parasite Nucleofector Kit 1, Lonza).
2. Medium: BHI.
3. Equipment: Nucleofector II-S®(Lonza) and Lonza-certified cuvettes.

2.6 Molecular Biology

1. The plasmid used here is the pLEXSY-I-neo3 (Jena Bioscience) with the neo marker gene allowing selection of recombinant LEXSY strains with G418, and designed for inducible expression of target genes in LEXSY host T7-TR.
2. Kits for small- (3–10 mg) and medium- (50–100 mg) scale plasmid (5–10 kbp) DNA preparations (NucleoSpin™ Plasmid, Macherey-Nagel).
3. Go Taq DNA polymerase (Promega).
4. NucleoSpin Extract II kit (Macherey-Nagel).
5. Restriction enzymes *Nco*I, *Xba*I, and *Swa*I.

2.7 SDS-PAGE

1. Separating buffer stock solution: 1.5 M Tris–HCl, pH 8.8.
2. Stacking buffer stock solution: 1 M Tris–HCl, pH 6.8.
3. 10% SDS.
4. Acrylamide/bis solution: 40%, 37.5:1 with 2.6% C (*see* **Note 3**).
5. 10% Ammonium persulfate (stored at 4 °C up to 15 days).
6. *N*,*N*,*N*,*N*-tetramethylethylenediamine (TEMED).
7. Running buffer: Dilute Tris-glycine-SDS 10× (Euromedex), can be stored at room temperature.
8. Laemmli-type loading buffer (5×) “5×U”: 100 mM Tris–HCl, pH 8.0, 8 M urea, 4% SDS, 1.4 M β-mercaptoethanol, 0.0025% bromophenol blue. The solution is stored at −20 °C and aliquoted to freeze/thaw ten times maximum [20].
9. Pre-stained molecular weight markers: Kaleidoscope markers (Bio-Rad).

10. Staining solution: Dissolve 1 g of Coomassie Brilliant Blue (Bio-Rad) in 1 L of 50% [v/v] ethanol, 10% [v/v] glacial acetic acid, 40% H_2O; stir the solution until complete solubilization and then filter through Whatman filter paper; store at room temperature; do not reuse.

2.8 Western Blotting

1. Transfer buffer: Dilute 10× Tris-glycine (Euromedex), 20% methanol (*see* **Note 4**). Prepare fresh and use cold, with a cooling ice bag during transfer.
2. Nitrocellulose membrane and 3 MM chromatography paper (Whatman).
3. Tris-buffered saline with tween and triton (TBS-TT): Dilute 10× TBS stock (Euromedex) with water, add 0.05% Tween-20 and 0.2% triton.
4. Blocking solution: 0.5% Blocking reagent (Qiagen) in TBS, 0.1% Tween-20.
5. Antibody anti-His HRP conjugated (Qiagen) is used 1/20,000 diluted in blocking solution.
6. Enhanced chemiluminescent (ECL) reagents and autoradiography films are used for revelation.

3 Methods

3.1 Molecular Biology

Cloning of *lb*ABCG6 and the truncated *lb*ABCG6ΔN into pLESXY plasmid (*see* restriction map in Fig. 1) is achieved using classical methods of molecular biology described in the LESXY kit and in [20]. *lb*ABCG6 was cloned between the *Nco*I and *Xba*I restriction sites inside the multiple cloning sites controlled by the T7 RNA polymerase promoter. Utr1, utr2, and utr3 are optimized non-translated gene-flanking regions providing the splicing signals for posttranscriptional mRNA processing for expression of target and marker genes in the LEXSY host. The following plasmids including the different constructs were generated: *pLEXSY-lbabcg6* and *pLEXSY-lbabcg6Δ N*. Once constructed, each plasmid was checked by sequencing and digested with *Swa*I to remove the *E. coli* fragment (Fig. 1). Each construction was extracted from agarose gel using the kit NucleoSpinExtract II and used in the nucleofection of *L. tarentolae*.

3.2 *L. tarentolae* Growth

1. Cells are grown at 26 °C in the dark, under the promastigote shape with flagella allowing them to swim in the medium.
2. Healthy cells tend to aggregate as cell density increases, forming larger aggregates at higher cell densities (*see* **Note 5**). Ideally, cells are amplified by dilution in fresh medium when they are in the exponential phase, $OD^{600} = 1.4–2$ ($6–8 \times 10^7$ cells/mL).

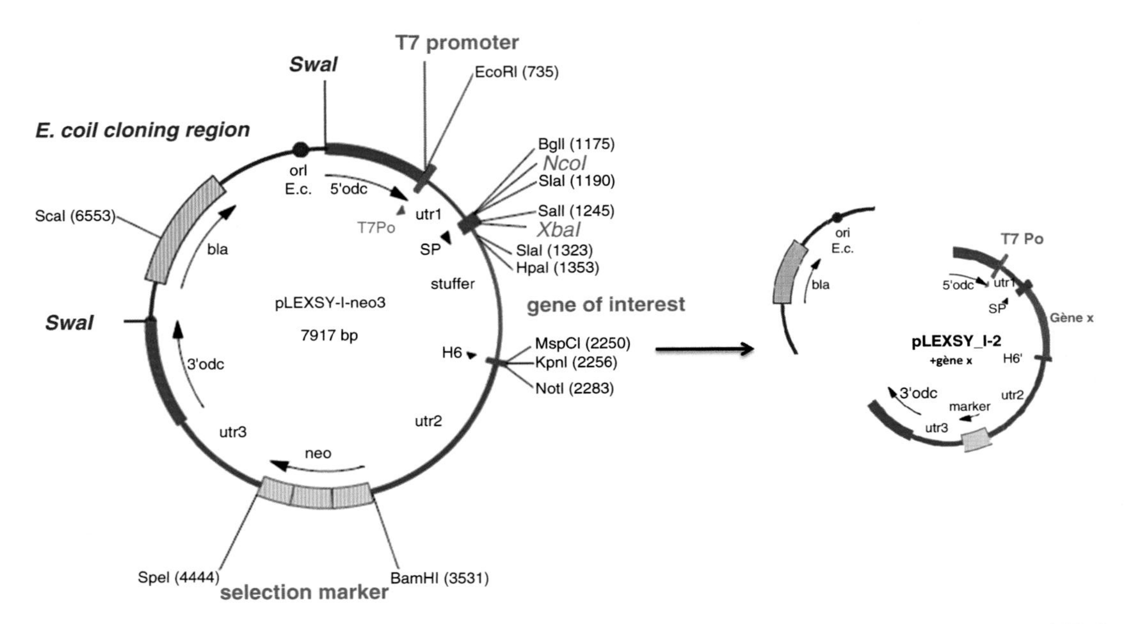

Fig. 1 Cloning strategy of ***Ib*ABCG6** based on pLEXSY system. ***Ib*ABCG6** was cloned between the *Nco*I and *Xba*I restriction sites inside the multiple cloning site. Then, *E. coli* fragment was removed by *Swa*I digestion

3. *L. tarentolae* is maintained in BHI medium in static T-25 flasks. Typically, a 50-fold dilution in 10 mL fresh BHI medium is carried out at time 0, and then cells reach a sufficient density for a 20-fold dilution 5 days later (*see* **Notes 6–8**).
4. Yeast extract medium is used for scale-up. A culture in BHI medium in exponential phase is diluted 20-fold in yeast extract medium (pre-culture). When the culture reaches the exponential phase, a baffled Fernbach flask of 500 mL of yeast extract medium is inoculated to a final OD^{600} of 0.1–0.2 and incubated in the dark, 26 °C and 75–90 rpm (higher cell densities can be obtained in agitated flasks compared to static cultures) (*see* **Notes 9** and **10**). The exponential phase will be reached after 36 h.

3.3 Cryo-Conservation of *L. tarentolae*

1. Cryo-conservation of *L. tarentolae* should be realized for a culture in exponential phase in BHI medium.
2. Add sterile glycerol to the cells in BHI medium, to a final concentration of 20 %, and aliquot by 1.6 mL in sterile cryotubes.
3. Incubate the cells for 10 min at room temperature, and then transfer the tubes to a precooled (4 °C) isopropanol cryobox for 10 min. Transfer the cryobox to −80 °C and incubate overnight. Store at −80 °C or in liquid nitrogen.

4. To reactivate frozen stocks, thaw a cryotube on ice, and then pour the content of the tube into 10 mL of fresh BHI medium in a T-25 flask. Check that the cells are vital by direct observation under a microscope. Incubate at 26 °C until $OD^{600} = 1.4$–2, which usually takes 2–3 days. Then dilute tenfold for allowing cells to fully recover from the freezing, and proceed to normal dilution.

3.4 *L. tarentolae* Nucleofection

1. The best efficiency of transfection is obtained for *L. tarentolae* in the exponential phase. Grow 10 mL of cells in BHI medium until $OD^{600} = 1.4$ and ensure by microscopy that they are vital and of drop-like shape grouping in aggregates.
2. Spin cells for 3 min, $2000 \times g$, at room temperature and suspend pellet in 100 μL of supplemented Nucleofector® solution. Add 4 μg of DNA and transfer to an electroporation cuvette. Electroporate according to the Basic Parasite Nucleofector Kit 1 (Lonza), using the program U-033.
3. Transfer electroporated cells to 10 mL of LEXSY BHI medium in a ventilated flask. Incubate for 24 h as static suspension culture (Fig. 2). Proceed to a clonal selection.

3.5 Monoclonal Selection of Recombinant Cells

1. 24 h after transfection, harvest 2 mL from the transfected 10 mL culture obtained by the electroporation protocol.
2. Pellet cells for 3 min at $2000 \times g$ at room temperature. Remove supernatant and suspend the cells in the residual medium left in the tube, approximately 50 μL.
3. Carefully spread the suspended cells onto freshly prepared BHI agar supplemented with the selective markers: G418, nourseothricin, hygromycin, and penicillin-streptomycin.
4. Seal plates with parafilm and incubate them covering up.
5. 5–7 days after plating, small defined colonies begin to appear. After these colonies have grown up to 1–2 mm diameter, they can be transferred to 0.2 mL of selective growth medium in a 96-well plate using a pipette tip.
6. After 24-h incubation at 26 °C, these clones must be expanded into 1 mL selective medium in a 12-well plate and incubated under agitation (140 rpm).
7. After 48-h incubation at 26 °C, the cultures are expanded into 10 mL selective medium in T-25 flasks and can be used for evaluation.

3.6 Verification of lbABCG6 Gene Integration into Leishmania Genome

After genomic DNA extraction from the cells, the integration of genes of interest into the *L. tarentolae* genome is verified by PCR using GoTaq polymerase (*see* Fig. 2e). The following primers were used, recognizing the pLEXY-I-neo3 sequence flanking the inserted gene so that the same primers can be used for all the clones:

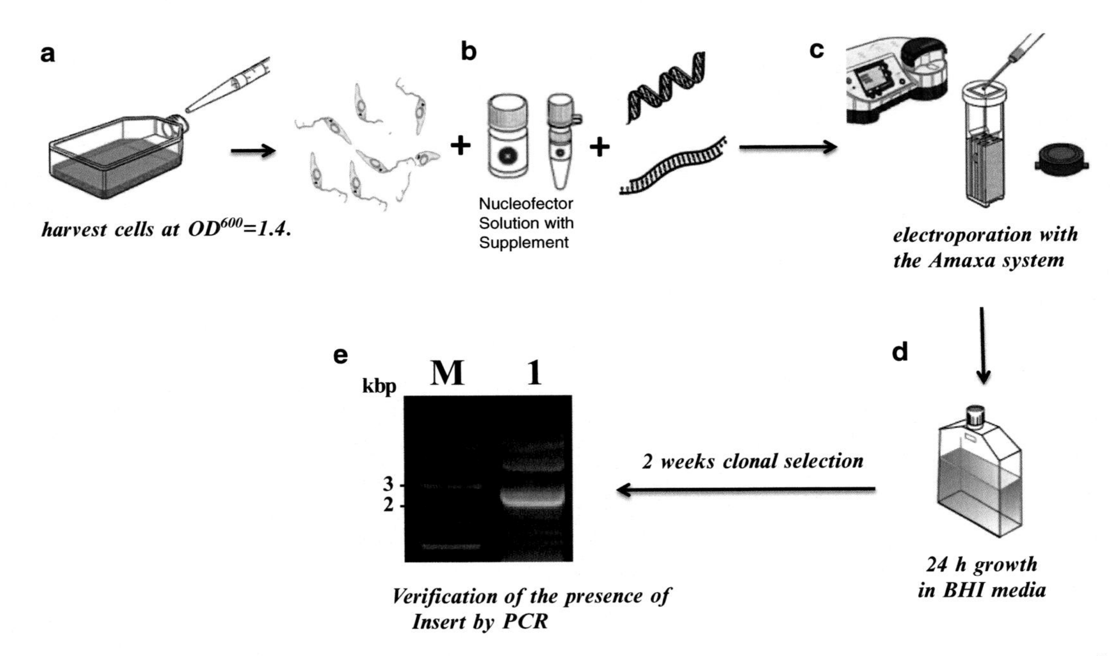

Fig. 2 Protein expression based on pLEXSY system. (**a**) Cells of the parasite are harvested and diluted to 6.10^7 cells/mL and OD^{600} ~1.4. (**b**) After spinning, cells are suspended in 100 μl of supplemented Nucleofector solution and mixed with 4 μg DNA. (**c**) All the mix is transferred to an electroporation cuvette and electroporated according to the Basic Parasite Nucleofector Kit 1 (Lonza), using the program U-033. (**d**) Electroporated cells are transferred to BHI media in a ventilated flask. After 24-h incubation, proceed with clonal selection. (**e**) Genomic DNA is extracted from cells and integration of *lb*ABCG6 is verified by PCR. *Lane M* molecular weight marker, *line 1* genomic DNA from *lb*ABCG6 transfected cells giving a PCR product of 2300 pb

Fwd 5′- CCGACTGCAACAAGGTGTAG and Rev 5′- GAGATGTTCCTGACCGACC.

3.7 Analysis of Protein Expression by SDS-PAGE and Western Blot

The expression of the protein can be checked rapidly after cell transfection. Cells grown to an OD^{600} = 1.4 in a T-25 flask and protein expression are induced with 10 μg/mL tetracycline for 24 h. One milliliter of culture is harvested by centrifugation and suspended in 50 μL of 50 mM Hepes-NaOH pH 7.5 and cells are broken by three cycles of freeze/thaw in liquid nitrogen/warm water. Fifteen microliters of broken cells are mixed with 5 μL of loading buffer 5×U, followed by analysis on SDS-PAGE and Western blot (Fig. 3).

1. Generate the separating gel (4 mL) of a 10% SDS-PAGE by mixing 1.92 mL of water, 1 mL of 40% acrylamide bisacrylamide solution, 1 mL of 1.5 M Tris–HCl pH 8.8, 40 μL of 10% SDS, 40 μL 10% ammonium persulfate, and 1.6 μL TEMED. Pour the Bio-Rad Mini-Protean 3 device to be 8 mm under the bottom of the wells. Add 200 μL of water at the surface of the gel for preventing the formation of waves (*see* **Note 11**). Polymerization occurs in 30 min at room temperature (22 °C).

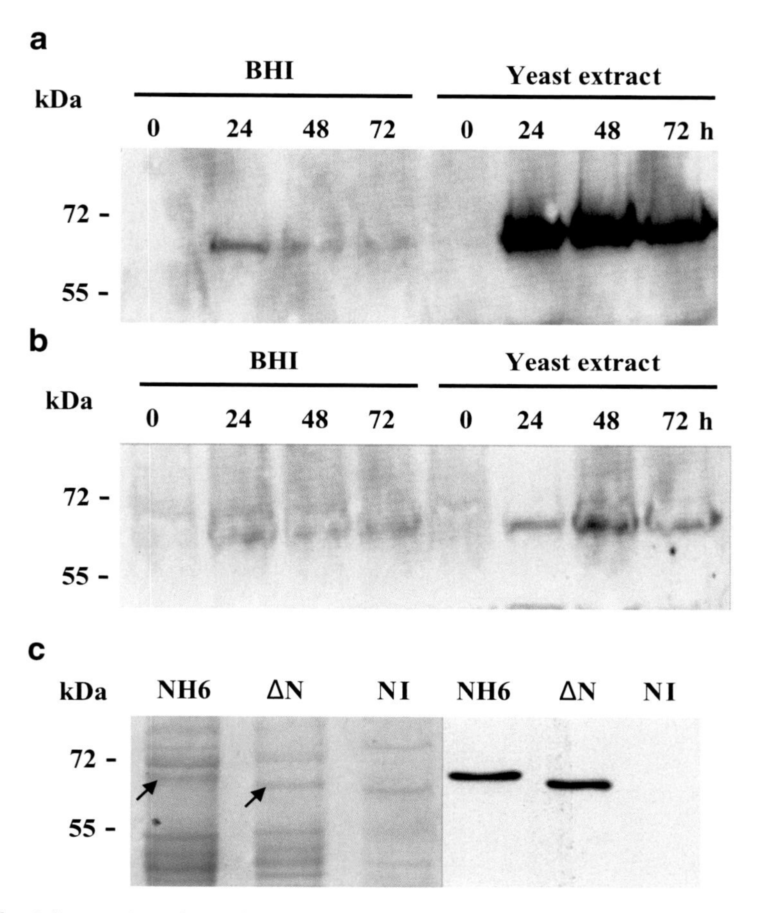

Fig. 3 Expression of *lb*ABCG6 (**a**) and *lb*ABCG6ΔN (**b**) in *L. tarentolae* cells as a function of time and culture media, under agitation. Expression of *lb*ABCG6/*lb*ABCG6ΔN is carried out as described in Subheading 3.5. Transfected cells were maintained under antibiotic selection for 2 weeks in BHI and yeast extract medium. Then, protein expression is induced by adding 10 μg/mL tetracycline and cells harvested after 24, 48, and 72 h. *lb*ABCG6/*lb*ABCG6ΔN expression was analyzed by SDS-PAGE and Western blot loading 15 μl of samples normalized to OD^{600} ~1.4 onto a 10 % SDS-PAGE. (**c**) Membrane expression of *lb*ABCG6 and *lb*ABCG6ΔN in *L. tarentolae* cells. Expression and membrane preparation of *lb*ABCG6/*lb*ABCG6ΔN are carried out as described in Subheading 3.8. Protein expression was analyzed by SDS-PAGE (Coomassie blue-stained Western blot) loading 20 μg and 10 μg of samples, respectively, onto a 10 % SDS-PAGE. *NI*, protein expression without induction as negative control

2. Generate the 5 % stacking gel by mixing 1.46 mL of water, 0.25 mL of 40 % acrylamide bisacrylamide solution, 0.25 mL of 1.5 M Tris–HCl pH 6.8, 20 μL of 10 % SDS, 20 μL 10 % ammonium persulfate, and 2 μL TEMED.
3. Load 20 μL samples onto the stacking gel and run for about 1.5 h at 120 V at room temperature.

4. After electrophoresis, proteins are either stained with Coomassie blue or transferred onto nitrocellulose membrane (*see* below).

For Coomassie Blue staining, incubate the gel for 30 min in 50 mL of staining solution, and then wash it three times in 10% [v/v] glacial acetic acid and 90% H_2O.

When proteins are transferred from the SDS-PAGE to the nitrocellulose membrane, proceed as follows:

1. Incubate the gel in 5 mL of cold transfer buffer for 5 min.
2. Wet the nitrocellulose membrane for 5 min in the cold transfer buffer.
3. Prepare the transfer sandwich built by superposing successively two paper sheets briefly wet in the transfer buffer, the acrylamide gel, the nitrocellulose membrane, and again two wet paper sheets.
4. Add the ice cube to the Mini-Protean 3 transfer device and a magnetic stirrer and transfer for 2 h at 100 V under agitation to optimize cooling.
5. After transfer, block the membrane for 1 h into 20 mL of TBS containing 0.5% blocking reagent and 0.1% Tween 20.
6. Add the primary antibody to the solution and incubate for an additional 1 h.
7. Wash three times with 20 mL of TBS-TT buffer.
8. Wash once with 20 mL of TBS buffer.
9. Withdraw the buffer and incubate with a 1:1 mix of 2 mL ECL solutions A and B for 5 min and expose onto a sensitive film for 1–20 min depending on the antibodies.

   A typical result is illustrated in Fig. 3.

3.8 Optimization of *lb*ABCG6 Expression

For optimization of culture medium and time of expression, proceed as follows:

1. To evaluate the type of medium and the influence of protein expression induction time, transfected *L. tarentolae* cells were grown in BHI or yeast extract media in a T-25 flask as described above.
2. When cells reach $OD^{600}=1.4$, add 10 μg/mL tetracycline to induce protein expression.
3. At 24, 48, and 72 h, harvest an aliquot of cell culture and check for *lb*ABCG6 expression by SDS-PAGE and Western blot as described in Subheading 3.7.

The result is illustrated in Fig. 3. It shows that the expression is higher with yeast extract medium and in both cases it is better at 48 h post-induction. This result permits the scale-up of cell culture, which will grant the quantities of protein required for structural studies.

Scale-up in Fernbach flasks to achieve the yields needed for protein production and purification for structural studies:

1. Grow cells in a T-25 flask (10 mL) used to seed a pre-culture in a T-75 flask (30 mL), both in BHI medium.
2. Once the pre-culture reached a suitable cell density, inoculate a 1 L baffled Fernbach flask filled with 500 mL of yeast extract medium to a final $OD^{600} = 0.2$.
3. When cells reach $OD^{600} = 1.4$, add 10 μg/mL tetracycline to induce protein expression. Protein expression is continued for 48 h, and cells reach an OD^{600} around 4.
4. Harvest the cells by centrifugation for 10 min at 7500 × *g*, 4 °C.
5. Take the pellet in 50 mL Hepes buffer per liter of culture and break the cells by two passages at 15,000 psi in a microfluidizer.
6. Centrifuge for 30 min at 15,000 × *g*, 4 °C, discard the pellet, and centrifuge the supernatant for 1 h at 180,000 × *g*, 4 °C, to collect the membranes.

   Routinely, 1 L of cell culture yields 0.5 g of dry membrane.

As observed in Fig. 3c, the first 30 residues of the protein are not very relevant in terms of expression. There is no remarkable difference in expression levels between *lb*ABCG6ΔN and *lb*ABCG6 constructs when analyzing membrane samples both by Western blot and Coommassie blue-stained SDS-PAGE. Differences shown in Fig. 3a, b, using whole cells, might be due to an artifact in Western blot.

The detailed methodology described above to express membrane proteins using the pLEXSY system in *L. tarentolae* represents an interesting alternative for structural studies. Indeed, culture medium optimization allows cost reduction while maintaining high protein expression levels. In addition, *L. tarentolae* cell culture does not require special biosafety requirements and cross-contamination with other cultures is very low.

4 Notes

1. *Leishmania tarentolae* is not infectious for human; its culture can be done in a class I culture room.
2. Ampicillin should be prepared fresh to a maximal efficiency.
3. Acrylamide is neurotoxic when non-polymerized; thus handle with gloves.

4. Methanol is neurotoxic.
5. Care should be taken not to dilute too much upon passages, as isolated cells do not divide well.
6. A 100-fold dilution is the limit of good growth and should be kept occasionally.
7. *L. tarentolae* can be kept in culture for up to 3 months, after which a new frozen stock should be used.
8. As the number of passages increases, *L. tarentolae* can reach higher cell densities.
9. To reduce costs, nourseothricin is not added in large cultures, without incidence on protein expression.
10. Cultures can also be carried out in Erlenmeyer flasks for volumes ranging from 50 to 500 mL, under agitation of 100–140 rpm.
11. Do not use organic solvent for this step as membrane proteins have a tendency to interact with it.

References

1. Robey RW, Honjo Y, Morisaki K, Nadjem TA, Runge S, Risbood M, Poruchynsky MS, Bates SE (2003) Mutations at amino-acid 482 in the ABCG2 gene affect substrate and antagonist specificity. Br J Cancer 89:1971–1978
2. Nakanishi T, Doyle LA, Hassel B, Wei Y, Bauer KS, Wu S, Pumplin DW, Fang HB, Ross DD (2003) Functional characterization of human breast cancer resistance protein (BCRP, ABCG2) expressed in the oocytes of *Xenopus laevis*. Mol Pharmacol 64:1452–1462
3. Béjà O, Bibi E (1996) Functional expression of mouse Mdr1 in an outer membrane permeability mutant of *Escherichia coli*. Proc Natl Acad Sci U S A 93:5969–5974
4. Trometer C, Falson P (2010) Mammalian membrane protein expression in baculovirus-infected insect cells. Methods Mol Biol 601:105–117
5. Sarkadi B, Price EM, Boucher RC, Germann UA, Scarborough GA (1992) Expression of the human multidrug resistance cDNA in insect cells generates a high activity drug-stimulated membrane ATPase. J Biol Chem 267:4854–4858
6. Mao Q, Conseil G, Gupta A, Cole SP, Unadkat JD (2004) Functional expression of the human breast cancer resistance protein in *Pichia pastoris*. Biochem Biophys Res Commun 320:730–737
7. Janvilisri T, Venter H, Shahi S, Reuter G, Balakrishnan L, van Veen HW (2003) Sterol transport by the human breast cancer resistance protein (ABCG2) expressed in *Lactococcus lactis*. J Biol Chem 278:20645–20651
8. Pozza A, Perez-Victoria JM, Di Pietro A (2010) Insect cell versus bacterial overexpressed membrane proteins: an example, the human ABCG2 transporter. Methods Mol Biol 654:47–75
9. Ozvegy C, Varadi A, Sarkadi B (2002) Characterization of drug transport, ATP hydrolysis, and nucleotide trapping by the human ABCG2 multidrug transporter. Modulation of substrate specificity by a point mutation. J Biol Chem 277:47980–47990
10. Telbisz A, Müller M, Ozvegy-Laczka C, Homolya L, Szente L, Váradi A, Sarkadi B (2007) Membrane cholesterol selectively modulates the activity of the human ABCG2 multidrug transporter. Biochim Biophys Acta 1768:2698–2713
11. Lerner-Marmarosh N, Gimi K, Urbatsch IL, Gros P, Senior AE (1999) Large scale purification of detergent-soluble P-glycoprotein from *Pichia pastoris* cells and characterization of nucleotide binding properties of wild-type, Walker A, and Walker B mutant proteins. J Biol Chem 274:34711–34718
12. Cai J, Daoud R, Georges E, Gros P (2001) Functional expression of multidrug resistance protein 1 in *Pichia pastoris*. Biochemistry 40:8307–8316
13. Niimi T (2012) Recombinant protein production in the eukaryotic protozoan parasite *Leishmania tarentolae*: a review. Methods Mol Biol 824:307–315

14. Kushnir S, Gase K, Breitling R, Alexandrov K (2005) Development of an inducible protein expression system based on the protozoan host *Leishmania tarentolæ*. Protein Expr Purif 42:37–46
15. Breitling R, Klingner S, Callewaert N, Pietrucha R, Geyer A, Ehrlich G, Hartung R, Müller A, Contreras R, Beverley SM, Alexandrov K (2002) Non-pathogenic trypanosomatid protozoa as a platform for protein research and production. Protein Expr Purif 25:209–218
16. Bolhassani A, Taheri T, Taslimi Y, Zamanilui S, Zahedifard F, Seyed N, Torkashvand F, Vaziri B, Rafati S (2011) Fluorescent *Leishmania* species: development of stable GFP expression and its application for in vitro and in vivo studies. Exp Parasitol 127:637–645
17. Castanys-Muñoz E, Pérez-Victoria JM, Gamarro F, Castanys S (2008) Characterization of an ABCG-like transporter from the protozoan parasite Leishmania with a role in drug resistance and transbilayer lipid movement. Antimicrob Agents Chemother 52:3573–3579
18. Litman T, Brangi M, Hudson E, Fetsch P, Abati A, Ross DD, Miyake K, Resau JH, Bates SE (2000) The multidrug-resistant phenotype associated with overexpression of the new ABC half-transporter, MXR (ABCG2). J Cell Sci 113:2011–2021
19. Liu XL, Tee HW, Go ML (2008) Functionalized chalcones as selective inhibitors of P-glycoprotein and breast cancer resistance protein. Bioorg Med Chem 16:171–180
20. Lenoir G, Menguy T, Corre F, Montigny C, Pedersen PA, Thinès D, le Maire M, Falson P (2002) Overproduction in yeast and rapid and efficient purification of the rabbit SERCA1a Ca(2+)-ATPase. Biochim Biophys Acta 1560: 67–83

Chapter 9

Overexpression, Membrane Preparation, and Purification of a Typical Multidrug ABC Transporter BmrA

Benjamin Wiseman and Jean-Michel Jault

Abstract

The production and purification is normally the first step in any biophysical or biochemical study of a new target protein. For membrane proteins, due to their generally low expression levels and hydrophobic properties this is often a major hurdle. Some multidrug transporters are members of one of the largest families of membrane proteins, the ABC ("ATP-binding cassette"), and are responsible for the uptake and export of a wide variety of molecules. This can lead to resistance when those molecules are antibiotics or chemotherapy drugs. To better understand their role in multidrug resistance pure and active protein is required. Here we outline a protocol to produce a highly pure and functionally active multidrug transporter BmrA that is suitable for use in biophysical and biochemical studies. We show that BmrA can be heterologously overexpressed in huge amount in *E. coli* and extracted from the membrane in a functionally active form.

Key words ATP-binding cassette, Transporter, Membrane protein, Multidrug resistance, Purification

1 Introduction

Multidrug transporters belong to one of the largest membrane protein families known as ATP-binding cassette (ABC) proteins. In this family, membrane transporters are responsible for either the import or the export of many different molecules across the membrane [1]. In general exporters efflux a wide array of molecules including anticancer drugs and antibiotics implicating them in multidrug resistance (MDR) phenotypes in human cancer cells and pathogenic bacteria, respectively [2]. These ABC transporters share a common architecture consisting of two soluble nucleotide-binding domains and two transmembrane domains that couple ATP binding and hydrolysis with the transport of molecules across a membrane against a concentration gradient [3, 4]. Despite the success of several reported structures of exporters [5–11] many open questions still remain about the exact mechanism of substrate recognition and multidrug efflux. BmrA ("bacillus multidrug resistance ATP") is one such transporter belonging to this

Isabelle Mus-Veteau (ed.), *Heterologous Expression of Membrane Proteins: Methods and Protocols*, Methods in Molecular Biology, vol. 1432, DOI 10.1007/978-1-4939-3637-3_9, © Springer Science+Business Media New York 2016

large family that can be overexpressed in *E. coli* [12] and purified to homogeneity in a functional state [13].

Membrane proteins in general make up approximately one-third of the proteome in all living organisms and are the target of the vast majority of the drugs on the market today. Despite their importance the hydrophobic nature and low expression levels of membrane proteins make them extremely difficult to study. In recent years however, great progress in their expression and purification has been made. Together with new detergents, advances in crystallization, powerful free electron lasers, and micro-focus synchrotron beams [14] these once notoriously difficult proteins are becoming a lot more attractive targets to study. Functional and structural studies of any protein normally require a highly pure, stable, and homogenous sample. Here we outline a protocol for the production and purification of a typical ABC transporter suitable for such studies.

2 Materials

2.1 Cell Culture and Overexpression

1. *E. coli* strain C41 (*see* **Note 1**) transformed with plasmid containing T7 inducible BmrA with a 6xHis *N*-terminal tag (*see* **Note 2**) plated on LB-ampicillin agar.
2. 3 × 5 L flasks containing 2 L of autoclaved 2TY media: 20 g/L LB base, 6 g/L tryptone, 5 g/L yeast extract, 5 g/L NaCl.
3. 100 mg/mL Ampicillin.
4. 1 M Isopropyl β-D-1-thiogalactopyranoside (IPTG).

2.2 Membrane Preparation

1. 200 mL Cell lysis buffer: 50 mM Tris–HCl pH 8.0, 5 mM $MgCl_2$, 1 mM phenylmethanesulfonylfluoride (PMSF) (*see* **Note 3**), complete EDTA-free protease inhibitor cocktail pills (Roche, 1 pill/50 mL).
2. 100 mL Membrane washing buffer: 50 mM Tris–HCl pH 8.0, 1 mM PMSF, anti-protease pills.
3. 50 mL Membrane storage buffer: 20 mM Tris–HCl pH 8.0, 300 mM sucrose.
4. 30 mL Glass homogenizer with pestle.

2.3 Membrane Solubilization and Protein Purification

1. 10 mL 10% Stock solution of *n*-Dodecyl-β-D-maltopyranoside (DDM, Anagrade, Affymetrix) (*see* **Note 4**).
2. 100 mL Solubilization buffer: 50 mM Tris–HCl pH 8.0, 10% glycerol, 100 mM NaCl, 10 mM imidazole, 1 mM PMSF, anti-protease pills.
3. 3 mL Ni-NTA resin.

4. 2 × 20 mL Plastic gravity flow columns with filters (*see* **Note 5**).
5. 100 mL Washing buffer A: 50 mM Tris–HCl pH 8.0, 10% glycerol, 100 mM NaCl, 25 mM imidazole, 1 mM PMSF, anti-protease pills, 0.04% DDM.
6. 100 mL Washing buffer B: 50 mM Tris–HCl pH 8.0, 10% glycerol, 100 mM NaCl, 50 mM imidazole, 1 mM PMSF, anti-protease pills, 0.04% DDM.
7. 50 mL Elution buffer: 50 mM Tris–HCl pH 8.0, 5% glycerol, 100 mM NaCl, 250 mM imidazole, 1 mM PMSF, anti-protease pills, 0.03% DDM.
8. 500 mL Gel filtration buffer: 25 mM Tris–HCl pH 8.0, 100 mM NaCl, 0.02% DDM.
9. SEC/FPLC HiLoad 16/60 Superdex 200 PG column and AKTA system with a 10 mL superloop (GE Healthcare).

3 Methods

Producing pure, stable, and active protein is often the first step in the biophysical, functional, and structural characterization of any protein. In the case of membrane proteins this can often be a difficult and tedious task. Although the exact conditions (e.g., pH, salt concentration, additives, detergent) will most likely need to be imperially optimized for each target protein, the method outlined below can act as a good base for the purification of any new ABC transporter and in fact any His-tagged membrane protein overproduced in *E. coli*. The method in particular is well suited to produce highly pure membrane protein samples suitable to commence crystallization trails with. It brings to light the often overlooked or rarely mentioned problem of trace contaminates present in membrane protein purifications and presents a simple solution to remove them [15].

Ultimately it is important to check if the final purified product is pure, stable, and active before proceeding with downstream applications. This can be easily accomplished with SDS-PAGE, SEC, and activity tests. An overloaded SDS-PAGE gel with 15 μg or more of purified protein can easily identify contaminates. Incubation of the purified, concentrated sample at room temperature and/or 4 °C and reinjection into the same SEC column used during purification can assess stability by visualizing the appearance of peaks that elute before and after the main peak representing large aggregates or degradation products, respectively, along with the disappearance of the main protein peak. These together with an ATPase activity assay [16] allow the quality of the final purified protein to be quickly determined.

3.1 Cell Culture and Overexpression

1. Once autoclaved media has cooled, add 2 mL of 100 mg/mL ampicillin to each flask containing 2 L of media to a final concentration of 100 μg/mL. Using a sterile 1 mL pipette tips pick 6–8 colonies of freshly transformed *E. coli* and expunge the entire tip into the media (it is possible to pick 3–4 colonies per pipette tip). Incubate flasks overnight at 25 °C with vigorous shaking (*see* **Note 6**).
2. When the OD_{600} reaches 1.0–1.2 induce expression with 0.7 mM IPTG (1.4 mL of 1 M IPTG stock per 2 L culture) and incubate for an additional 3–4 h at 25 °C with vigorous shaking. At harvest the OD_{600} will be approximately 2.5.
3. Centrifuge the culture at 5000 × *g* for 15 min to collect cells. The cells can then be washed with 200 mL of 50 mM Tris–HCl pH 8.0 by pipetting up and down with a 25 mL disposable pipette until they reach a homogenous mixture. Collect the cells by centrifuging again at 5000 × *g* for 15 min and transfer to a 250 mL plastic beaker with a spatula. Cover with tinfoil, label, and store at −20 °C until membranes can be prepared.

3.2 Membrane Preparation

1. Thaw cells and suspend in an approximate final volume of 130 mL of cell lysis buffer (*see* **Note 7**). Place at 4 °C and stir until the mixture is completely homogenous with no visible clumps of cells remaining. From this point on keep everything at 4 °C or on ice.
2. Lyse cells with a high-pressure cell disrupter such as a French press, Emulsiflex (Avestin), or Microfluidizer (Microfluidics). Follow the manufacturer's guidelines for operation of the cell disrupter but normally three passages through the chosen disrupter are more than sufficient for lysis of almost 100 % of the cells.
3. Centrifuge at 15,000 × *g* for 30 min at 4 °C to pellet cell debris and unlysed cells. Carefully decant supernatant to a clean beaker (*see* **Note 8**).
4. Add fresh PMSF to a final concentration of 1 mM.
5. Ultracentrifuge at 150,000 × *g* (37,000 rpm, rotor type 45Ti) for 1 h at 4 °C.
6. Pour off the supernatant and add approximately 10 mL of washing buffer to each centrifuge tube containing the pelleted membranes. *E. coli* membranes are normally brown in color. Using a metal spatula scrape the membranes off the side of the centrifuge tubes and pour them into a 30 mL glass homogenizer (*see* **Note 9**). Add another 10–20 mL of washing buffer and homogenize the membranes with the pestle until they are completely homogenous.

7. Pour the suspended membranes into a clean 45Ti ultracentrifuge tube, fill to the top with washing buffer, and ultracentrifuge again at 150,000 × *g* for 1 h at 4 °C.
8. Similar to **step 6** above, suspend the pelleted membranes in membrane storage buffer to a final volume of 20 mL. Take a 10 μL aliquot to determine concentration (modified Lowery method) and visualization with SDS-PAGE. Flash freeze with liquid nitrogen in 5 mL aliquots in 15 mL Flacon tubes and store at −80 °C.

3.3 Membrane Solubilization and Protein Purification

1. Prepare a 10 mL 10% stock solution of DDM. This should be made fresh the day before or in the morning of the day of the purification.
2. Equilibrate Superdex200 16/60 column and 10 mL superloop with 1 volume of gel filtration buffer (approximately 150 mL) (*see* **Note 10**).
3. Thaw membranes and dilute to 10 mg/mL or less with cold solubilization buffer. Once the membranes are thawed and diluted in buffer add DDM to a final concentration of 1%. It is convenient to dilute to a final volume of about 65 mL (including detergent) (*see* **Note 7**). Incubate with gentle stirring at 4 °C for 90–120 min. Try to avoid as much as possible the formation of detergent bubbles during incubation.
4. To remove any unsolubilized membranes ultracentrifuge solubilized membrane mixture at 150,000 × *g* (37,000 rpm, rotor type 45Ti) for 1 h (*see* **Note 11**).
5. During ultracentrifugation wash two plastic gravity flow columns and filters with water and 20% ethanol (*see* **Note 12**).
6. Add 1.5 mL of Ni-NTA agarose to each clean column and wash with copious amounts of water. Add DDM to a final concentration of 0.05% to the remaining solubilization buffer and equilibrate the Ni-NTA agarose by adding 5 mL buffer to each column. Once the 5 mL has passed through add an additional 10 mL or more of buffer, swirl to resuspend the agarose, and pour the contents of each column into a separate 50 mL Falcon tube. Store at 4 °C until the solubilized membranes are ready.
7. Five to ten minutes before the solubilized membranes are ready, centrifuge the equilibrated Ni-NTA agarose for 5 min at 4000 × *g* and carefully decant as much supernatant as possible without loosing any Ni-NTA.
8. Pour half of the supernatant of the ultracentrifuged solubilized membranes into each 50 mL Falcon tube containing the equilibrated Ni-NTA agarose and incubate for 90 min at 4 °C on a rotating wheel.
9. After incubation collect the Ni-NTA agarose by pouring into each plastic column. Let the flow-through run until there is

no more liquid on the top of the agarose. There should be approximately 1.5 mL of agarose in each column.

10. Wash each column with 50 mL of washing buffer A (25 mM imidazole) at 4 °C (the following steps are also performed at the same temperature). Wash by filling column to the top with buffer and let it run dry before adding more washing buffer. Continue in this way until approximately 50 mL has passed through each column.
11. Wash each column with 50 mL of washing buffer B (50 mM imidazole) (*see* **Note 13**). Wash by filling column to the top with buffer and let it run dry before adding more washing buffer. Continue in this way until approximately 50 mL has passed through each column.
12. Elute the bound protein by adding 6 mL of elution buffer to each column and collect in a 15 mL Falcon tube. Add to each column one at a time to be able to collect each column elution in the same 15 mL Falcon tube. At this point there should be an elution of approximately 12 mL.
13. Concentrate elution to 5 mL or less using a 100 kDa MWCO filter (*see* **Note 14**). Centrifuge in 5-min intervals with mixing in between by pipetting up and down using a 1 mL pipette.
14. When the volume reaches 5 mL or less transfer elution to microcentrifuge tubes and centrifuge at maximum speed at 4 °C for 15 min. Transfer supernatant to clean tubes using a pipette or syringe being careful not to disturb any pellet.
15. Inject the concentrated elution into the equilibrated SEC column. Run at a flow rate of 1 mL/min and collect 1 mL fractions. Here is a good break point. The fractions could be left at 4 °C overnight to be analyzed in the morning.
16. Pool the appropriate fractions. In this case the BmrA dimer in a DDM micelle elutes at about 60 mL (Fig. 1). Try to avoid pooling aggregates as much as possible.
17. Concentrate the pooled fractions using a 100 kDa MWCO concentrator (*see* **Note 14**) to the desired concentration. Similar to **step 13**, centrifuge in 5-min intervals with mixing in between.
18. When the desired volume is reached transfer to microcentrifuge tube and centrifuge at maximum speed at 4 °C for 15 min. Transfer supernatant to clean tubes using a pipette or syringe being careful not to disturb any pellet.
19. Determine concentration using the Bradford method and check purity of final concentrated sample by visualizing on a 12 % SDS-PAGE (Fig. 2) (*see* **Note 15**).
20. Aliquot into 50 μL aliquots, flash freeze with liquid nitrogen, and store at −80 °C if not used immediately.

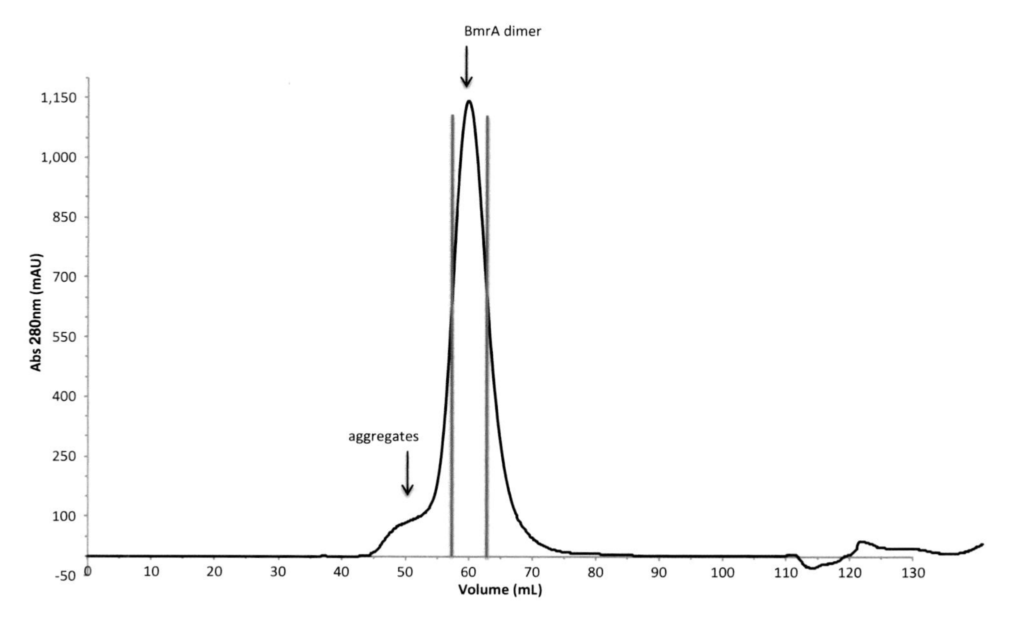

Fig. 1 Size-exclusion chromatography of Ni-NTA-purified BmrA. The red bars (57 mL to 63 mL) represent the pooled fractions as described in the text

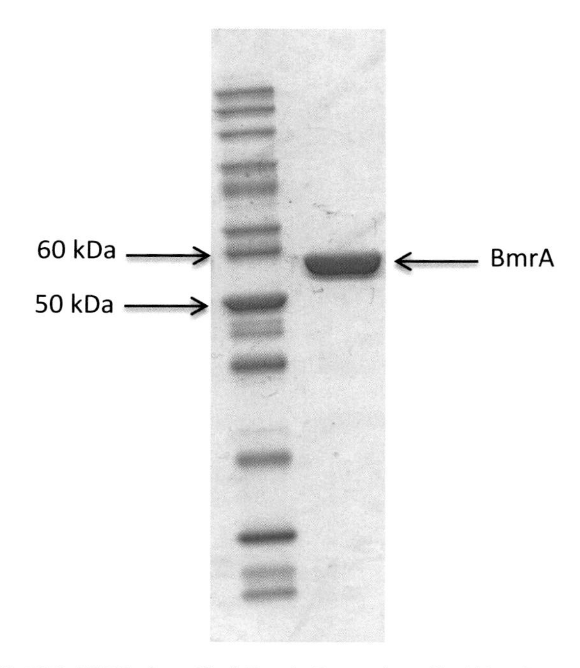

Fig. 2 12% SDS-PAGE of purified BmrA. 5 μg of purified BmrA was visualized with Coomassie brilliant blue. The sample was mixed with loading buffer and incubated at 37 °C for 5 min prior to loading. Note that the expected molar mass of BmrA is 65 kDa, but it migrates to an apparent mass of 55 kDa

4 Notes

1. It is highly recommended to use a strain deficient in AcrB if possible. Due to a histidine-rich cluster this menacing protein tends to bind to Ni-NTA resin and therefore contaminate Ni-NTA-based purifications. This is especially problematic if the downstream application is crystallization trails as AcrB is easily crystallizable and can crystallize at concentrations in the picomolar range. It has been repeatedly crystallized by accident [17–19].
2. Creating a cleavable His-tag version is also an option to eliminate AcrB and any other contaminating proteins that tend to stick to Ni-NTA resins. By running the cleaved protein through a second Ni-NTA column only the contaminating proteins will bind.
3. PMSF is a serine protease inhibitor that rapidly degrades in aqueous solutions. A 100 mM stock solution in 100% isopropanol is stable for months at 4 °C. Add PMSF from the stock solution to cold buffers just before use to ensure that it has its full inhibitory effect.
4. Although it is a good idea to screen many detergents before finally selecting a detergent for purification, DDM is a good and relatively safe detergent for an initial trail. It has been used successfully many times for the purification and crystallization of ABC transporters and many other α-helical membrane proteins [20].
5. We have noticed that using two gravity flow columns with 1.5 mL of Ni-NTA-resin each opposed to one column with 3 mL of Ni-NTA resin results in faster dripping columns and thus a faster purification and purer final product.
6. Inoculating the large overexpression flasks directly with colonies opposed to an overnight preculture has the advantage of saving one day and very often results in greater overexpression of the target protein. Care should be taken however as not to inoculate with too many colonies that would result in an overgrown culture in the morning. If unsure it is advisable to inoculate with fewer colonies; if the OD_{600} is too low in the morning the incubator could always be turned up to 37 °C.
7. It is convenient to dilute to a final volume of about 130 mL (multiples of 65 mL) since the type 45Ti ultracentrifuge tubes hold 70 mL and must be filled to the top to prevent implosion during ultracentrifugation. If using a different ultracentrifuge rotor adjust the volume accordingly.
8. It is also possible to filter the supernatant through a 0.2 μm filter to ensure having a clean supernatant free of contaminating unlysed cells and/or outermembrane debris although this can

be very slow as the filter tends to get clogged very quickly with cell debris.

9. *E. coli* membranes are normally brown in color. If a small amount of cell debris is accidently decanted with the cell lysate in **step 3** there will be a small white button visible. When scrapping the membranes off the centrifuge tube try to avoid this button as much as possible.
10. It is convenient to have the SEC column equilibrated, so the Ni-NTA elution can be injected immediately after elution and thus the protein spends as little time as possible in a high concentration of imidazole.
11. Here it is recommended to suspend the pelleted membranes in the same starting volume (in this case 20 mL) to ensure the same concentration as the starting membranes. Check the amount of unsolubilized protein that remains in the membrane by running an SDS-PAGE with an equal volume of membranes before and after solubilization. Do not be alarmed to see target protein in pelleted membranes. We have found that doing a second round of solubilization does not extract more of the target protein.
12. Check that the filters are not blocked. If the filters are blocked or dripping slowly washing with 95% ethanol can sometimes clear them. If washing with ethanol does not help, replace filters. Slow-dripping filters will not only make for a very long day but could also result in a less pure final product.
13. We have noticed that a 50 mM imidazole wash is sufficient to remove AcrB contamination from Ni-NTA resins. However, it should be tested that the target protein is not washed off the Ni-NTA before washing with 50 mM imidazole. Even if using an AcrB-deficient strain during overexpression it is still recommended to include a second wash with higher imidazole concentration.
14. It is important here to use the highest possible MWCO filter to avoid concentrating detergent as much as possible. We have found that empty DDM micelles will concentrate in the presence of protein even when using a 100 kDa MWCO filter. For this reason it is important to record the starting and final volumes before and after this concentrating step. Knowing this, future purifications can be concentrated the same number of times ensuring that they will have a similar amount of detergent in the final concentrated purified sample.
15. In general, unlike soluble proteins, membrane protein samples should not be boiled prior to loading onto SDS-PAGE. Boiling often results in large aggregates; thus instead of visualizing the target protein at the correct mass all of the protein will be stuck at the top of the gel. Incubating at room temperature or 37 °C for a few minutes is normally sufficient.

References

1. Sk N (2015) What do drug transporters really do? Nat Rev Drug Discov 14:29–44
2. Rees DC, Johnson E, Lewinson O (2009) ABC transporters: the power to change. Nat Rev Mol Cell Biol 10:218–227
3. Kos V, Ford RC (2009) The ATP-binding cassette family: a structural perspective. Cell Mol Life Sci 66:3111–3126
4. Wilkens S (2015) Structure and Mechanism of ABC Transporters. F1000Prime Rep. 7: 14.
5. Dawson RJ, Locher KP (2006) Structure of a bacterial multidrug ABC transporter. Nature 443:180–185
6. Ward A, Reyes CL, Yu J, Roth CB, Chang G (2007) Flexibility in the ABC transporter MsbA: Alternating access with a twist. Proc Natl Acad Sci U S A 104:19005–19010
7. Aller SG, Yu J, Ward A, Weng Y, Chittaboina S, Zhuo R, Harrell PM, Trinh YT, Zhang Q, Urbatsch IL, Chang G (2009) Structure of P-glycoprotein reveals a molecular basis for poly-specific drug binding. Science 323: 1718–1722
8. Jin MS, Oldham ML, Zhang Q, Chen J (2012) Crystal structure of the multidrug transporter P-glycoprotein from *Caenorhabditis elegans*. Nature 490:566–569
9. Hohl M, Briand C, Grutter MG, Seeger MA (2012) Crystal structure of a heterodimeric ABC transporter in its inward-facing conformation. Nat Struct Mol Biol 19:395–402
10. Fribourg PF, Chami M, Sorzano CO, Gubellini F, Marabini R, Marco S, Jault JM, Lévy D (2014) 3D cryo-electron reconstruction of BmrA, a bacterial multidrug ABC transporter in an inward-facing conformation and in a lipidic environment. J Mol Biol 426:2059–2069
11. Kim J, Wu S, Tomasiak TM, Mergel C, Winter MB, Stiller SB, Robles-Colmanares Y, Stroud RM, Tampé R, Craik CS, Cheng Y (2015) Subnanometre-resolution electron cryomicroscopy structure of a heterodimeric ABC exporter. Nature 517:396–400
12. Steinfels E, Orelle C, Dalmas O, Penin F, Miroux B, Di Pietro A, Jault JM (2002) Highly efficient over-production in *E. coli* of YvcC, a multidrug-like ATP-binding cassette transporter from *Bacillus subtilis*. Biochim Biophys Acta 1565:1–5
13. Steinfels E, Orelle C, Fantino JR, Dalmas O, Rigaud JL, Denizot F, Pietro A, Jault JM (2004) Characterization of YvcC (BmrA), a multidrug ABC transporter constitutively expressed in *Bacillus subtilis*. Biochemistry 43:7491–7502
14. Loll PJ (2014) Membrane proteins, detergents and crystals: what is the state of the art. Acta Crysallogr F Struct Biol Commun 70: 1576–1583
15. Wiseman B, Kilburg A, Chaptal V, Reyes-Mejia GC, Sarwan J, Falson P, Jault JM (2014) Stubborn Contaminants: Influence of detergents on the purity of the multidrug ABC transporter BmrA. PLoS One 9, e114864
16. Ravaud S, Do Cao MA, Jidenko M, Ebel C, Le Maire M, Jault JM, Di Pietro A, Haser R, Aghajari N (2006) The ABC transporter BmrA from *Bacillus subtilis* is a functional dimer when in a detergent-solublized state. Biochem J 395: 345–353
17. Glover CA, Postis VL, Charalambous K, Tzokov SB, Booth WI, Deacon SE, Wallace BA, Baldwin SA, Bullough PA (2011) AcrB contamination in 2-D crystallization of membrane proteins: lessons from a sodium channel and a putative monovalent cation/proton antiporter. J Struct Biol 176:419–424
18. Psakis G, Polaczek J, Essen LO et al (2009) AcrB: Obstinate contaminants in a picogram scale. One more bottleneck in the membrane protein structure pipeline. J Struct Biol 166: 107–111
19. Veesler D, Blangy S, Cambillau C, Sciara G (2008) There is a baby in the bath water: AcrB contamination is a major problem in membrane-protein crystallization. Acta Crystallogr Sect F Struct Biol Cryst Commun 64:880–885
20. Parker JL, Newstead S (2012) Current trends in α-helical membrane protein crystallization: an update. Protein Sci 9:1358–1365

Chapter 10

Expression of Eukaryotic Membrane Proteins in *Pichia pastoris*

Lucie Hartmann, Valérie Kugler, and Renaud Wagner

Abstract

A key point when it comes to heterologous expression of eukaryotic membrane proteins (EMPs) is the choice of the best-suited expression platform. The yeast *Pichia pastoris* has proven to be a very versatile system showing promising results in a growing number of cases. Indeed, its particular methylotrophic characteristics combined to the very simple handling of a eukaryotic microorganism that possesses the majority of mammalian-like machineries make it a very competitive expression system for various complex proteins, in amounts compatible with functional and structural studies. This chapter describes a set of robust methodologies routinely used for the successful expression of a variety of EMPs, going from yeast transformation with the recombinant plasmid to the analysis of the quality and quantity of the proteins produced.

Key words *Pichia pastoris*, Yeast, Eukaryotic membrane protein, GPCR, Heterologous expression

1 Introduction

Pichia pastoris is one of the few budding yeasts that have developed a specific methanol utilization metabolism. Upon deprivation of preferential carbohydrates (i.e., sugars, glycerol), the presence of methanol as the unique carbon source actually strongly stimulates the expression of genes involved in its degradation for energy and biomass production [1]. The exceptional expression levels triggered by these tightly regulated methanol-dependent promoters have thus been exploited to constitute one of the most efficient eukaryotic systems used for bioproduction purposes. Up to now indeed, hundreds of proteins have been successfully produced with this host, including more than 70 commercial products (http://www.pichia.com/science-center/commercialized-products).

Non-surprisingly, this system has also proven to be very powerful for the recombinant expression of a large number of eukaryotic membrane proteins (EMPs) from various key membrane functions (transport, signaling, enzymatic activities) and diverse

Isabelle Mus-Veteau (ed.), *Heterologous Expression of Membrane Proteins: Methods and Protocols,* Methods in Molecular Biology, vol. 1432, DOI 10.1007/978-1-4939-3637-3_10, © Springer Science+Business Media New York 2016

structural organizations (representative membrane-spanning topologies and oligomeric states involving various membrane and cytosoluble subunits). As a strong illustration of its versatility and competitiveness in this domain, the most diverse panel of structure-solved EMPs has been produced with *P. pastoris*, encompassing P-glycoprotein ABC transporters, monoamine oxidases, G-protein-coupled receptors (GPCRs), several aquaporins, and various types of ion channels (http://blanco.biomol.uci.edu/mpstruc/, as of June 2015).

A full description of the system, its advantages and drawbacks, and how it performs for the production of EMPs in comparison with other expression systems can be found in a number of very complete and authoritative reviews [2–6].

We present here a series of robust and straightforward methodologies and techniques that we routinely apply for the overexpression of EMPs that are intended to be purified for different biochemical and biophysical purposes [7–10]. Exemplified with a panel of representative EMPs, this chapter covers all the procedures needed to handle the expression system, from yeast transformation and clone selection steps down to the inducible expression phases in different formats. Associated analytical experiments performed to evaluate the expression levels and, when possible, the activity of the recombinant EMP, are also described. The following extraction and purification steps are not addressed in this chapter since, from our experience, these procedures are quite systematically protein dependent and request to be set up in a tailored-made fashion.

2 Materials

2.1 Yeast Transformation

1. The recombinant expression vector of your choice that you carefully selected (*see* Subheading 3.1).
2. Restriction enzyme *Pme*I and its specific buffer.
3. Sterile water.
4. Nucleic acid extraction and purification kit (e.g., NucleoSpin kit, Machery-Nagel).
5. Loading dye (e.g., 6× DNA gel loading dye, Thermo Scientific).
6. 1% (w/v) agarose gels supplemented with 0.5 μg/mL ethidium bromide.
7. Additional equipment for agarose gel electrophoresis.
8. 24:24:1 (v/v/v) chloroform:phenol:isoamyl alcohol.
9. Chloroform.
10. 100% (v/v) ethanol, ice cold.

11. 3 M Sodium acetate, pH 4.8.
12. 70 % (v/v) ethanol, ice cold.
13. *Pichia pastoris* strain SMD1163 streaked on a YPD plate.
14. YPD agar plates: 1 % (w/v) yeast extract, 2 % (w/v) peptone, 2 % (w/v) dextrose, 2 % (w/v) agar. Heat sterilize in autoclave.
15. Spectrophotometer .
16. 1 M 4-(2-Hydroxyethyl)piperazine-1-ethanesulfonic acid (HEPES), pH 8.
17. 1 M Dithiothreitol (DTT).
18. 1 M sorbitol, ice-cold.
19. 30 °C Shaking incubator.
20. Electroporation instrument (e.g., Gene Pulser system, BioRad).
21. Sterile 0.2 cm electroporation cuvettes.
22. YNB plates (minimal medium): 1.34 % (w/v) yeast nitrogen base without amino acids, 2 % (w/v) dextrose, 2 % (w/v) agar. Heat sterilize in autoclave.
23. All materials and solutions in contact with the cells must be sterile.

2.2 Screening of Recombinant Clones

1. YPD liquid medium and agar plates: 1 % (w/v) yeast extract, 2 % (w/v) peptone, 2 % (w/v) dextrose. For YPD agar plates, add 2 % (w/v) agar. Heat sterilize in autoclave. When needed, supplement with appropriate concentrations of antibiotic (typically 50–250 μg/mL geneticin).
2. Spectrophotometer.
3. 100 % (v/v) glycerol, autoclave sterilized.
4. 96-Well plate.
5. Sterile toothpicks or inoculating loop.
6. 30 °C Shaking incubator.
7. YEP agar medium (to be dissolved in 700 mL): 10 g Yeast extract, 20 g meat peptone, 20 g agar. Heat sterilize in autoclave.
8. 1 M Phosphate buffer, pH 6: 3.1 % (w/v) dipotassium hydrogen phosphate trihydrate (K_2HPO_4, 3 H_2O), 11.81 % (w/v) potassium dihydrogen phosphate (KH_2PO_4). Filter sterilize.
9. BMGY agar medium: 700 mL YEP agar medium (still liquid and warm), 100 mL 13.4 % (w/v) yeast nitrogen base without amino acid (10× solution, filter sterilized), 100 mL 10 % (v/v) glycerol (10× solution, filter sterilized), 100 mL 1 M phosphate buffer, pH 6 (10× solution, filter sterilized).

10. BMMY agar medium: 700 mL YEP agar medium (still liquid and warm), 100 mL 13.4% (w/v) yeast nitrogen base without amino acid (10× solution, filter sterilized), 100 mL 5% (v/v) methanol (10× solution, filter sterilized), 100 mL 1 M phosphate buffer, pH 6 (10× solution, filter sterilized).
11. 14 cm diameter Petri dishes.
12. 0.45 μm Nitrocellulose blotting membrane of a 96-well plate dimension.
13. Lysis buffer: 100 mM Tris–HCl pH 7.4, 150 mM β-mercaptoethanol, 20 mM ethylenediaminetetraacetic acid (EDTA), 2% (v/v) SDS, and 9 M urea. Dissolve the components at 65 °C.
14. Whatman paper of a 96-well plate dimension.
15. 65 °C Incubator.

2.3 Yeast Culturing

1. Recombinant clones freshly streaked on a YPD plate supplemented with appropriate concentrations of antibiotic (typically 50 μg/mL geneticin)
2. YEP medium (to be dissolved in 700 mL): 10 g Yeast extract, 20 g meat peptone. Heat sterilize in autoclave.
3. 1 M Phosphate buffer, pH 6.
4. BMGY liquid medium: 700 mL YEP medium, 100 mL 13.4% (w/v) yeast nitrogen base without amino acid (10× solution, filter sterilized), 100 mL 10% (v/v) glycerol (10× solution, filter sterilized), 100 mL 1 M phosphate buffer pH 6 (10× solution, filter sterilized).
5. BMMY liquid medium: 700 mL YEP medium, 100 mL 13.4% (w/v) yeast nitrogen base without amino acid (10× solution, filter sterilized), 100 mL 5% (v/v) methanol (10× solution, filter sterilized), 100 mL 1 M phosphate buffer pH 6 (10× solution, filter sterilized).
6. Phosphate-buffered saline (PBS): 137 mM NaCl, 2.7 mM KCl, 10 mM Na_2HPO_4, 1.76 mM KH_2PO_4, pH 7–7.4.
7. Baffled flasks: 250 mL (small-scale culturing), 1 and 2 L (upscale).
8. 30 °C Shaking incubator.
9. Spectrophotometer.

2.4 Yeast Cell Lysis and Membrane Preparation

1. Yeast cell pellet.
2. TNG buffer: 50 mM Tris–HCl pH 7.4, 0.5 M NaCl, 10% (v/v) glycerol, 1 mM PMSF (added extemporaneously).
3. TNGE buffer: 50 mM Tris–HCl pH 7.4, 0.5 M NaCl, 10% (v/v) glycerol, 1 mM PMSF (added extemporaneously), 1 mM EDTA.

4. Acid-washed glass beads (425 to 600 μm diameter, Sigma-Aldrich).
5. High-speed benchtop homogenizer (e.g., FastPrep 24, MP Biomedicals).
6. Ultracentrifuge equipped with an appropriate fixed-angle rotor and adapted polycarbonate bottles.
7. Potter homogenizer.
8. Protein assay kit for the determination of protein concentration (e.g., Pierce BCA Protein Assay Kit, Thermo Scientific).

2.5 Immunodetection

1. 40% Acrylamide/Bis-acrylamide solution, 19:1.
2. 3 M Tris–HCl pH 8.45, 0.3% (w/v) SDS.
3. 80% (v/v) glycerol.
4. 10% (w/v) ammonium persulfate (APS).
5. Tetramethylethylenediamine (TEMED).
6. Gel-casting stand and electrophoresis chamber (e.g., Mini-PROTEAN system, Bio-Rad).
7. Membrane preparation samples.
8. Tris-tricine-SDS cathode running buffer: 1 M Tris–HCl pH 8.2, 1 M tricine, 1% (w/v) SDS.
9. Tris anode running buffer: 1 M Tris–HCl pH 8.9.
10. 2× Tricine sample buffer (SB 2×): 100 mM Tris–HCl pH 6.8, 25% (v/v) glycerol, 8% (w/v) SDS, 0.02% (w/v) Coomassie blue G250, 200 mM DTT.
11. Tris-glycine transfer buffer: 25 mM Tris base, 200 mM glycine, 0.02% (w/v) SDS, 20% (v/v) ethanol.
12. 0.45 μm Nitrocellulose blotting membrane.
13. Whatman paper.
14. Electroblotting system (e.g., Mini Trans-Blot Cell, Bio-Rad).
15. Phosphate-buffered saline (PBS).
16. PBS containing 0.02% (v/v) Tween 80 (PBST).
17. Blocking buffer: PBST with 5% (w/v) nonfat dry milk.
18. Primary anti-tag or anti-protein antibody (e.g., monoclonal anti-FLAG antibody from mouse, Sigma).
19. Secondary anti-mouse IgG antibody linked to a reporter system (traditionally HRP-conjugated antibody, here an IRD800-coupled antibody).
20. Reagents and detection device adapted to the reporter system selected.
21. Orbital shaker.
22. Microfiltration blotting device (e.g., Bio-Dot apparatus, Bio-Rad).

2.6 Radioligand Binding Assay

1. Membrane preparation samples.
2. EMP-specific binding buffer (here: 50 mM Tris–HCl pH 7.4, 5 mM $MgCl_2$, 1 mM EDTA).
3. Filter preincubation buffer: 50 mM Tris–HCl pH 7.4, 0.3% (v/v) polyethylenimine.
4. Washing buffer: 50 mM Tris–HCl pH 7.4.
5. Nonradioactive, EMP-specific ligand, here melatonin.
6. Radiolabeled ligand, here [^{3}H]-O-methyl-melatonin (Perkin Elmer).
7. Scintillation cocktail (e.g., Microscint-O scintillation fluid, Perkin Elmer).
8. Low-protein-binding 96-well plate.
9. Shaking incubator.
10. GF/B-grade glass-fiber Unifilters (Perkin Elmer).
11. Scintillation counter.
12. Manifold vacuum filtration apparatus, here a Unifilter-96 harvester (Perkin Elmer).
13. Analysis software (e.g., Prism4, GraphPad Software).

3 Methods

3.1 *P. pastoris* Vector Design and Cloning Procedure

P. pastoris expression vectors are built on a classical *E. coli*/yeast shuttle model with components required for *E. coli* amplification (classically one origin of replication and one antibiotic selection marker) and specific elements for heterologous gene expression in *P. pastoris*. These typically include a selectable auxotrophy marker and/or an antibiotic resistance bacterial gene, as well as a promoter and a terminator sequences surrounding a cloning cassette.

The *P. pastoris* system offers a wide range of plasmid backbones, selection markers, promoters, and fusion sequences that can be combined to obtain the best-suited vector for a given protein. For an exhaustive description of the different elements available and guidelines to choose their assembly, we recommend the reader a chapter from a previous volume of the Methods in Molecular Biology series [11] that is fully dedicated to these aspects.

The examples illustrating the methods presented here are based on the utilization of modified pPIC9K vectors (Life Technologies) designed for the large-scale production of EMPs as described in [7] and [10] notably. Briefly, this vector comprises the gene coding for histidinol dehydrogenase (*HIS4*) as an auxotrophy marker, as well as a bacterial gene (*Kan*) conferring *P. pastoris* resistance to geneticin. The gene coding for the protein of interest is expressed under the control of the strong P_{AOX1} promoter, which is induced by methanol in the absence of other preferential carbohydrate source. The recombinant proteins are flanked at their N- and/or C-termini by a panel of tags (Flag, decahistidine, c-myc, biotinylation domain), which are used for analysis and/or purification purposes. If the fusion sequences need to be eliminated during or after the purification process, tobacco etch virus (TEV) protease sites may be inserted on both sides of the protein to be expressed.

3.2 Preparation of the Expression Vector

Contrary to other yeast systems, no autonomously replicating vectors are available for *P. pastoris*, so they are designed to be integrated in the yeast genome. This is achieved by homologous recombination events that naturally occur between linearized sequences carried by the plasmids (typically *HIS4* or P_{AOX1}) and their homologous counterparts present on the genome, leading to the targeted insertion of the expression vector. Moreover, such plasmid insertions frequently occur in tandem in yeasts and thus lead to multiple integration of the gene of interest with an associated impact on its expression levels.

We describe here a transformation protocol based on the electroporation of the *P. pastoris* SMD1163 strain (*his4, pep4, prb1*) (*see* **Note 1**).

1. Digest 5–7 μg of the purified expression vector with *Pme*I restriction enzyme (*see* **Note 2**) according to the manufacturer's instructions (typically mix 5–7 μg of the expression vector with 25 U of *Pme*I, 20 μL of 10× corresponding buffer, and sterile water to a final volume of 200 μL; incubate the reaction for 2 h at 37 °C).
2. Purify the DNA using the NucleoSpin kit. Alternatively to the use of a commercial kit for the purification of restriction-digested plasmids, we routinely perform this phenol-chloroform extraction procedure that yields a high-quality DNA leading to optimal transformation efficiencies (*see* protocol detailed in **steps 3–10**).
3. Add 400 μL of 24:24:1 (v/v/v) chloroform:phenol:isoamyl alcohol to 200 μL of digestion mixture.
4. Centrifuge for 5 min at 18,000 × *g*, room temperature, and transfer the upper aqueous phase to a new 1.5 mL microcentrifuge tube.
5. Add 400 μL of chloroform and vortex thoroughly for about 20 s.
6. Centrifuge for 5 min at 18,000 × *g*, room temperature, and transfer the upper aqueous phase to a new 1.5 mL microcentrifuge tube.

7. Add 1 mL of 100% ethanol and 50 μL of 3 M sodium acetate and incubate for at least 1 h at −20 °C to precipitate the DNA.
8. Centrifuge for 30 min 18,000×*g*, 4 °C. Discard the supernatant.
9. Wash the pellet with 100 μL of 70% (v/v) ethanol and centrifuge for 5 min at 18,000×*g*, 4 °C. Discard the supernatant.
10. Air-dry the pellet for 15 min and then resuspend in 15 μL sterile H_2O.
11. Check the DNA linearization by loading 1 μL of the reaction mix on a 1% (w/v) agarose gel (*see* **Note 3**).

3.3 Preparation of P. pastoris Electrocompetent Cells (See Note 4)

1. To prepare about 500 μL of electrocompetent *P. pastoris* cells, inoculate 100 mL YPD medium with a fresh SMD1163 colony and incubate overnight at 30 °C in a shaking incubator.
2. Measure the OD_{600} of the culture with a spectrophotometer, dilute the culture with 400 mL fresh YPD to obtain an OD_{600} of 0.25, and incubate at 30 °C (*see* **Note 5**).
3. When the culture reaches an OD_{600} of 1 (approximately after 4 h), harvest the cells by centrifugation in sterile tubes for 5 min at 4000×g, 4 °C.
4. Discard the supernatant and resuspend the cells in 100 mL YPD, 20 mL of 1 M HEPES pH 8, and 2.5 mL of 1 M DTT. Mix gently until the pellet is resuspended.
5. Incubate for 15 min at 30 °C.
6. Transfer onto ice and add ice-cold sterile H_2O to a final volume of 500 mL.
7. Pellet the cells by centrifuging for 5 min at 4000×*g*, 4 °C.
8. Discard the supernatant and wash the cell pellet with 250 mL ice-cold sterile H_2O.
9. Pellet the cells by centrifuging for 5 min at 4000×*g*, 4 °C.
10. Discard the supernatant and resuspend the cell pellet in 20 mL ice-cold 1 M sorbitol by gently mixing.
11. Pellet the cells by centrifuging for 5 min at 4000×*g*, 4 °C.
12. Discard the supernatant and resuspend the cell pellet in 500 μL ice-cold 1 M sorbitol by gently mixing.

3.4 Electrotransformation and Selection of the Recombinant Clones

1. Place an electroporation cuvette on ice at least 10–15 min before performing the transformation.
2. Mix gently 40 μL competent cells with 7.5 μL of the linearized DNA in the cuvette and incubate for 5 min on ice.
3. Adjust the electroporation settings as follows: 1500 V, 25 μF, and 400 Ω.

4. Place the cuvette in the electroporator chamber and apply the electric pulse.
5. Immediately resuspend the electroporated mixture in 1 mL ice-cold 1 M sorbitol and transfer into a sterile tube.
6. Allow the cells to recover for about 1 h at 30 °C, and then pellet the cells by centrifuging for 10 min at 4000 × *g*, room temperature.
7. Discard the supernatant and resuspend the pellet in 500 μL of 1 M sorbitol.
8. Spread, respectively, 10 % (50 μL) and 90 % (450 μL) of electrotransformed cells on each of the two YNB plates and incubate for 2–3 days at 30 °C (*see* **Note 6**).

3.5 Screening of Recombinant Clones

When using expression vectors comprising the *HIS4* auxotrophy selection marker (e.g., vectors from the pPIC9K series), recombinant yeast clones are selected in two consecutive steps. The first one, performed after the electrotransformation, is based on the recovery of histidine prototrophy through growth ability on minimal medium. The second one is dedicated to the identification of multicopy transformants producing high yields of recombinant proteins. Indeed, a high number of integrated copies often (but not always) correlates with higher expression levels, and it is necessary to screen several His + transformants in order to identify a high producer.

To this end, we propose two different procedures: either a classical one in which His + transformants are selected on YPD plates containing a range of geneticin concentrations or a more high-throughput immunostaining method named Yeastern blot in which histidine-prototroph clones are grown in expression conditions and lysed directly on a nitrocellulose membranes in a 96-well plate format.

The goal of screening on increasing geneticin concentration is to select clones with geneticin resistance phenotypes representative of the number of expression vector-integrated copies (the higher number of integrated copies, the potentially higher resistance level), before evaluating their performance in terms of EMP expression (*see* **Note** 7).

1. Harvest the His + transformants with 1 mL of YPD medium poured onto the YNB plates and scrape off all the clones using a sterile scraper.
2. Perform 10× and 100× dilutions and measure the OD_{600} for each.
3. Spread an equivalent of 10^5 cells/plate (OD_{600} of 1 is equivalent to approximately 5×10^7 cells/mL) on YPD plates supplemented with increasing geneticin concentrations ranging from 50 to 250 μg/mL (*see* **Notes 8** and **9**).

4. Incubate for 2–3 days at 30 °C.
5. Pick 6–12 representative colonies from the different geneticin concentration plates, streak them onto a fresh YPD plate supplemented with 50 μg/mL geneticin, and let them grow for 1–2 days at 30 °C.
6. Directly use these colonies for further expression tests and/or resuspend them in 1 ml of sterile YPD supplemented with 20% glycerol and store them at −80 °C.

The goal of the high-throughput expression-level screening assay or yeastern blotting that we developed is to screen at once the largest number of EMP-expressing clones with the fewest and simplest handling steps. Therefore, no selection based on geneticin resistance phenotypes is required (*see* **Note 10**).

Steps 1–8 must be performed in sterile conditions.

1. Fill each well of a 96-well plates with 100 μL YPD supplemented with 20 μg/mL geneticin.
2. Inoculate each well with a single His+colony (from the YNB plate) with a sterile toothpick or inoculation loop.
3. Incubate overnight on a shaker at 250 rpm, 30 °C.
4. On the next day, place a sterile nitrocellulose membrane on a BMGY plate (14 cm diameter) avoiding air bubbles.
5. With a multichannel pipette, spot 5 μL of each preculture on the membrane. Add 20 μL sterile glycerol in each well and conserve the plate at −20 or −80 °C.
6. Incubate for about 12 h at 30 °C (plate lid on the top).
7. After incubation, transfer the membrane on a BMMY plate avoiding air bubbles.
8. Incubate for about 20 h at 30 °C (plate lid on the top).
9. In a petri dish, soak 12 Whatman paper sheets in lysis buffer. The liquid level should not recover the last sheet.
10. Transfer the nitrocellulose membrane on top of the Whatman paper pile avoiding air bubbles.
11. Incubate for 4 h at 65 °C.
12. After lysis, rinse generously the membrane with water to get rid of all visible yeast cell traces and proceed to standard immunodetection of the protein of interest as described in Subheading 3.8. Figure 1 presents two examples of GPCR-expressing colonies assayed with this method that are representative of the typical patterns that can be obtained. This immunoblot clearly highlights clones that exhibit high expression levels of the proteins of interest.
13. Streak the high-expressing clones onto a fresh YPD plate supplemented with 50 μg/mL geneticin and let them grow for 1–2 days at 30 °C.

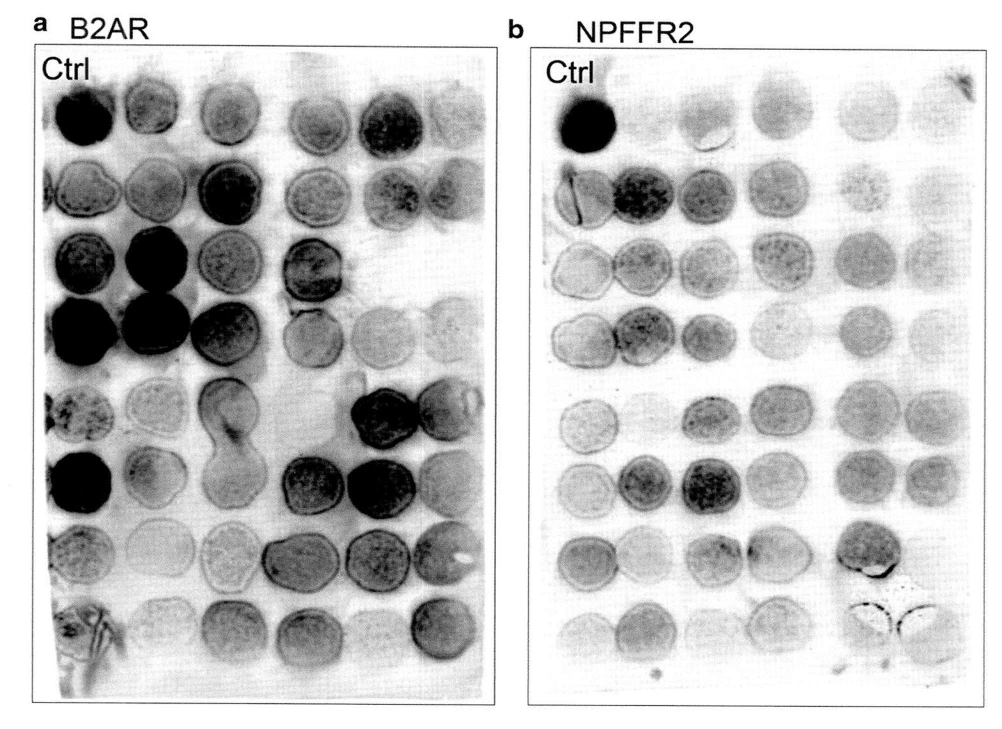

Fig. 1 Yeastern blot analysis of ca. 96 recombinant *P. pastoris* clones expressing class A GPCRs. (**a**) Flag-tagged human B2AR receptor; (**b**) flag-tagged human NPFFR2 receptor. *Ctrl*: colony expressing the AA2A receptor used as a positive control. After expression induction and direct lysis of the colonies grown on the nitrocellulose membrane, the tagged receptors were immunodetected with an M2 anti-flag antibody

14. Directly use these colonies for further expression tests and/or resuspend them in 1 ml of sterile YPD supplemented with 20% glycerol and store them at −80 °C.

3.6 Yeast Culturing for Membrane Protein Expression

Once interesting transformants have been identified, their expression abilities should be further explored in liquid culture. For expression based on P_{AOX1}-dependent vectors, yeasts are first cultured in a glycerol-containing medium to an appropriate cell density and growth phase. Transferring the cells into a methanol-containing medium then induces protein expression.

The procedure presented below describes small-scale culturing using shaken baffled flasks, which allows the parallel screening of several clones. It can easily be up-scaled to bigger baffled flasks for the production of larger amounts of EMP (*see* **Note 11**).

1. Inoculate 50 mL of freshly prepared BMGY medium in a 250 mL baffled flask with a fresh recombinant colony isolated on a YPD agar plate. Incubate on a shaker at 250 rpm, 30 °C, ON.
2. On the next day, measure the OD_{600} of the culture. Dilute the cells into 50 mL of fresh BMGY medium in a 250 mL flask to achieve an OD_{600} of 1 (about 5×10^7 cells/mL) and incubate

on a shaker at 250 rpm, 30 °C. This step usually corresponds to a tenfold dilution of the pre-culture.

3. When the culture reaches ≈5 OD_{600} (about 5 h later), pellet the cells by centrifuging in sterile tubes for 5 min at 4000 × *g*, room temperature.
4. Discard the supernatant and resuspend the cell pellet with 50 mL fresh BMMY medium (*see* **Note 12**). Incubate for 18–24 h in a shaker at 30 °C, 250 rpm.
5. After induction, harvest the cells by centrifuging for 5 min at 4000 × *g*, 4 °C.
6. Discard the supernatant and wash the cell pellet with 50 mL PBS. Pellet the cells by centrifuging for 5 min at 4000 × *g*, 4 °C.
7. Discard the supernatant and weigh the cell pellet. At this stage, the cell pellet can either be snap-frozen in liquid nitrogen and stored at −80 °C, or kept on ice to be directly used for membrane preparation.

3.7 Yeast Cell Lysis and Membrane Preparation Procedure

P. pastoris cells are surrounded by a thick protective cell wall, resulting in the need to use a robust cell lysis method. The protocol described below involves glass microbeads associated with vigorous mechanical shaking. Programmable equipment such as the FastPrep 24 from MP Biomedicals employed here can be used to achieve reliable and reproducible results. It can also be adapted to various sample volumes and formats (from 1.5 mL microcentrifuge tubes to 50 mL conical centrifuge tubes). However, similar results can be obtained using a basic vortex apparatus in 4–8 cycles of alternating shaking and ice-cooling phases.

1. Resuspend the yeast pellet obtained (about 1 g wet cells) with 10 mL ice-cold TNGE buffer in a 50 mL conical centrifuge tube.
2. Add 5 mL of acid-washed glass beads.
3. Place the tubes on the cell breaker device and proceed to cell lysis by alternating shaking and cooling steps on ice (3 cycles of 40 s each at 6.5 m/s on a FastPrep 24).
4. Centrifuge the samples for 5 min at 4000 × *g*, 4 °C, collect the supernatant, and store at 4 °C.
5. Dissolve the remaining pellet in 10 mL ice-cold TNGE buffer.
6. Repeat steps 3–4 times (or until the supernatant is clear).
7. Centrifuge the collected supernatants for an additional 5 min at 4000 × *g*, 4 °C.
8. Proceed to supernatant ultracentrifugation for 30 min at 100,000 × *g*, 4 °C.

9. Discard the supernatant and resuspend the membrane pellet with a Potter homogenizer in 3 mL ice-cold TNG buffer until a homogeneous suspension is obtained.
10. Determine the protein concentration using the BCA assay kit following the manufacturer's recommendations.
11. Use directly the membrane preparation for expression-level analysis or purification, or store them at −80 °C.

3.8 Membrane Analysis Procedure: Immunodetection

Immunodetection methods allow the determination of the overall amount of the recombinant EMP contained in the membrane preparations. These analyses can be performed using specific antibodies targeted against the EMP of interest or against the tags fused to the protein.

The expression level can either be estimated on membrane samples previously separated by a denaturing SDS-PAGE or directly by spotting membrane preparation on a nitrocellulose membrane with a dot-blot device. Both methods are presented below.

The following proportions are given for casting two gels of 1 mm thickness in a standard Mini-Protean system from Bio-Rad.

1. Prepare the separating gel by mixing 2.5 mL of the acrylamide solution, 3.3 mL Tris–HCl SDS buffer, 1.25 mL 80% (v/v) glycerol, and 2.9 mL H_2O. Add 90 µL APS and 6 µL TEMED, mix, and immediately cast the gel. Allow space to cast the stacking gel.
2. Prepare the stacking gel by mixing 0.6 mL of the acrylamide solution, 1.6 mL Tris–HCl SDS buffer, and 4.1 mL H_2O. Add 90 µL APS and 6 µL TEMED, mix, and cast carefully over the separating gel. The presence of glycerol in the separating part avoids the need to wait for its complete polymerization. Insert a 10-well gel comb immediately without introducing air bubbles.
3. Let polymerize for about 30 min.
4. Preincubate 10 µg of membrane preparation in SB 2× for about 10 min at room temperature (*see* **Note 13**).
5. Load the sample in a well of a 10% SDS-polyacrylamide gel. Proceed to electrophoresis using Tris-tricine-SDS cathode running buffer and Tris anode running buffer in a tank unit for about 1 h 30 min at 100 V.
6. Transfer the proteins from the gel to a nitrocellulose membrane by electroblotting in Tris-glycine transfer buffer for about 1 h 30 min at 100 V.
7. Incubate the membrane in 50 mL blocking buffer for 1 h at room temperature on an orbital shaker. Alternatively, incubate the membrane overnight at 4 °C.
8. Remove the blocking solution and incubate the membrane with the selected antibody diluted in blocking buffer (for

instance a monoclonal anti-flag antibody at a final concentration of 0.1 μg/mL) for 1 h at room temperature on an orbital shaker.

9. Wash the membrane three times, each time with 50 mL PBST on an orbital shaker for 5 min at room temperature.
10. Remove the PBST and incubate the membrane with the adapted anti-IgG antibody diluted in blocking buffer (typical final concentration of 0.1 μg/mL) for 1 h at room temperature on an orbital shaker.
11. Wash the membrane three times, each time with 50 mL PBST on an orbital shaker for 5 min at room temperature.
12. Remove the PBST and wash the membrane with 50 mL PBS on an orbital shaker for 5 min at room temperature.
13. Store the membrane in PBS until revelation.
14. Proceed to membrane revelation according to the reporter system selected and following the manufacturer's recommendations. Figure 2 exemplifies a typical Western blot profile obtained for membrane samples prepared from different clones expressing the same EMP construct (here a 1 TM human enzyme). This example illustrates the variation of expression levels that can be observed between clones, from no (clone #7) to low (clone #3) and relatively high (clone #4) amounts of protein produced.

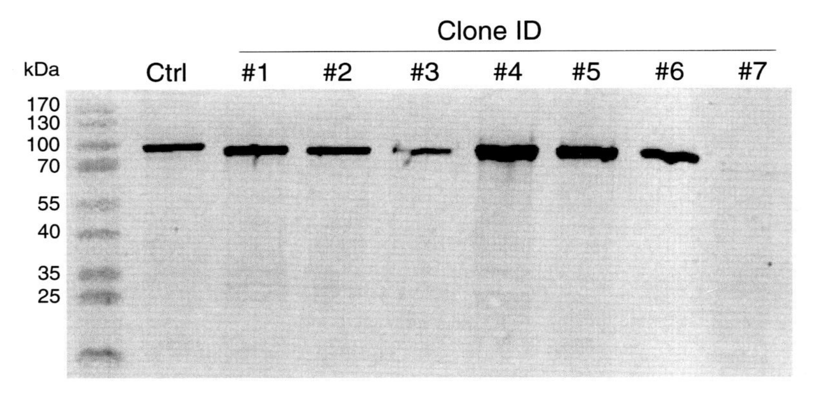

Fig. 2 Western blot analysis of membrane samples from eight different recombinant *P. pastoris* clones expressing a native 1TM, ER-located, human enzyme. *Ctrl*: membrane sample of a clone used as a positive control, expressing a tagged version of the ER enzyme. Equal amounts of membrane proteins were loaded in each lane (10 μg). Proteins were separated by a 10 % SDS-PAGE and immunoblotted with a protein-specific monoclonal antibody. Molecular weights are indicated in kilodalton (kDa) on the *left*

For Dot-Blot Immunodetection

1. Pre-soak with PBS a piece of nitrocellulose membrane and fix it tightly into the microfiltration-blotting device following the manufacturer's instructions.
2. Pipet 100 μL of PBS into each well of the blotting device and let it flow through the nitrocellulose membrane by applying the vacuum.
3. Turn the vacuum off and pipet 5–10 μg of membrane protein samples into the appropriate wells. Incubate for 5 min at room temperature before applying the vacuum.
4. Turn the vacuum on and wash the wells three times, each time with 100 μL PBS.
5. Remove the nitrocellulose membrane from the blotting device and proceed to the blocking and immunodetection reactions (as already described). Figure 3 provides an illustration of how this approach could be helpful in two different applications.

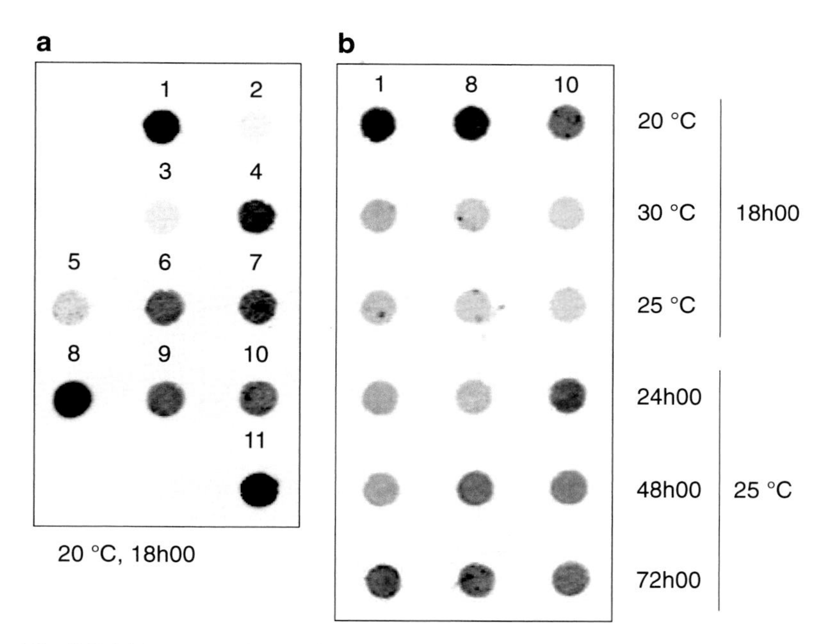

Fig. 3 Dot blot comparisons of membrane samples from independent *P. pastoris* clones induced in different conditions for the expression of a human class C GPCR. (**a**) Membrane samples from 11 different clones (#1 to #11) induced with methanol for 18 h at 20 °C. (**b**) Membranes from clones #1, #8, and #10 induced at various temperatures (20, 25, and 30 °C) and for different periods of time (18, 24, 48, and 72 h). Equal amounts of corresponding membrane proteins (10 μg) were loaded in each well of the blotting device and immunoblotted with an M2 anti-flag antibody (0.1 μg/mL)

With a higher throughput than a Western blot, the first application (panel A) is the ranking of several individual clones on their capacity to produce the recombinant EMP of interest. In a second direction, this method allows to screen relatively easily a number of parameters that may influence the yields of the expressed protein. The present example notably highlights the significant effect of induction time and temperature on the expression levels of a class C GPCR, but several other parameters may also be influential as shown in a number of studies (*see* **Note 12**).

3.9 Membrane Analysis Procedure: Radioligand Binding Assay

In the case of GPCRs and other ligand-regulated EMPs, ligand binding assays are highly valuable procedures to infer the amounts of active receptors (B_{max}) in a membrane preparation and their affinity (K_d) for the ligand tested. These very sensitive and reliable techniques however suppose the availability of protein-specific radiolabeled ligands and an access to a lab facility where radioactive material can be handled.

In the following protocol, a saturation ligand binding procedure is exemplified with a class A GPCR, the MT1 melatonin receptor, assayed with the [^{3}H]-O-methyl-melatonin used as the tracer radioligand, and the agonist melatonin used to determine the nonspecific binding.

1. Thaw and homogenize membrane preparations on ice.
2. For each concentration of radiolabeled ligand assayed (e.g., 0.5, 1, 2, 5, 10, 20, and 50 nM), ligand-binding measurements are performed in triplicate for total (T) binding and in parallel for nonspecific (NS) binding (i.e., 6-point measurement for each concentration of radioligand).
3. Fill each T well with 10 μg membrane proteins diluted in 90 μL binding buffer.
4. Fill each NS well with 10 μg membrane proteins diluted in 80 μL binding buffer. Add 10 μL of 100 μM nonradioactive ligand (for a final cold ligand concentration of 10 μM).
5. Add to each well 10 μL of tenfold concentrated radiolabeled ligand to achieve the concentration range of 0.5, 1, 2, 5, 10, 20, and 50 nM in a final volume of 100 μL.
6. Incubate at room temperature on a shaker for 2 h to achieve ligand-binding equilibrium.
7. During the incubation or at least 15 min prior to filtration, pre-soak the GF/B filters in filter buffer.
8. Terminate the reactions by a rapid filtration of the samples through the pre-soaked GF/B filters using a vacuum manifold according to the manufacturer's instructions.
9. Wash the filters three times with ice-cold washing buffer.

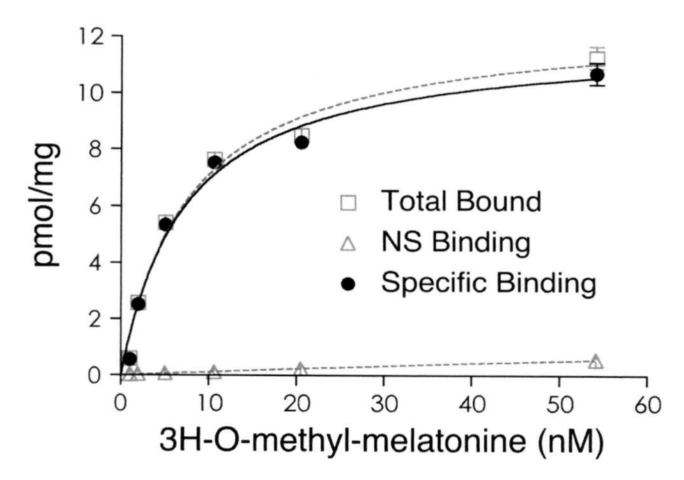

Fig. 4 Radioligand binding saturation curve determined on membranes of a *P. pastoris* clone expressing the human MT1 melatonin receptor. The specific binding curve (*black circles*) is calculated by subtracting the nonspecific (*triangles*) to the total (*squares*) binding values. Data are fitted using the one-site saturation binding model

10. Let the 96-well plate filter dry and add 40 μL scintillation cocktail.
11. Incubate the vials for 2 h in the dark before proceeding to counting.
12. Subtract the NS from the T values to determine specific (S) binding values. Analyze the data by nonlinear regression using appropriate software.

Figure 4 represents a typical saturation binding curve obtained with membranes of a *P. pastoris* clone expressing the MT1 receptor. In the present case, the GraphPad Prism software determined from these data a B_{max} value of 11.9 ± 0.4 pmol/mg and a K_D of 7.2 ± 0.8 nM. According to the conversion factors detailed in [3], the total number of ligand-binding sites (B_{max}) present in this membrane sample approximately corresponds to about 0.5 mg of active receptor produced per 1 l of culture. On the other side, the measured K_D value suggests that the affinity of MT1 for its agonist ligand melatonin is decreased about tenfold in *P. pastoris* membranes compared to the receptor expressed in mammalian cells [10].

4 Notes

1. A number of *P. pastoris* strains that are frequently used for EMP expression present an auxotrophic behavior for histidine due to a histidinol dehydrogenase deficiency (*his4* genotype). The prototrophy for histidine can be restored upon transfor-

mation with a series of pPIC vectors bearing the *HIS4* gene, thereby allowing positive selection of the transformants on minimum medium. Strains from the SMD series bear additional deficiencies in endogenous proteases (*pep4*, *prb1* for SMD1163) and are often preferred for this convenient phenotype (*see* ref. 10 for more details).

2. *Pme*I is used to linearize the expression vector in the P_{AOX1} promoter, thus favoring its integration at the homologous P_{AOX1} genomic locus. The absence of the *Pme*I site in the gene to be expressed should be checked; otherwise several fragments would be generated instead of a linearized vector. In case *Pme*I is present, another restriction enzyme that cut only once in the P_{AOX1} or *HIS4* sequences of the vector can be selected instead (*Sac*I or *Sal*I for instance).
3. When the plasmid is correctly linearized, one single DNA band of the corresponding size should be observed. If additional bands of nonlinearized plasmid are present, proceed back to **step 1** and double the amount of *PmeI* restriction enzyme.
4. Unlike bacteria, yeast cells cannot conserve their competence properties when stored at −80 °C. Electrocompetent *P. pastoris* cells should then be prepared extemporaneously before each transformation experiment.
5. One OD_{600} unit measured with an Eppendorf Biophotometer approximately corresponds to 5×10^7 cells/mL. In order to keep a proper OD_{600}/cell density proportionality, be also aware that it is important to dilute the cell culture before the spectrophotometer measurement so that OD_{600} values do not exceed *ca.* 0.3 (higher values are underestimating the actual cell density).
6. Commercial vectors from the pPICZ series (Life Technologies) do not contain the *HIS4* auxotrophy marker and the geneticin resistance gene present on the pPIC9K vectors. They comprise instead a single resistance marker to zeocin antibiotic. In this case, positive transformants are isolated on YPD plates containing a low zeocin concentration (e.g., 25 μg/mL) before proceeding to the screening of expressing clones.
7. Direct selection of transformants on geneticin-containing medium is not recommended. Indeed, the level of resistance is dependent on the cell density and false-positive clones may be isolated. In addition, high concentration of antibiotics applied directly after transformation may eliminate potentially valuable clones that have not fully recovered from electroporation nor achieved homologous recombination.
8. Because of its mutations conferring protease deficiencies, the strain SMD1163 exhibits a relatively high susceptibility to

geneticin and the concentration range of antibiotic should be reduced to 50–250 μg/mL. When using other strains, recombinant clones may be selected on YPD plates supplemented with up to 2 mg/mL geneticin.

9. When using vectors from the pPICZ series, zeocin concentrations ranging from 25 to 1500 μg/mL may be applied.
10. This second procedure enables to screen a high number of transformants within a shorter time frame. However, for difficult-to-express membrane proteins, a combination of both methods can be beneficial (e.g., a Yeastern Blot performed with clones selected on high geneticin concentrations).
11. This protocol can be applied to larger culturing format. It is however recommended to maintain a 1:5 ratio between the volume of the culture and the total volume of the selected baffled flasks for an optimal aeration of the culture. Robust procedures for large-scale culturing in biorectors are also available [12] but are less straightforward to handle as they necessitate specific equipment and a succession of culturing conditions before the induction step.
12. BMMY induction medium can be supplemented with different components depending on the protein expressed. In particular, dimethyl sulfoxide (DMSO) supplemented at 2.5 % (v/v) has been shown to increase remarkably the production yield of ligand-binding active GPCRs [7]. Similarly, adding a ligand specific to the receptor to be produced has been highly beneficial for a large majority of the GPCRs tested [7], probably playing the role of a pharmacological chaperone [13].
13. It is usually not recommended to boil membrane protein samples prior to electrophoresis. When compact and highly hydrophobic proteins such as GPCRs are boiled, they usually aggregate and keep stuck in the stacking gel.

References

1. Gellissen G (2000) Heterologous protein production in methylotrophic yeasts. Appl Microbiol Biotechnol 54(6):741–750
2. Cereghino JL, Cregg JM (2000) Heterologous protein expression in the methylotrophic yeast *Pichia pastoris*. FEMS Microbiol Rev 24:45–66
3. Sarramegna V, Talmont F, Demange P, Milon A (2003) Heterologous expression of G-protein-coupled receptors: comparison of expression systems from the standpoint of large-scale production and purification. Cell Mol Life Sci 60(8):1529–1546
4. Alkhalfioui F, Logez C, Bornert O, Wagner R (2011) Expression systems: *Pichia pastoris*. In: Robinson AS (ed) Production of membrane proteins–strategies for expression and isolation. Wiley-VCH, Weinheim. doi:10.1002/9783527634521
5. Bill RM (2014) Playing catch-up with *Escherichia coli*: using yeast to increase success rates in recombinant protein production experiments. Front Microbiol 5:85
6. Bertheleme N, Singh S, Dowell S, Byrne B (2015) Heterologous expression of G-protein-coupled receptors in yeast. Methods Enzymol 556:141–164
7. André N, Cherouati N, Prual C, Steffan T, Zeder-Lutz G, Magnin T, Pattus F, Michel H,

Wagner R, Reinhart C (2006) Enhancing functional production of G protein-coupled receptors in *Pichia pastoris* to levels required for structural studies via a single expression screen. Protein Sci 15:1115–1126

8. Magnin T, Fiez-Vandal C, Potier N, Coquard A, Leray I, Steffan T, Logez C, Alkhalfioui F, Pattus F, Wagner R (2009) A novel, generic and effective method for the rapid purification of G protein-coupled receptors. Protein Expr Purif 64(1):1–7
9. Bornert O, Møller TC, Boeuf J, Candusso MP, Wagner R, Martinez KL, Simonin F (2013) Identification of a novel protein-protein interaction motif mediating interaction of GPCR-associated sorting proteins with G protein-coupled receptors. PLoS One 8(2):e56336
10. Logez C, Berger S, Legros C, Banères JL, Cohen W, Delagrange P, Nosjean O, Boutin JA, Ferry G, Simonin F, Wagner R (2014) Recombinant human melatonin receptor MT1 isolated in mixed detergents shows pharmacology similar to that in mammalian cell membranes. PLoS One 9(6):e100616
11. Logez C, Alkhalfioui F, Byrne B, Wagner R (2012) Preparation of *Pichia pastoris* expression plasmids. Methods Mol Biol 866:25–40
12. Singh S, Gras A, Fiez-Vandal C, Martinez M, Wagner R, Byrne B (2012) Large-scale production of membrane proteins in *Pichia pastoris*: the production of G protein-coupled receptors as a case study. Methods Mol Biol 866:197–207
13. Bernier V, Bichet DG, Bouvier M (2004) Pharmacological chaperone action on G-protein-coupled receptors. Curr Opin Pharmacol 4(5):528–533

Chapter 11

Integral Membrane Protein Expression in *Saccharomyces cerevisiae*

Rebba C. Boswell-Casteel, Jennifer M. Johnson, Robert M. Stroud, and Franklin A. Hays

Abstract

Eukaryotic integral membrane proteins are challenging targets for crystallography or functional characterization in a purified state. Since expression is often a limiting factor when studying this difficult class of biological macromolecules, the intent of this chapter is to focus on the expression of eukaryotic integral membrane proteins (IMPs) using the model organism *Saccharomyces cerevisiae*. *S. cerevisiae* is a prime candidate for the expression of eukaryotic IMPs because it offers the convenience of using episomal expression plasmids, selection of positive transformants, posttranslational modifications, and it can properly fold and target IMPs. Here we present a generalized protocol and insights based on our collective knowledge as an aid to overcoming the challenges faced when expressing eukaryotic IMPs in *S. cerevisiae*.

Key words Integral membrane protein, *Saccharomyces cerevisiae*, Protein expression, Protein overproduction, Yeast

1 Introduction

Saccharomyces cerevisiae is a well-characterized eukaryotic model organism for recombinant protein expression, especially for integral membrane proteins [1–4], because it combines the advantages of unicellular organisms (e.g., rapid growth and genetic material is easily manipulated) with the capacity to perform eukaryotic posttranslational modifications. In contrast to more multifarious eukaryotic organisms, *S. cerevisiae* expression systems are cost-effective, are capable of rapidly reaching high cell densities, can produce high protein yields, and *S. cerevisiae* is generally regarded as safe (GRAS). These advantages position *S. cerevisiae* as a leading expression system for the overproduction of eukaryotic integral membrane proteins (IMPs).

Procuring sufficient quantities of IMPs for downstream studies can be a formidable task. In this chapter we present our approach to

Isabelle Mus-Veteau (ed.), *Heterologous Expression of Membrane Proteins: Methods and Protocols*, Methods in Molecular Biology, vol. 1432, DOI 10.1007/978-1-4939-3637-3_11, © Springer Science+Business Media New York 2016

the over-expression of IMPs using the budding yeast *S. cerevisiae*. However, at almost every step throughout this protocol, an alternative method, vector, buffer, etc. could be substituted to further optimize expression or meet the specific requirements of your IMP of interest. The primary intent of this chapter is to provide a generalized approach to the expression of eukaryotic IMPs and, when possible, provide alternative strategies or important considerations that will aid in the overproduction of target IMPs (Fig. 1).

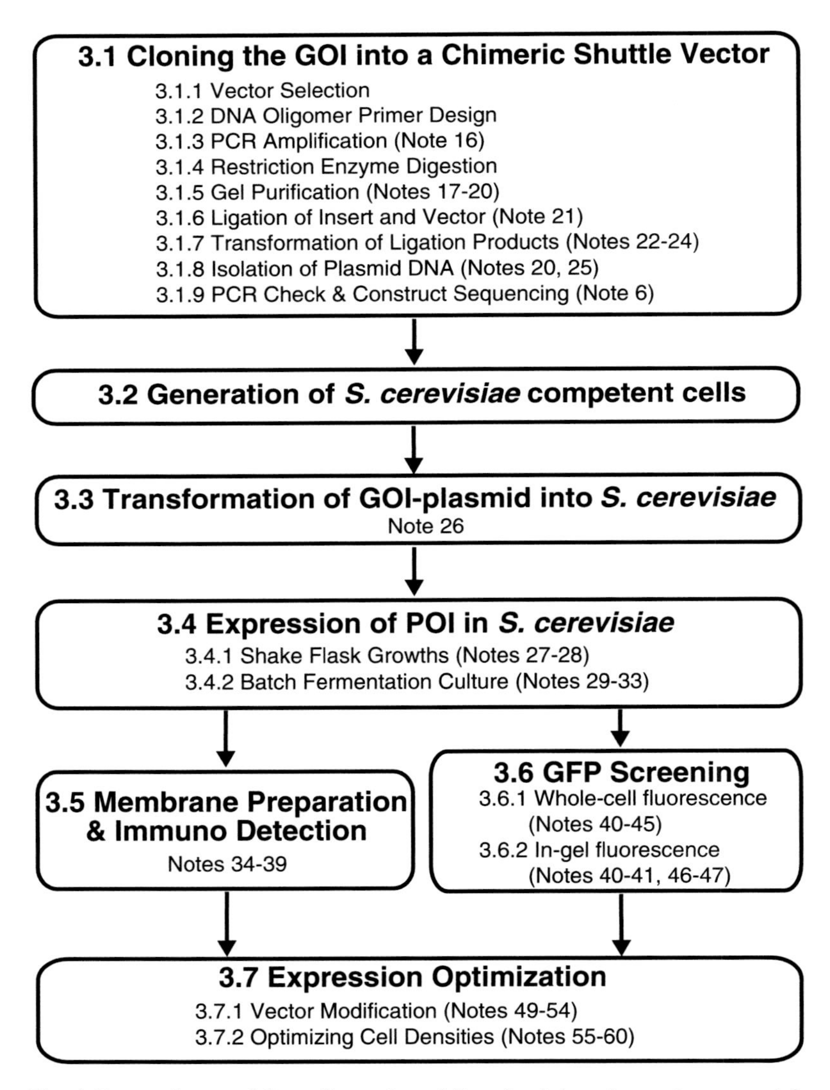

Fig. 1 Expression workflow. General workflow for integral membrane protein expression in *S. cerevisiae* with relevant Notes indicated for each section

2 Materials

Prepare all solutions using ultrapure water (simply referred to as ddH_2O, prepared by purifying deionized water to attain a sensitivity of 18 MΩ at 25 °C). Prepare and store all reagents at room temperature (RT, unless indicated otherwise). All listed pH values were determined at room temperature (unless indicated otherwise).

2.1 Cloning Your Gene of Interest into the "83Xi" Vector

1. DNA oligomer primers of less than 40 nucleotides in length can be ordered as "desalted" quality (*see* **Note 1**).
2. Phusion® High-Fidelity DNA Polymerase (New England BioLabs) kit: 50 mM $MgCl_2$, 5× Phusion Reaction Buffer, DMSO, and Phusion® High-Fidelity DNA Polymerase.
3. dNTP solution mix (New England BioLabs) was obtained separately. Stored −20 °C.
4. QIAquick® PCR purification kit (Qiagen).
5. Ethidium bromide (EtBr, 2 mg/ml). It is highly carcinogenic and light sensitive. Keep wrapped in aluminum foil and store at RT (*see* **Note 2**).
6. 50× TAE (2 M Tris, 1 M acetic acid, 50 mM EDTA).
7. Molecular biology grade ethanol.
8. Molecular biology grade agarose.
9. Restriction enzymes and corresponding 10× buffer stocks (New England BioLabs). Stored at −20 °C. (Specific buffer compositions can be found in the appendix of the NEB catalogue) (*see* **Note 3**).
10. 10× BlueJuice™ DNA gel loading buffer (Invitrogen).
11. Quick-Load® 2-Log DNA Ladder (0.1–10.0 kb, New England BioLabs).
12. QIAquick® gel extraction kit (Qiagen).
13. T4 DNA ligase and ligation buffer.
14. SOC Media: autoclave 1 l of SOB broth and add 10 mM $MgCl_2$ (sterile solution) and 20 mM glucose (sterile solution).
15. *Escherichia coli* strain XL2-Blue. Store at −80 °C.
16. Luria–Bertani (LB) medium and or on LB agar (*see* **Note 4**).
17. Ampicillin (200 mg/ml) (*see* **Note 5**).
18. QIAprep® spin miniprep kit (Qiagen).
19. GAL1 promoter and Cyc terminator sequencing primers (*see* **Note 6**).

2.2 Generation of S. cerevisiae Competent Cells

1. Yeast strain W303-Δpep4 (*leu2-3, 112 trp1-1, can1-100, ade2-1, his3-11,15, Δpep4, MATα*).
2. YPD broth (Research Products International) 0.2 μm sterile filtered.
3. YPD agar. Plates are poured following a 15 min autoclave cycle, and stored at 4 °C.
4. YPD broth plus 15% (v/v) glycerol. 0.2 μm sterile filtered.

2.3 Transformation of S. cerevisiae Competent Cells

1. PLATE solution: 40% PEG 3350 (w/v), 0.1 M lithium acetate, 10 mM Tris–HCl, pH 7, and 1 mM EDTA. Solution is brought to volume with ddH_2O and 0.2 μm sterile filtered (*see* **Note 7**).
2. TE Buffer: 10 mM Tris–HCl, pH 9.0 and 150 mM EDTA.
3. Molecular biology grade sterile DMSO.
4. Sheared salmon sperm DNA (10 mg/ml). Stored at −20 °C.
5. CSM-His plates (500 ml = 20 plates): of 0.77 g/L CSM-His (Sunrise Science Products), 20 g/L low melting agar, 10 g/L glucose, 3.0 g/L ammonium sulfate, 1.7 g/L yeast nitrogen base without amino acids and ammonium sulfate, and this is brought to volume with ddH_2O. Autoclave for 15 min and store plates at 4 °C.

2.4 Protein Expression in S. cerevisiae

1. Glucose solution, 40% (w/v). Autoclave for 15 min and store at room temperature.
2. 20× galactose solution, 40% (w/v). Add galactose to autoclaved, sterile water. Once dissolved and cooled to room temperature filter-sterilize using 0.45 μm filter. Store solution at room temperature (*see* **Note 8**).
3. 10× raffinose solution, 10% (w/v). Filter-sterilize using 0.22 μm filter and store at room temperature (*see* **Note 9**).
4. 10× CSM-His solution, 7.9 g/L. Filter-sterilize using 0.22 μm filter and store at 4 °C (*see* **Note 10**).
5. 20× YNB solution. 13.4% (w/v) yeast nitrogen base without amino acids. Filter-sterilize using 0.22 μm filter and store at 4 °C.
6. 4× YPG solution is used as the inductant. Yeast extract 8% (w/v), peptone 16% (w/v), and galactose 8% (w/v). Add yeast extract and peptone to hot water to dissolve. Once dissolved autoclave for 15 min. After solution cools to room temperature add galactose using the sterile 20× galactose stock, and stir (*see* **Note 11**).
7. Antifoam 204.
8. Resuspension buffer Y, pH 7.5 : 50 mM Tris–HCl, pH 7.5, 10% (v/v) glycerol, 150 mM NaCl, and 10 mM EDTA, pH 8.0. Store at 4 °C.

2.5 Membrane Preparation and Immunodetection

1. 2× Solubilization Buffer, pH 7.4: 100 mM Tris–HCl, pH 7.4, 800 mM NaCl, and 10% (v/v) glycerol. Store at 4 °C.
2. Coomassie Protein Staining Solution.
3. SDS Polyacrylamide Gel Components.
   (a) Resolving gel buffer: 1.5 M Tris–HCl, pH 8.8
   (b) Stacking gel buffer: 0.5 M Tris–HCl, pH 6.8
   (c) 30% Acrylamide/Bis Solution, 29:1. Store at 4 °C (*see* **Note 12**).
   (d) Ammonium persulfate : 10% (w/v) solution in water. Store at −20 °C
   (e) *N,N,N,N′*-tetramethyl-ethylenediamine (TEMED)
   (f) SDS-PAGE Running Buffer: 25 mM Tris–HCl, pH 8.3, 0.192 M glycine, 0.1% (w/v) SDS (*see* **Note 13**).
   (g) 4× Laemmli Sample Buffer: 200 mM Tris–HCl, pH 6.8, 8% (w/v) SDS,) 40% (v/v) glycerol, 50 mM EDTA, pH 8.0, and 0.08% (w/v) bromophenol blue. Mix thoroughly and make 960 μl aliquots in 1.5 ml microfuge tubes. *Prior to use add 40 μl β-mercaptoethanol to the microfuge tube.* Store at −20 °C (*see* **Note 14**).
   (h) Precision Plus Protein ™ Kaleidoscop™ (Bio-Rad).
4. Immunoblotting Components.
   (a) PVDF transfer membrane (0.2 μm, Thermo Scientific)
   (b) Western blot transfer buffer: 25 mM Tris–HCl, 192 mM glycine, and 20% (v/v) methanol.
   (c) Tris buffered saline (TBS, 10× stock): 1.5 M NaCl, 0.1 M Tris–HCl, pH 7.4
   (d) 1× TBS containing 0.05% (v/v) Tween-20 (TBST)
   (e) Blocking solution: 5% (w/v) milk in TBST (*see* **Note 15**).
   (f) Wash buffer: 1× TBS
   (g) Thick blotting paper (Bio-Rad)
   (h) Thin blotting paper (Bio-Rad)
   (i) Anti-histidine antibody, Penta-His Alexa Fluor® 647 conjugate (Qiagen), or an antibody specific for your POI.

2.6 GFP Detection Components

1. Black Nunc 96 well plates with an optical bottom (Thermo Scientific).

2.7 Instruments and Useful Apparatus

1. PCR machine.
2. Agarose gel tank.
3. Cell disrupter or bead beater.
4. Glass beads 0.5 mm diameter (for use in bead beater).

5. Polyacrylamide gel pouring apparatus.
6. Polyacrylamide gel tank.
7. Various centrifuges (micro, low speed, high speed, etc.) and corresponding centrifuge tubes.
8. Trans-Blot SD semidry electrophoretic transfer cell (Bio-Rad).
9. Water bath.
10. UV transilluminator with camera.
11. Shaking incubators.
12. Fermenter.
13. Fluorescent microplate reader.

3 Methods

The following steps are necessary for the functional overexpression of your Protein of Interest (POI) in *S. cerevisiae* (Fig. 1): (1) cloning of the corresponding gene of interest (GOI) into a chimeric shuttle vector, (2) generation of *S. cerevisiae* competent cells, (3) transformation of GOI plasmid into *S. cerevisiae* competent cells, (4) cell growth (shake flasks or batch fermentation), (5) membrane preparations of *S.cerevisiae*, (6) expression confirmation of your POI via western blotting or GFP detection, and (7) expression optimization.

3.1 Cloning of GOI into a Chimeric Shuttle Vector

There are multiple approaches for cloning your GOI into a shuttle vector. These include the high-throughput ligase independent cloning [4] and GAP repair cloning [5, 6], methods previously described with detailed protocols. However, this chapter is organized around the expression of target IMPs and not high-throughput screening methods. For detailed approaches to various high-throughput screening methods for IMPs expressed in *S. cerevisiae*, we refer the readers to previously published detailed protocols [2, 4]. Cloning for our purpose will focus on traditional *E. coli* based approaches utilized for targeted GOIs.

3.1.1 Vector Selection

The vector we selected to use is the "83Xi" designed by the Membrane Protein Expression Center (MPEC.ucsf.edu, Fig. 2). Briefly, this is a GAP compatible expression plasmid that is based on a 2 μ plasmid backbone containing an N-terminal 10× Histidine tag followed by a thrombin protease cleavage site. This plasmid contains a HIS3 selection marker and is driven by a galactose inducible (GAL1) promoter.

3.1.2 DNA Oligomer Primer Design

Primers were designed for the 5′ and 3′ ends of our GOI. These primers included the palindromic sequences for XmaI (N-terminus of GOI) and XhoI (C-terminus of GOI) restriction enzymes, each containing a four basepair overhang (we choose XmaI and XhoI,

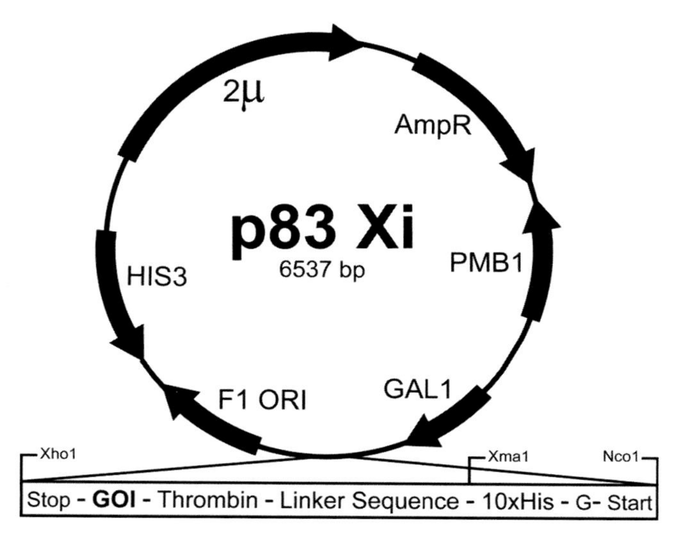

Fig. 2 p83 Xi Plasmid Map: The p83 Xi plasmid is a chimeric shuttle vector based on a 2 μ backbone containing both an ampicillin and HIS3 selection marker. Gene expression is driven by the GAL1 promoter and terminates with the Cyc terminator. Specifically, the p83 Xi plasmid contains an NcoI restriction site with an internal start codon followed by a 10× histidine tag, linker region, thrombin protease, and the multiple cloning site (XmaI, SmaI, and XhoI)

but any unique restriction sites within the multiple cloning site may be used), and the first 12–18 base pairs of either the N- or C-terminal of the GOI. A start codon was not included as it is already designed into the vector. Ensure stop codons (we suggest two) are placed at the end of the gene sequence. Before ordering primers for your GOI, make certain quality checks have been made for melting temperature, self-complementarity, and dimer formation.

3.1.3 PCR Amplifications of GOI from cDNA

1. To amplify your GOI use a high-fidelity proof reading DNA polymerase such as Phusion®.
2. Set up a 50 μl PCR reaction in a 0.5 ml PCR tube by mixing the following components in the given order:

36.5 μl	ddH_2O
10 μl	5× HF buffer
1.0 μl	dNTPs, 10 mM
0.5 μl	100 μM Forward Primer
0.5 μl	100 μM Reverse Primer
1.0 μl	template DNA
0.5 μl	Phusion® polymerase

3. Mix reaction mixture thoroughly. Make sure that the tubes are closed properly and put them into the PCR Machine.

4. Close the lid to the PCR machine and start the following protocol:
   (a) 98 °C for 30 s
   (b) 98 °C for 10 s
   (c) T_m of lowest T_m primer + 3 °C for 30 s
   (d) 72 °C for 30 s per kb
   (e) 72 °C for 10 min
   (f) Hold reaction at 4 °C
   (g) Repeat steps b through d for 35 cycles
5. Run 1–5 μl of the above PCR reaction (plus 1 μl of 10× BlueJuice™ gel loading dye and bring to a 10 μl volume with ddH_2O) on a 0.7% (w/v) agarose gel containing ethidium bromide to check the quality and amount of amplified DNA fragment (*see* **Note 16**).
6. Clean the PCR products using the QIAquick® PCR purification kit. To column purify the remaining PCR products follow the instructions of the kit manufacturer.
7. In short, mix the PCR products with 5 volumes (5 × 49 μl = 245 μl) of PB buffer (binding buffer) and load onto a spin column. The PCR products will bind to the column by spinning the solution through the column at maximum speed in a microfuge.
8. Discard the flow-through.
9. Wash the column with 750 μl PE buffer (contains ethanol), and spin again discarding the flow-through.
10. Spin the column once more to ensure that all residual buffer has been removed.
11. Elute the bound DNA from the spin column into a 1.5 ml microfuge tube by adding 30 μl of EB buffer (elution buffer) to the spin column. Incubate column with buffer for 1–2 min before spinning the column at maximum speed in a microfuge.
12. Purified PCR products will be in the flow-through.

3.1.4 Digestion of GOI and Vector with Restriction Enzymes

1. Digest the purified GOI fragment and the "83 Xi" vector with XmaI and XhoI (or the preferred enzymes you designed within the PCR primers). To do so, add the following items to 25 μl of the purified PCR product and 25 μl of vector (5 μl will be used as an undigested control)

(a) 18 μl	ddH_2O
(b) 5 μl	10× NEB Cut Smart Buffer
(c) 1 μl	XmaI
(d) 1 μl	XhoI

2. Mix the reaction thoroughly and incubate at 37 °C for 2 h.

3.1.5 Gel Purification of DNA Fragments

In order to successfully clone your GOI into the shuttle vector, it is absolutely necessary to gel purify the digested GOI and vector fragments (*see* **Note 17**).

1. After the double restriction enzyme digest is complete add 6 μl of 10× BlueJuice™ gel loading dye (2 μl 10× BlueJuice™ gel loading dye and 13 μl ddH_2O to undigested controls) and mix thoroughly.
2. Load onto a 0.7% (w/v) agarose gel. Run the gel at 100 V for 1–1.5 h for separation.
3. Isolate the fragments of interest by excising them from the gel using a UV transilluminator that visualizes ethidium bromide strained DNA bands. *Warning:* UV light is a potent mutagen—do not over expose DNA (or bare skin) to UV light. To avoid over exposure to the UV light, cut the agarose gel into strips and only expose one lane at a time. Remove the band of interest using a scalpel. Make sure to cut as closely to the band as possible (*see* **Note 18**).
4. Collect the excised bands into pre-weighed 1.5 ml microfuge tubes and calculate weights of the agarose gel slices (*see* **Note 19**).
5. Extract DNA from gel slices using the QIAquick® gel extraction kit, following the manufacturer's instructions.
6. In short, add 3 volumes of QG buffer to one volume of gel slice (0.7 g × 3 volumes = 210 μl) and dissolve the agarose gel by placing the closed microfuge tube at 50 °C for 10 min. Make sure the gel slice is fully dissolved before moving on to the next step.
7. Add one volume of isopropanol to the dissolved gel solution, and mix by inverting the tube.
8. Bind the DNA to the spin column by loading and spinning the solution through the column at maximum speed in a microfuge for 1 min. Repeat until all the DNA solution is bound to the column. Discard the flow-through
9. Wash with 500 μl of QG buffer, spin at maximum speed in a microfuge for 1 min, and discard the flow-through.
10. Wash with 750 μl of PE buffer, spin at maximum speed in a microfuge for 1 min, and discard the flow-through.
11. Spin once more at maximum speed to remove any remaining buffer.
12. Elute the bound DNA from the spin column into a 1.5 ml microfuge tube by adding 30 μl of EB buffer (elution buffer) to the spin column. Incubate column with buffer for 1–2 min before spinning the column at maximum speed in a microfuge (*see* **Note 20**).
13. Purified DNA products will be in the flow-through.

3.1.6 Ligation of DNA Fragments

To ligate the GOI DNA fragment into shuttle vector everything needs to be completely digested with XmaI and XhoI restriction enzymes (or the preferred enzymes you designed within the PCR primers) to prevent the plasmids from self-ligating in the subsequent DNA ligation reaction. For an optimum ligation, choose a ~10× molar excess of insert over plasmid (*see* **Note 21**).

1. Set up the following ligation reactions (20 μl):

(a) 17 − (*x* + *y*) μl	ddH_2O
(b) *x*μl	Insert (not added to the control reaction)
(c) *y*μl	Plasmid
(d) 2 μl	10× T4 ligase buffer
(e) 1 μl	T4 Ligase

2. Incubate at room temperature overnight.
3. Store the ligation reactions at 4 °C until needed.

3.1.7 Transformation of Ligation Products into E. coli

The ligation reactions can now be used to transform *E. coli* XL2 Blue competent cells for the propagation of individual ligated plasmids. Once the *E. coli* cells have been plated on a selective medium, only the *E. coli* cells that have been successfully transformed will grow on the selective medium.

1. Remove a tube of ultra or super competent *E. coli* XL2 Blue competent cells from the −80 °C storage (*see* **Note 22**).
2. Aliquot 200 μl of ice-cold competent cells into two separate 1.5 ml microfuge tubes.
3. Add 5 μl of the ligation mixture and 5 μl of ligation control to individual 1.5 ml microfuge tubes (*see* **Note 23**).
4. Gently mix each tube by flicking the tube several times.
5. Incubate the cells on ice for 10 min.
6. Heat-shock the cells for 30 s in a 42 °C water bath.
7. Place on ice for an additional 5 min following the heat-shock.
8. Add 800 μl of SOC medium to the transformed cells and incubate at 37 °C for 1 h in shaking incubator (~250 RPM).
9. Pipette ~25 μl of the transformed cells onto LBAmp plates, spread the cells evenly using a cell spreader or glass beads.
10. Incubate at 37 °C overnight for the transformed *E. coli* colonies to appear.
11. Remove the plates from the incubator and compare the ligation plate (vector plus insert) to the control plate (vector only). The control plate should have significantly fewer colonies (*see* **Note 24**).

3.1.8 Isolation of Plasmid DNA from E. coli Transformants

Plasmid preparations are made from individual *E. coli* colonies to identify those that contain the desired GOI containing plasmid. If significantly more colonies are located on the ligation plate (vector plus insert) when compared to the control plate (vector only), proceed with isolating the plasmid DNA. If the ligation step appears to have failed, determine which of the previous DNA manipulation steps failed and repeat the protocol starting from the failed step (*see* **Note 25**). Isolate plasmid DNA from 5–10 *E. coli* colonies.

1. Pick a single *E. coli* colony, using sterile technique, and inoculate 8 ml portions of LB-Amp broth in sterile plastic tubes.
2. Grow cells at 37 °C in a shaking incubator at 250 RPM overnight.
3. Harvest the cells by centrifugation, 3000 × *g* for 10 min at 4 °C.
4. Use the QIAprep® spin miniprep kit to isolate the plasmid DNA following the manufacturer's instructions.
5. In short, resuspend the cells in 250 μl of buffer P1, transfer the cells to a 1.5 ml microfuge tube.
6. Add 250 μl of buffer P2 to the cells and mix by inverting the tube.
7. Add 350 μl of buffer N3 to the cells and mix my inverting the tube.
8. Spin at maximum speed in a microfuge for 10 min to sediment the cellular debris.
9. Load the supernatant into a spin column and spin at maximum speed in a microfuge for 1 min. Discard flow-through, the DNA will be bound the spin column.
10. Wash the spin column with 500 μl of buffer PB, spin for 1 min and discard the flow-through.
11. Wash the spin column with 750 μl of buffer PE, spin for 1 min, and discard the flow-through.
12. Re-spin the column to remove any buffer.
13. Place the column into a clean 1.5 ml microfuge tube.
14. Add 30 μl of buffer EB to the column and incubate at room temperature for 1–2 min. Spin the column to elute the DNA. DNA is stored at −20 °C (*see* **Note 20**).

3.1.9 PCR Check, Sequencing, and Generation of GOI-Fusion Plasmid Stocks

1. Instead of sending the isolated plasmid DNA to sequencing, perform a PCR check to identify plasmid stocks that contain the GOI. Set up the PCR reactions according to the parameters used in Subheading 3.1.3. The presence of amplification will signify that the plasmid sample has the target GOI sequence.
2. Send plasmids with positive amplification to a sequencing facility for sequence confirmation.

3. Send sequencing primers together with the column-purified plasmids to a DNA sequencing facility for confirmation of sequence (*see* **Note 6**).
4. Keep additional unsequenced plasmids as reserves in case the first plasmid DNA sequences are not 100% correct.
5. Repeat the steps for transformation into *E. coli* cells and the isolation of plasmid DNA using the QIAprep ® spin miniprep kit to generate sequence confirmed stocks of the GOI-fusion construct.

3.2 Generation of S. cerevisiae Competent Cells

1. Inoculate 10 ml of YPD broth and place in a shaking incubator at 30 °C, 220 RPM overnight.
2. Streak a YPD plate for colony isolation and incubate at 30 °C for 48 h.
3. Place 5 ml of YPD broth in five aerated culture tubes (you may increase/decrease this number as desired) and inoculate each tube with a single colony of yeast. Grow at 30 °C, 220 RPM, for 24 h.
4. Spin down growths at 3000 × *g* for 10 min, and discard the supernatant.
5. Resuspend cell pellets in 5 ml of YPD + 15% (v/v) glycerol broth. Aliquot 500 μl into each microfuge tube. Store at −80 °C.

3.3 Transformation of GOI-Fusion Plasmid into S. cerevisiae Competent Cells

1. Remove a vial of competent cells from the −80 °C freezer. Thaw and pellet the cells.
2. Resuspend cells in 150 μl PLATE solution.
3. Add 5 μl of sheared salmon sperm (carrier DNA, 10 μg) plus ~0.1 μg of plasmid DNA of your GOI-fusion construct and vortex well.
4. Add 10 μl if DMSO and vortex briefly.
5. Incubate at room temperature for 15 min.
6. Heat shock at 42 °C for 20 min.
7. Pellet cells in a microfuge for 30 s and remove supernatant.
8. Add 200 μl TE to the cell pellet and gently resuspend cells by aspirating up-and-down with a pipette tip.
9. Pipette 50 and 150 μl of the resuspended cells onto selective CSM-His plates.
10. Incubate at 30 °C for 2 days (*see* **Note 26**).

3.4 Expression the POI in S. cerevisiae

3.4.1 Shake Flask Growths

Preculture outgrowth

1. Inoculate 5 ml SC-His media with a single colony of the transformant from the CSM-His selective plate containing your GOI-fusion plasmid.

2. Incubate 4 h at 30 °C and at 220 RPM.
3. Autoclave 270 ml ddH_2O in a 1 l size baffled flask. Allow flask to cool to room temperature.
4. Add 37.5 ml 10× CSM-His, 18.8 ml of 20× YNB and 40% (w/v) glucose.
5. Inoculate the flask with 5 ml of the starter culture.
6. Incubate for 24 h at 30 °C and at 220 RPM.

Scale up shake flask cultures (7.5 l)

7. Autoclave 15 × 270 ml ddH_2O in 1 l baffled flasks.
8. Add 37.5 ml 10× CSM-His and 10% (w/v) raffinose, 18.8 ml of 20× YNB, and 9.4 ml 40% (w/v) glucose to each flask.
9. Inoculate with 10 ml of preculture.
10. Incubate for 24 h at 30 °C and at 220 RPM (*see* **Note 27**).

Induction of shake flask cultures

11. Following this growth period the optical density at 600 nm ranged between 15 and 20 for most cultures with glucose concentrations generally <0.1%. Induce the cultures using 125 ml of the 4× YPG stock medium to each flask (*see* **Note 28**).
12. Incubate for 16 h at 30 °C and at 220 RPM.

3.4.2 Batch Fermentation Culture

Preparation of the fermenter

1. Add 5.1 l of ddH_2O and 500 μl of antifoam 204 (Sigma) to the fermentation vessel.
2. Calibrate the pH probe using pH standards of 4 and 7.
3. Prepare the DO (dissolved oxygen) probe
   (a) Clean probe
   (b) Pour out old electrolyte solution
   (c) Refill with new electrolyte solution
4. Ensure all tubing is disconnected, clamps are shut, all protective caps are in place (motor, pH probe, and DO probe). Foil all non-sterile connections and place sterile lines in 50 ml conical tubes sealed with foil. Use autoclave tape where necessary (*see* **Note 29**).
5. Autoclave for 1 h (*see* **Note 30**).
6. Allow fermenter to cool to room temperature (*see* **Note 31**).
7. Add 175 ml of 40% (v/v) glucose, 350 ml of 20× YNB, and 700 ml of 10× CSM-His and 10% (w/v) raffinose to the fermentation vessel.
8. Calibrate the DO probe at 0%.
9. Set temperature at 30 °C.

10. Set the agitation rate from 200 to 350 RPM based on a DO scale of 20–90 %.
11. Flow air at 2.5 l/min.
12. Calibrate DO probe at 100 % (*see* **Note 32**).

Batch fermentation growth

1. Inoculate with 375 ml of overnight growth as previously described in Subheading 3.4.1 (**steps 1–6**).
2. Grow at 30 °C for 24 h

Inoculation of batch fermentation growth

1. Induce the culture by adding 2.5 l of the 4× YPG stock media to the vessel.
2. Increase air flow to 5.0 l/min (*see* **Note 33**).
3. Incubate for 16 h.

3.4.3 Cell Harvest

1. Pellet cells via centrifugation.
2. Resuspend cells in 60 ml of Resuspension Buffer Y per 80 g of cell pellet.
3. Lyse cells immediately or store at −20 °C.

3.5 Membrane Preparation and Immunodetection

1. 12 % SDS-PAGE Gels (makes four mini gels): First, prepare and pour the resolving gel using the following components in the order listed (6.6 ml ddH_2O, 8.0 ml 30 % acrylamide mix, 5.0 ml 1.5 M Tris–HCl, pH 8.8, 0.2 ml 10 % (w/v) SDS, 0.2 ml 10 % (w/v) ammonium persulfate, and 0.008 ml TEMED). Once gel has been poured, top gel off with water to prevent drying. Once the resolving gel has set, pour off any remaining water. Next, prepare and pour the stacking gel (5 %) using the following components in the order listed (3.4 ml ddH_2O, 0.83 ml 30 % acrylamide mix, 0.63 ml 1.5 M Tris–HCl, pH 8.8, 0.05 ml 10 % (w/v) SDS, 0.05 ml 10 % (w/v) ammonium persulfate, and 0.005 ml TEMED). Insert the comb, taking care not to trap bubbles in the wells. Once the stacking gel has set remove from gel apparatus and store at 4 °C (*see* **Note 12**).
2. If frozen cells are to be used thaw them first.
3. Lyse cells using a C3 Emulisflex (~28,000 psi, 3 passes) or a bead beater (blend at maximum speed for 60 s, stop for 60 s. Repeat for a total of five cycles of beating and cooling.) (*see* **Note 34**).
4. Spin down the crude cell extract at 7500 ×*g* for 1 h at 4 °C.
5. Collect the supernatant of the previous low speed spin. Save sample for SDS-PAGE gel (*see* **Note 35**).

6. Resuspend the post lysis pellet in ddH_2O to the original volume (pre centrifugation). Save sample for SDS-PAGE gel.
7. Spin supernatant at 101,000 × *g* for 1 h at 4 °C.
8. Separate the supernatant from the high-speed membrane pellet. Save sample of the supernatant for SDS-PAGE gels.
9. Either scrape out the membrane pellet from each tube and store at −20 °C or resuspend each pellet with 1× solubilization buffer (16 ml buffer per g membrane) then store at −20 °C. Save sample for SDS-PAGE gels (*see* **Note 36**).
10. Locate the samples that were saved for SDS-PAGE gels. Normalize the volume of sample added based on the original starting volumes. Bring to 30 μl using Resuspension Buffer Y. Add 10 μl of 4× Laemmli Sample Buffer for a total volume of 40 μl (*see* **Note 37**).
11. For single construct screening: load 5 μl of protein standard in lanes 1 and 6 of a 10 lane 12 % SDS-PAGE mini gel. Load 20 μl of the cleared cell lysate, resuspended cell pellet, supernatant from the high-speed spin and solubilized membrane pellet in lanes 2–5 and in lanes 7–10 of the gel.
12. For multiple construct screening: load 5 μl of protein standard in lanes 1 or 6 of a 10 lane 12 % SDS-PAGE mini gel. Load 20 μl of the cleared cell lysate, resuspended cell pellet, supernatant from the high-speed spin and solubilized membrane pellet in lanes 2–5 for the first construct and in lanes 7–10 of the gel for the second construct. Prepare two gels loaded identically.
13. Run SDS-PAGE gels in the electrophoresis unit (~140 V for 1 h, or until dye front has reached the base of the gel).
14. Prewet two pieces of thick and thin filter papers (per gel, cut to size) in 1× western transfer buffer, and pre-wet enough 0.2 μm PVDF membrane in 100 % (v/v) methanol to adequately cover the gel. Rinse the membrane with 1× western transfer buffer immediately before use.
15. Prepare the semidry transfer blot apparatus. In brief, make a sandwich of thick blot paper, thin blot paper, PVDF membrane, SDS-PAGE gel, thin blot paper, and thick blot paper. (Use ½ the gel for single constructs and one of the full duplicate gels for multiple constructs.)
16. Transfer proteins to the PVDF membrane for 30 min, 25 V, and 0.33 amps (per gel).
17. Stain the remaining SDS-PAGE gels with a Coomassie Protein staining solution.
18. Block the membranes with a solution of 5 % (w/v) milk in 1× TBST for 1 h at room temperature (*see* **Note 38**).
19. Make working solutions of the desired antibody. For the Penta-His Alexa Fluor® 647 conjugate antibody use a 1:10,000

dilution in 5% (w/v) milk 1× TBST. *Make sure the antibody solution is protected from light.*

20. Incubate at room temperature for 1 h (*see* **Note 38**).
21. Wash the membrane(s) with TBS for 15 min at room temperature. Repeat for a total of three times (*see* **Note 39**).
22. Image membranes, and evaluate POI localization and expression level.

3.6 GFP-Based Optimization Screen as Described by David Drew et al. [2]

This method is applicable if the expression construct has a GFP fluorescent fusion partner. Expression is determined by using whole-cell and in-gel fluorescence via this method.

1. Follow steps outlined previously for cell growths (Subheading 3.4.1 or 3.4.2).
2. Prepare one culture without galactose addition (protein expression not induced) (*see* **Note 40**).
3. Harvest 10 ml of cells that have been cultured with and without galactose via centrifugation and resuspend in 200 μl Resuspension Buffer Y (*see* **Note 41**).
4. Harvest and resuspend the remaining cells according the steps outlined in Subheading 3.4.3.

3.6.1 Expression Quantification Using Whole-Cell Fluorescence

1. Transfer 200 μl of the resuspended cells to a black Nunc 96-well optical bottom plate (*see* **Note 42**).
2. Measure fluorescence emission in a microplate spectrofluorometer (GFP has an excitation wavelength of 488 nm and emission wavelength of 512 nm) (*see* **Note 43**).
3. Estimate membrane protein expression levels (mg/L) using the following methods:
   (a) Aliquot 200 μl of cell suspension from induced (MP-I, RFU) and non-induced (galactose not added, MP-NI, RFU) cultures.
   (b) Measure fluorescence as described in **steps 1–2** in Subheading 3.6.1
   (c) Measure the fluorescence of a known concentration of the GFP fusion partner (STD, RFU) (*see* **Note 44**).
   (d) Calculate the concentration of the GFP fusion partner in mg/L as follows:

$$\frac{(\mathrm{MP}-\mathrm{I})-(\mathrm{MP}-\mathrm{NI})}{\mathrm{STD}}\times(\mathit{conc.}\,\mathrm{STD})=\mathit{conc.}\,\mathrm{protein}$$

For example:

$$\frac{(30{,}000\mathrm{RFU}-2000\mathrm{RFU})}{10{,}000\mathrm{RFU}}\times 0.02\mathrm{mg}/\mathrm{ml}=0.056\mathrm{mg}/\mathrm{ml}\ \text{of protein}$$

4. Next divide the calculated GFP concentration from whole cells by 40 (8000 μl cell culture/200 μl resuspended cells) in order to determine the GFP concentration in 200 μl of resuspended cell culture (*see* **Note 45**).
5. According to David Drew et al. "the typical recovery of GFP counts from a 1 l culture into membranes is 60% or 0.6." Multiply the calculated GFP concentration by 0.6.

$$1.4\ \text{mg} / \text{L} \times 0.6 = 0.84\ \text{mg} / \text{L GFP fusion protein}$$

6. Calculate the amount of membrane protein (MP) expression as follows:

$$\frac{\text{Molecular Mass of MP}\left(\text{kDa}\right)}{\text{Molecular Mass of GFP}\left(\text{kDa}\right)} \times \text{amount of GFP}\left(\text{mg} / \text{l}\right)$$
$$= \text{Membrane Protein Expression}\left(\text{mg} / \text{l}\right)$$

For Example:

$$\frac{50\ \text{kDa}\left(\text{POI}\right)}{28\ \text{kDa}\left(\text{GFP}\right)} \times 0.84 = 1.5\ \text{mg} / \text{l}$$

3.6.2 Expression Screening Using In-Gel Fluorescence

1. Lyse the resuspended cells obtained in Subheading 3.6 (Subheading 3.5, **step 2**).
2. Follow **steps 1–12** as outlined in Subheading 3.5 to prepare, resuspend the membrane pellet, and run the SDS-PAGE gel.
3. Once the gel has finished running, rinse the gel with ddH_2O and detect the fluorescent bands with a CCD camera system. Gel bands are visualized via exposure to blue light (EPI source) set at 460 nm and cut-off filter of 515 nm (*see* **Note 46**).
4. The gel may then be stained with a Coomassie staining solution and destained if desired (*see* **Note 47**).

3.7 Expression Optimization

If the initial cloning efforts result in less than optimal expression levels there are several options available to increase protein yields. The two most prevalent options are altering the expression vector or modifying the growth conditions (*see* **Note 48**).

3.7.1 Vector Modifications

There are several options available for modifying the expression vector to improve overall yields. Firstly, the "83Xi" vector could be modified to express a C-terminal tag (*see* **Note 49**). Next, the choice of tag can affect expression levels. While we prefer the general ease of a His tag for downstream applications, an alternative tag may improve protein production (*see* **Note 50**). Another available option is to choose a different promoter. The current vector is driven by the inducible GAL1 promoter. Changing to a different tightly regulated promoter with a strong transcriptional start signal, such as ADH2, is a viable option (*see* **Notes 51** and **52**). Additionally, modifying the selection marker could assist in improving protein expression as different media formulations could then be tried (*see* **Note 53**). Finally, your gene of interest can be codon optimized using synthetic gene redesign (*see* **Note 54**).

3.7.2 Optimizing Cell Densities

In addition to optimizing the expression vector, growth conditions can be modified to maximize the amount of biomass being produced and the protein content within each cell. The media conditions may not be optimal to produce your POI, in this case changing to an alternative minimal media (*see* **Note 55**), semidefined medium, or a rich complex medium may be beneficial (*see* **Note 56**). Altering the dissolved oxygen content can also be beneficial (*see* **Note 57**). Oxygen availability can effect growth rates and energy availability [7], which may alter plasmid replication or partitioning. Studies have also shown that reducing growth temperatures can improve yields if toxicity is suspected [8], but it can also be detrimental to the expression of IMPs [1] (*see* **Note 58**). Several reports have shown chemical chaperones can increase protein expression levels by improving the folding of IMPs [9, 10]. These chemical chaperones include: (1) DMSO (2.5 % v/v), glycerol (10 % v/v), or histidine (0.04 mg/ml) (*see* **Note 59**). You may also alter the media by changing the composition of the carbon sources. We opt to use 1 % (w/v) glucose and 1 % (w/v) raffinose for a final carbohydrate concentration of 2 % (w/v) because it ensures the absence of glucose repression once the galactose inductant has been added. Variations to this include increasing or decreasing the final carbohydrate concentration, using only glucose, or substituting with alternative secondary carbon sources (e.g., lactose, maltose, and raffinose). Finally more mechanical approaches can be taken. These include switching from shake flask growths to a fermentation process, or increasing the complexity of your fermentation scheme (*see* **Note 60**).

4 Notes

1. No further purification is needed if the oligos are ordered in the salt free format. Ensure that the start codon from the target gene has been removed, as it is already incorporated into the "83Xi" plasmid.
2. Ethidium bromide is highly carcinogenic and poses a reproductive hazard. Use proper personal protective equipment (e.g., gloves and lab coat) when handling ethidium bromide and thoroughly rinse with water all equipment that comes into contact with ethidium bromide. Dispose of ethidium bromide waste according to local regulations.

3. The majority of restriction enzymes are stable for at least 1 year and up to 5 years when stored at −20 °C. Use a portable −20 °C freezer box when handling the restriction enzyme or any other enzyme (e.g., DNA polymerase and DNA ligase) used for molecular biology purposes.
4. Autoclaved sterile stocks of LB can be stored at room temperature and will remain sterile for long periods of time if the stocks remain unopened. LB plates made with ampicillin (LBAmp) are stable for 1 month when protected from light at stored at 4 °C.
5. Ampicillin is light sensitive. Ampicillin stocks stored at 4 °C will remain viable for approximately 1 month. Store at −20 °C for time frames extending past 1 month.
6. The sequences for the sequencing primers are as follows: GAL1 forward primer—5′ CTT TCA ACA TTT TCG GTT TG-3′ and Cyc reverse primer—5′ GGG GGG AGG GCG TGA ATG TAA-3′. When ordering oligos, obtain them in the desalted form if possible.
7. This solution will acidify overtime causing transformation efficiency to decline.
8. Do NOT autoclave galactose. Galactose isomerizes at elevated temperature. Galactose stored at room temperature and 0.45 μm sterile filtered is stable for approximately 4 months.
9. Dissolve using warm water. Do NOT autoclave. Raffinose will isomerize if exposed to elevated temperature for extended periods of time.
10. Dissolve using warm water.
11. Dissolve yeast extract and peptone using warm water. Do not autoclave in the presence of galactose. Galactose isomerizes at elevated temperatures and this will result in unacceptable levels of repressive glucose.
12. Caution: unpolymerized acrylamide is a neurotoxin and proper protective equipment should be used.
13. This solution can be made up as a 10× concentrated stock. Add SDS last, as it causes bubbles.
14. Add SDS last as it causes bubbles. SDS precipitates at 4 °C. Warm the Laemmli Sample Buffer prior to use.
15. Make this solution fresh and store at 4 °C. Once hydrated, the milk will spoil if left out at room temperature.
16. To visualize the DNA fragments we add 5 μl of ethidium bromide (2 mg/ml) to melted agarose (70 ml of 0.7% (w/v) agarose in 1× TAE) prior to pouring the gel (*see* **Note 2**).
17. Gel purification is critical for successful cloning, PCR clean-up alone is not sufficient.

18. If the double digestion appears to have failed, repeat the digestion using a single enzyme at a time.
19. Large pieces of agarose (>400 mg) will result in lower DNA recovery.
20. Minimizing the elution volumes or eluting more than one column (identically prepared) with the same elution buffer will increase the final concentrations of recovered DNA.
21.

$$\text{Insert Mass (ng)} = \text{Desired ratio} \times \left[\frac{\text{Insert Length (base pairs)}}{\text{Vector Length (base pairs)}}\right] \times \text{Vector mass (ng)}$$

22. Commercially available high efficiency competent cells work best. However, in-house competent cells will also work but transformation efficiency may be diminished.
23. It is important to include the ligation control containing only the digested vector. This allows you to determine if a successful double digestion and gel purification has occurred—because the vector will remain linear and thus will not impart any antibiotic resistance to the transformed cells.
24. If the control plate does not have significantly fewer colonies, repeat the cloning process starting with the restriction enzyme digestion. Modify by digesting with each enzyme individually. If the enzymes are compatible in the same buffer, heat-inactivate the first enzyme.
25. The cloning process can be considered to have failed if the background number of transformants on the ligation control plate in higher than the number of transformants on the ligation plate containing vector and insert.
26. If colony size is small, incubate for an additional 24 h. If the transformation efficiency is low or absent, remake the PLATE solution—it will acidify over time.
27. After 24 h of growth the glucose concentration should be less than 0.1 % (w/v). This yeast strain will turn pink due to the *ade2-1* mutation.
28. Glucose represses the GAL1 promoter. If the glucose concentrations have not been sufficiently depleted during growth, induction will not occur.
29. Ensure that at least one line of the fermenter is opened to allow for venting, during autoclaving.
30. An hour autoclave time should be sufficient. However, a dummy growth to check for adequate sterility is recommended if the reader is not familiar with the operation of the fermenter. If contamination occurs, increase the time of the autoclave cycle.

31. The fermenter can be left at room temperature to cool or it can be attached to a cooling unit.
32. It is important to calibrate the DO probe at 100% in the exact growth conditions that will be used (e.g., agitation rate, temperature, and airflow), as these conditions can affect the dissolved oxygen content.
33. The addition of the 4× YPG will decrease the dissolved oxygen content because it has not been pre-equilibrated in order to maintain sterility. Increasing the airflow will limit exposure to an anoxic environment.
34. Yeast cells will NOT sufficiently lyse using sonication or freeze fracture methods.
35. There are two ways determine the lysis efficiency: (1) visually inspect cells under a microscope or (2) the cell debris pellet typically has two layers: bottom darker colored layer of unlysed cells, and a top lighter colored layer consisting or organelles and lysed cells. The estimated ratio of the top layer of lysed cells verses the bottom layer of intact cells will give you an estimate of the lysing efficiency.
36. Solubilizing the membrane pellet using a defined ratio of buffer will ease expression comparisons between various constructs.
37. Normalizing the loading volumes for the SDS-PAGE gel is essential for comparing expression levels between various constructs or comparing various growth conditions during optimization.
38. It can be blocked overnight at 4 °C.
39. Increasing the number of washes or the length of each wash can help reduce the amount of background.
40. This culture will be used to estimate the amount of background fluorescence.
41. It is important to remove all of the supernatant, it can affect whole-cell fluorescence measurements by altering the final volume of the resuspended cells.
42. Yeast cells will settle to the bottom of the 96-well plate. Immediately measure fluorescence to ensure accurate readings.
43. Choose the bottom read option on the plate reader if available.
44. This is needed in order to correlate the whole-cell fluorescence with the amount of GFP produced.
45. Although the initial cell volume was 10 ml there is an effective 2 ml loss by only transferring 200 μl of the resuspended cells (approximately 250 μl total volume, 200 μl buffer + cell pellet = 250 μl).
46. Blue light is closer to the excitation wavelength of GFP, therefore it is preferred over UV light. Detection of GFP expression

using western blotting is not recommended because the transfer of GFP-fusion IMPs can be inconsistent between samples.

47. Coomassie staining is a poor indication of the expression level, because some IMPs bind Coomassie better than others.
48. Another option would be to change the cell line being used, but this may require additional changes with the expression vector or growth conditions.
49. Although our suggested plasmid has an N-terminal tag, N-terminal tags can interfere with the processing of signal peptides. If the POI does not contain a signal peptide, then N-terminal tags provide greater flexibility for the development of expression constructs. However, if a signal peptide is suspected or present a C-terminal tag is preferable.
50. There is a wide variety of expression tags available including, but not limited too: GFP, FLAG, galectin, or maltose binding protein (MBP). Our experience has shown that MBP or galectin tags may help improve expression levels and aid in processing properly folded protein.
51. Like the GAL1 promoter, ADH2 is also subject to glucose repression. The ADH2 promoter is induced in the presence of ethanol, and gene expression increases as glucose is consumed. Two advantages of this promoter include: (1) expression is turned on when the culture is in stationary phase (high biomass), and (2) no inductant is required, and therefore, no disruptions to the growth process occur [11, 12].
52. Constitutive promoters are generally not well suited for the expression of IMPs. Toxicity issues can arise from the inability to control onset of expression. Additionally, constitutive promoters usually results in lower expression levels for IMPs [1, 4].
53. The vector we suggest currently has a HIS3 selection marker. However, the yeast strain that we use is compatible with vectors containing at least one of the following selection markers: *leu2-3, 112 trp1-1, can1-100, ura3-1, ade2-1,* and *his3-11.* Since our chosen yeast strain is compatible with numerous selection methods, substituting the current selection markers may provide additional optimal growth conditions.
54. Synthetic gene redesign can either be performed in house based upon your extensive knowledge of your gene of interest, or it can be out-sourced to companies specializing in synthetic gene redesign/production.
55. Changing to alternative minimal medias may require altering the selection marker on the expression vector.
56. Generally, minimal media is preferred as it favors plasmid stability [7]. However this is not the case for plasmids containing an ADH2 promoter [11, 12].

57. Oxygenation can be controlled in shake flask cultures by increasing the speed at which they are shaking and changing flask size (e.g., liter vs. 2.5 l) or shape (e.g., baffled vs. non-baffled). For growth utilizing fermentation, the agitation speed and air supply rate may be altered or the air source can be changed (e.g., nitrogen, air, pure oxygen).
58. The general starting point for temperature reduction screening is 20 °C, but any temperature below 30 °C may be tried.
59. In order to use histidine as a chemical chaperone with the system we describe, the selection marker and current media configuration will need to be altered. Our systems employ a histidine selection marker and use a synthetic complete histidine dropout media. The URA3 marker is common for growths utilizing a histidine chemical chaperone.
60. Increasing the cell biomass is one crude option for increasing the amount of recoverable protein. This can be done through using advanced fermentation techniques (continuous or fed-batch process) to reach higher cell densities. However, these fermentation processes will have to be designed around each particular POI-vector-yeast strain combination used. Feb-batch process are the most commonly used and designing optimized feeding protocols has two major general points of consideration: (1) is the host strain Crabtree positive or negative, and (2) is protein expression constitutive or regulated [13]?

Acknowledgement

This work was supported by an Institutional Development Award (IDeA) from the National Institute of General Medical Sciences (NIGMS) of the National Institutes of Health under grant number P20GM103639, Oklahoma Center for the Advancement of Science grant HR11-046 (to F.A.H.), American Heart Association predoctoral fellowship 13PRE17040024 (to R.C.B-C.), and NIGMS grant GM24485 (to RMS).

References

1. Newstead S, Kim H, von Heijne G, Iwata S, Drew D (2007) High-throughput fluorescent-based optimization of eukaryotic membrane protein overexpression and purification in *Saccharomyces cerevisiae*. Proc Natl Acad Sci U S A 104:13936–13941
2. Drew D, Newstead S, Sonoda Y, Kim H, von Heijne G, Iwata S (2008) GFP-based optimization scheme for the overexpression and purification of eukaryotic membrane proteins in *Saccharomyces cerevisiae*. Nat Protoc 3:784–798
3. Li M, Hays FA, Roe-Zurz Z, Vuong L, Kelly L, Ho CM, Robbins RM, Pieper U, O'Connell JD III, Miercke LJ, Giacomini KM, Sali A, Stroud RM (2009) Selecting optimum eukaryotic integral membrane proteins for structure determination by rapid expression and solubilization screening. J Mol Biol 385:820–830
4. Hays FA, Roe-Zurz Z, Stroud RM (2010) Overexpression and purification of integral membrane proteins in yeast. Methods Enzymol 470:695–707

5. Lundblad V, Hartzog G, Moqtaderi Z (2001) Manipulation of cloned yeast DNA. Curr Protoc Mol Biol/ edited by Frederick M. Ausubel ... [et al.] Chapter 13: Unit13 10
6. Bessa D, Pereira F, Moreira R, Johansson B, Queiros O (2012) Improved gap repair cloning in yeast: treatment of the gapped vector with Taq DNA polymerase avoids vector self-ligation. Yeast 29:419–423
7. Zhang Z, Moo-Young M, Chisti Y (1996) Plasmid stability in recombinant *Saccharomyces cerevisiae*. Biotechnol Adv 14:401–435
8. Dragosits M, Frascotti G, Bernard-Granger L, Vazquez F, Giuliani M, Baumann K, Rodriguez-Carmona E, Tokkanen J, Parrilli E, Wiebe MG, Kunert R, Maurer M, Gasser B, Sauer M, Branduardi P, Pakula T, Saloheimo M, Penttila M, Ferrer P, Luisa Tutino M, Villaverde A, Porro D, Mattanovich D (2011) Influence of growth temperature on the production of antibody Fab fragments in different microbes: a host comparative analysis. Biotechnol Prog 27:38–46
9. Andre N, Cherouati N, Prual C, Steffan T, Zeder-Lutz G, Magnin T, Pattus F, Michel H, Wagner R, Reinhart C (2006) Enhancing functional production of G protein-coupled receptors in *Pichia pastoris* to levels required for structural studies via a single expression screen. Protein Sci 15:1115–1126
10. Figler RA, Omote H, Nakamoto RK, Al-Shawi MK (2000) Use of chemical chaperones in the yeast Saccharomyces cerevisiae to enhance heterologous membrane protein expression: high-yield expression and purification of human P-glycoprotein. Arch Biochem Biophys 376:34–46
11. Lee KM, DaSilva NA (2005) Evaluation of the *Saccharomyces cerevisiae* ADH2 promoter for protein synthesis. Yeast 22:431–440
12. Price VL, Taylor WE, Clevenger W, Worthington M, Young ET (1990) Expression of heterologous proteins in *Saccharomyces cerevisiae* using the ADH2 promoter. Methods Enzymol 185:308–318
13. Mattanovich D, Branduardi P, Dato L, Gasser B, Sauer M, Porro D (2012) Recombinant protein production in yeasts. Methods Mol Biol 824:329–358

Chapter 12

High-Throughput Baculovirus Expression System for Membrane Protein Production

Ravi C. Kalathur, Marinela Panganiban, and Renato Bruni

Abstract

The ease of use, robustness, cost-effectiveness, and posttranslational machinery make baculovirus expression system a popular choice for production of eukaryotic membrane proteins. This system can be readily adapted for high-throughput operations. This chapter outlines the techniques and procedures for cloning, transfection, small-scale production, and purification of membrane protein samples in a high-throughput manner.

Key words Sf9, Insect cell, Baculovirus, Bacmid, Membrane protein, High-throughput, Expression, Purification, Nickel affinity chromatography

1 Introduction

Membrane proteins are of particular biological interest due to their involvement in various cellular functions and the fact that their genes make up between 20 and 30% of a genome. Some membrane proteins span the membrane (integral or intrinsic membrane proteins) while others are exposed on the outside or the inside (extrinsic or peripheral membrane proteins). Membrane proteins are gateways of the cell and this feature makes them attractive targets for drug discovery [1]. Structural biology is an integral part of drug discovery process and this truly is the golden era for structural biologists. Thanks to the advent of high-throughput technologies and structural genomics (SG) consortia (such as the NIH-funded Protein Structure Initiative) one can easily screen several thousand constructs for a given target. Another major recent development has been the emergence of single-particle cryo-electron microscopy for determining three-dimensional structures with high resolution [2, 3] although X-ray crystallography still dominates the field when it comes to structure determination.

Isabelle Mus-Veteau (ed.), *Heterologous Expression of Membrane Proteins: Methods and Protocols,* Methods in Molecular Biology, vol. 1432, DOI 10.1007/978-1-4939-3637-3_12, © Springer Science+Business Media New York 2016

E. coli is the most dominant expression host for production of membrane proteins, but many eukaryotic membrane proteins cannot be produced in bacterial systems in a functional form, particularly if the protein of interest requires some type of post-translational modification, such as glycosylation, for proper folding and function. However yeast, insect cell and mammalian expression systems can be used to overcome (to various degrees) the deficiencies of *E. coli*. Among these systems, the insect cell baculovirus expression vector system (BEVS) is the most attractive option not only for soluble protein production but also for eukaryotic membrane protein production. It is also gaining popularity in commercial manufacturing for vaccines and gene therapy vectors [4].

Structural studies on membrane proteins are notoriously difficult [5, 6]. In most cases only little amount of protein is obtained. Even when large amounts of membrane protein are expressed, proper folding and insertion into the host membrane are roadblocks still to be overcome. Further during the purification step the membrane proteins may get destabilized due to the inadequacy of the detergents used to extract the proteins from the membranes. Even if all this works out, many membrane proteins do not crystallize as most of them commonly lack surfaces sufficiently large for the strong protein–protein interactions needed for crystal formation.

Our experience with high-throughput screening of membrane proteins in baculovirus expression system shows that about 20–30 % of eukaryotic membrane protein targets can be expressed (NYCOMPS unpublished data). It was also observed that in cases where a target protein could be expressed most of its orthologs also showed appreciable expression. The inverse was also true; that is, orthologs of target proteins that failed to express also could not be expressed.

In order to improve success rate, a multi-thronged approach is needed covering all aspects of the workflow including expression, extraction, purification, and crystallization. Achieving success in all these aspects is time consuming and incremental. The other route is a brute-force approach of structural genomics [7–10], where large numbers of constructs (different tags, homologs, truncations, and mutations) for a given target protein are screened in different detergents, and look for the well-behaved constructs for carrying out structural studies.

Our high-throughput strategy utilizes ligation-independent cloning (LIC) [11, 12], the BAC-to-BAC system for production of recombinant baculovirus DNA molecules [13], and use of robotics wherever necessary.) LIC was chosen over other cloning methods for the reasons of cost, ease, and simultaneous cloning of the same amplified product into several other LIC vectors (with same overhangs). BAC-to-BAC was preferred over co-transfection method for reasons of cost. Considerable research effort has gone into increasing the productivity of the BEVS [14]. Detergent choice is very tricky but important for the success of the membrane protein isolation. We chose n-dodecyl-β-d-maltoside (DDM) as our primary detergent, even though it forms large micelles, since this has been shown to maintain many membrane proteins in a stable state over the prolonged periods of time (2–4 day) required for sample preparation. Lastly, SDS-PAGE rather than western blot has been used to assess the expression and recovery of membrane protein following solubilization, since a band on Coomassie-stained SDS-PAGE gel would be a better yardstick and indicator for the scale-up of these positive hits. We have designed this high-throughput protocol keeping in mind the cost, time, ease of use, and reproducibility (*see* Fig. 1).

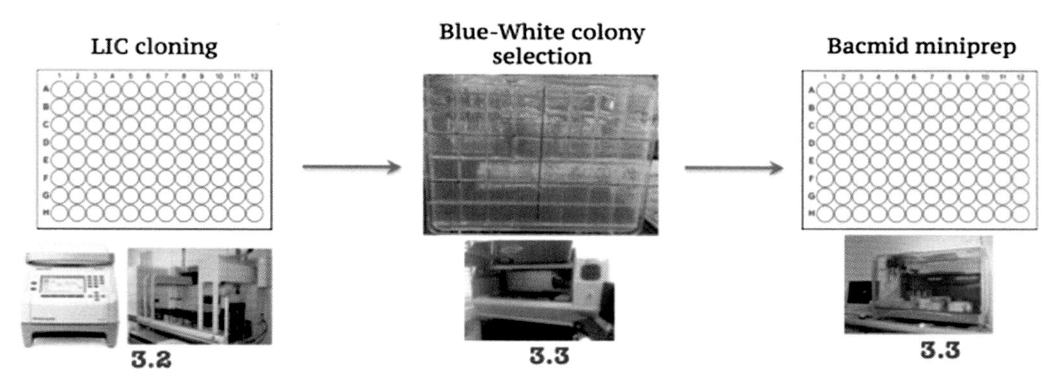

High throughput workflow for Baculovirus expression system

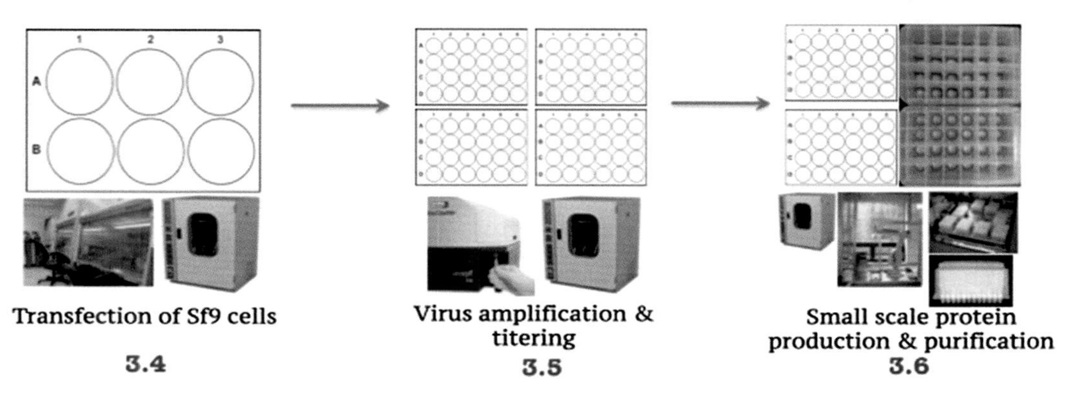

Fig. 1 The representation of the workflow of baculovirus expression system used to produce membrane proteins. The plate format and the instruments used for a particular step in the protocol are showed in the panel

2 Materials

2.1 Insect Cell Culture

1. SF9 and Hi5 cells (Expression Systems/Invitrogen).
2. ESF 921 Insect Cell Culture Medium: (Product No. 96-001, Expression Systems).
3. Cell countess and slides (Invitrogen).
4. Erlenmeyer flasks, polycarbonate, sterile.
5. Incubator shaker.

2.2 High-Throughput Cloning into pFastBac Vector

1. Expression vector.
2. Restriction enzyme BfuAI and 10× buffer.
3. PCR purification kit.
4. 100× Bovine serum albumin (BSA) (10 mg/ml).
5. Deoxynucleotides (100 mM).
6. T4 DNA polymerase.
7. 0.8–1.0 % Agarose gel and agarose gel electrophoresis system.
8. 37 °C, 42 °C, and 50 °C incubators.
9. Thermal cycler.
10. 25 mM EDTA.
11. *E. coli* DH10B-T1^R competent cells.
12. SOC medium.
13. LB agar.
14. Antibiotics.
15. 96-Well PCR plates.
16. Multichannel pipettor.
17. Adhesive foil seals (VWR).
18. Adhesive porous seals (VWR).
19. 37 °C Shaking incubator.
20. 96-Well deep-well blocks.
21. Centrifuge with plate rotor.
22. Liquid-handling robot (Beckman Coulter Biomek®).
23. 96-Well round-bottom plates.
24. Sterile toothpicks.
25. 2 × YT.
26. CosMCPrep plasmid DNA isolation kit (Beckman Coulter).
27. 24-Well blocks
28. Glass beads (Sigma, cat. no. Z273627).

2.3 Isolation of Recombinant Bacmid

1. Sequence-verified miniprep plasmid DNA.
2. *E. coli* DH10Bac competent cells.
3. Antibiotics: Kanamycin sulfate, tetracycline, and gentamycin.
4. Genetix Qpix 2 XT Automated Arraying Bacterial Colony Picker.
5. Genetix vented QTray with 48-well divider.
6. Glycerol: Prepare 65% (w/v) stock with 25 mM Tris, pH 8.0 and 100 mM $MgSO_4$, autoclave, and store at room temperature.
7. Eppendorf epMotion 5075 Vac.
8. 5 Prime Perfectprep Bac 96 kit.
9. KOD Hot-Start DNA polymerase kit (Novagen) containing:
   KOD Hot-Start DNA polymerase
   10× PCR buffer
   25 mM $MgSO_4$
   dNTP mix (2 mM each)
10. pUC/M13 forward and reverse primers (Integrated DNA Technologies, custom synthesized).

2.4 Transfection of Sf9 Cells with Bacmid

1. Cellfectin® II Reagent (Invitrogen).
2. Grace's Insect Cell Medium, unsupplemented (Gibco™).
3. 6-Well tissue culture plates, sterile.
4. 96-Well microplates.
5. ESF 921 Insect Cell Culture Medium (Expression Systems).
6. 25 ml Reagent reservoirs, polystyrene, sterile.
7. Light microscope.
8. Shaker incubator.
9. Centrifuge, with adaptors for centrifuging deep-well plates.

2.5 Baculovirus Titering and Amplification

1. 96-Well deep-well plates, sterile.
2. 24-Well deep-well blocks.
3. Adhesive foil seals.
4. Virus counter (ViroCyt).
5. Sample dilution buffer; Combo Dye solution (Virocyt).
6. Autoclaved 96-well deep-well blocks.
7. Aluminum foil.
8. Shaker incubator.
9. Centrifuge, with adaptors for centrifuging deep-well blocks.

2.6 Small-Scale Expression and Purification

1. Resuspension buffer: 50 mM HEPES pH 7.8, 300 mM NaCl, 20 mM imidazole pH 7.8, 5 % (v/v) glycerol, 0.1 mM TCEP (*tris*(2-carboxyethyl)phosphine), 1 mM $MgCl_2$.
2. Resuspension buffer containing 12 % (w/v) *N*-dodecyl-β-d-maltopyranoside (DDM).
3. Benzonase nuclease (EMD Millipore).
4. Protease inhibitor.
5. Titramax 1000 shaking platform (Heidolph Instruments), 4 °C.
6. Sonicator robot-small probe (ST Robotics) or handheld sonicator.
7. 50 ml centrifuge tubes.
8. 96-Well, 2 ml filter plates (Thompson Instrument Company).
9. 96-Well bottom plate seal (Thompson Instrument Company).
10. Large orifice pipet tips.
11. Nickel-chelating resin.
12. Wash buffer: 25 mM HEPES pH 7.8, 500 mM NaCl, 75 mM imidazole pH 7.8, 5 % (v/v) glycerol, 0.1 mM TCEP, 0.05 % DDM.
13. Elution buffer: 25 mM HEPES pH 7.8, 200 mM NaCl, 500 mM imidazole pH 7.8, 5 % (v/v) glycerol, 0.1 mM TCEP, 0.05 % DDM.
14. 5× SDS-PAGE loading buffer: 200 mM Tris-Cl pH 6.8, 10 % (w/v) SDS, 25 % (v/v) glycerol, 0.05 % (w/v) bromophenol blue, 5 % (v/v) β-mercaptoethanol (added fresh).
15. Criterion Precast 5 to 20 % (w/v) acrylamide-Tris-Cl gels, 26 wells (Bio-Rad).
16. Coomassie Brilliant Blue R-250 Dye.

3 Methods

3.1 Insect Cell Culture: Sf9 and Hi5

1. Use the sterile, disposable 125 ml and/or 500 ml flasks without baffles at the bottom.
2. Passage the SF9 cells in 200 ml of ESF 921 medium at a cell density of 0.5×10^6 viable cells per ml.
3. Shake the flask in an orbital shaker at 140 rpm and 27 °C.
4. Allow the culture to reach a density of $3–5 \times 10^6$ viable cells per ml. These cells can be used for both adherent and suspension culture without any adaptation procedures.

The abovementioned procedure can be used for the maintenance of Hi5 cells except that, as Hi5 cells grow faster, they can be split to a lower cell density (0.3×10^6 cells/ml) compared to SF9 cells. These cells can be maintained for approximately 40 passages. Rather than strictly going by the passage number, cells have to be constantly checked for any morphological changes and doubling time. Sf9 cells were used for transfection, virus propagation, and protein production, whereas Hi5 cells were used only for protein production.

3.2 High-Throughput Cloning into pFastBac Vector

The generation of recombinant baculoviruses is based on the Bac-to-Bac expression system (Invitrogen). The pFastBac transfer vector is modified to include overhangs to carry out ligation-independent cloning (LIC). The LIC pFastBac vectors come in many versions with affinity tags either introduced at the N- or at C-terminus of the target gene. Other constructs also fuse the GFP gene along with affinity tags to the gene of interest. High-throughput cloning is performed on a 96-well format. The procedure described below is for the pFastBac-C vector with c-terminally tagged 10× His and FLAG epitope. The vector also contains the SacB gene for negative selection of parental plasmid.

1. Digest 30–40 μg of plasmid pFastBac-C vector with BfuAI. Incubate at 50 °C for 2.5 h.

BfuAI digestion:	x μl plasmid DNA (30–40 μg)
	10 μl of NEB 10× buffer 3
	6 μl BfuAI (NEB, 30 units total)
	ddH_2O to 100 μl final

2. Purify entire digest using the Qiagen QiaQuick PCR purification kit. Use 3 columns per digestion reaction. Elute each column twice: first with 50 μl EB, and then with 30 μl EB. Pool all eluates (approximately 240 μl) and measure the DNA concentration.
3. Run a 2 μl sample on an agarose gel to verify the completion of the digestion reaction.
4. Assemble the vector LIC reaction in one or two rows of a 96-well plate.

LIC treatment:	x μl of BfuAI-digested DNA (10 μg)
	40 μl of NEB 10× buffer 2
	4 μl of 100× BSA (NEB)
	10 μl 100 mM dCTP (2.5 mM final)
	5 μl of T4 DNA polymerase (NEB, 15 units total)
	ddH_2O to 400 μl final

The final DNA concentration should be 25 ng/μl.

5. Incubate at 22 °C for 1 h and then 20 min at 75 °C to inactivate the polymerase (programmed on thermal cycler). Keep at –20 °C for long-term storage.
6. Amplify target gene using primers with LIC overhang for selected vector (*see* **Notes 1** and **2**).
7. Purify amplified products using standard protocols (*see* **Note 3**).
8. Prepare LIC reaction mixture for purified PCR products.

LIC treatment (120 inserts):	120 μl of NEB 10× Buffer 2
	12 μl of 100× BSA (NEB)
	30 μl 100 mM dGTP (2.5 mM final)
	15 μl T4 DNA polymerase (0.0375 units/μl)
	785 μl ddH_2O

Combine 8 μl of the above mixture with 2 μl of each purified PCR product. Incubate at 22 °C for 60 min. Heat inactivate enzyme at 75 °C for 20 min (programmed on thermal cycler).

9. Assemble the annealing reaction by combining 2 μl of vector (*see* **step 2**) with 4 μl of each insert (**step 5**) and incubate at 22 °C for 60 min.
10. Add 2 μl of 25 mM EDTA and incubate at 22 °C for 5 min (annealed products can be stored at –20 °C if necessary).
11. *Transformation:* Use a multichannel pipettor to combine 2 μl of each LIC reaction to 20 μl of competent DH10B-T1^R cells (96-well PCR plate of frozen competent cells, *see* **Note 4**) and leave on ice for 30 min. Heat shock for 45 s at 42 °C (performed on thermal cycler). Add 80 μl pre-warmed (37 °C) SOC medium. Cover with porous seal and recover cells for 1 h in a 37 °C shaking incubator at 400–500 rpm. Plate transformations onto LB-agar-containing carbenicillin and 5 % sucrose in 24-well blocks with two to four sterile glass beads per well (*see* **Notes 5** and **6**). Shake blocks at 37 °C shaker until all liquid is absorbed into LB agar (approximately 45 min). Incubate inverted blocks overnight at 37 °C.
12. *Plasmid isolation:* Plasmid DNA is isolated using the CosMCPrep plasmid purification system and a liquid-handling robot. For non-automated, 96-well plasmid purification, the PureLink 96 HQ mini plasmid DNA purification kit (Invitrogen) can be used. Use sterile toothpicks to pick a single colony from each well of the 24-well blocks and inoculate 1.2 ml of 2× YT medium containing the appropriate antibiotic in a 96-well deep-well block. Incubate overnight in a 37 °C shaking incubator at 800 rpm (a small amount of the overnight

culture should be used to inoculate another culture for sequencing purposes (*see* **Note 7**)). Recover cells by centrifuging for 10 min at 4500×*g*, 4 °C. Decant medium. Use the CosMCPrep plasmid DNA isolation kit according to the manufacturer's protocol. Elute DNA with 100 μl buffer RE1 (or water) into a 96-well round-bottom plate (*see* **Note 8**). DNA can be stored at −20 °C for up to 1 year.

3.3 Isolation of Recombinant Bacmid

The next step is the transfer of the cloned fragment (from the pFastBac vector) into the DH10Bac *E. coli* strain. DH10Bac cells contain a baculovirus shuttle vector (i.e., bacmid) with a mini-*att*Tn7 target site and a helper plasmid. In this system, the recombinant baculovirus genome is generated in bacteria through a transposition reaction, which occurs between the mini-Tn7 site on the pFastBac vector and the mini-*att*Tn7 site on the bacmid.

1. Use 1–2 μl of the above sequence-verified plasmids to transform 20 μl of competent DH10-Bac cells (96-well PCR plate of frozen competent cells).
2. Meanwhile, add 1 ml of pre-warmed SOC into the Costar 96-well assay block (2 ml capacity).
3. Incubate the 96-well PCR plate containing transformation mix in ice for 30 min.
4. Heat shock at 42 °C for 45 s.
5. Incubate in ice for 2 min.
6. Transfer the contents to the deep-well block containing SOC media. Grow the cells in a shaking incubator for 5 h at 37 °C with medium at 600 rpm.
7. Prepare LB agar Q-plates containing kanamycin (50 μg/ml), tetracycline (10 μg/ml), gentamicin (7 μg/ml), IPTG (40 μg/ml), and Bluo gal (100 μg/ml). Label four Q-plates as A1-H6 and A7-H12 for plating 150 μl of undiluted and 150 μl of (1:10) diluted samples, respectively. The Q-plating protocol on Genetix robotic colony picker was used to spread the colonies. If the above instrument is unavailable, plating can be done manually using glass beads. Incubate the plates at 37 °C for 48 h (*see* **Note 9**).
8. White colonies for each bacmid sample are picked and grown in 1.5 ml of 2× YT media containing kanamycin, gentamicin, and tetracycline at 37 °C for 12–18 h (*see* **Note 10**). Some of the sample (300 μl) can be kept aside to make glycerol stocks of these bacmids (*see* **Note 11**).
9. Isolation of transfection-grade bacmid DNA from 96-well culture plates can be completed in approximately 75 min using the Perfectprep BAC 96 Kit (5 PRIME) on the Eppendorf epMotion 5075 Vac. The Perfectprep BAC 96 Kit and the

Eppendorf epMotion 5075 Vac workstation are integrated to provide a complete "walk-away" protocol. This protocol is optimized to yield 0.5–1.0 μg of bacmid DNA that is immediately ready to use for multiple downstream applications. If the above instrumentation is unavailable, the 5 PRIME Perfectprep BAC 96 Kit has alternative protocols suited for either a vacuum manifold compatible with 96-well plates or a centrifuge with a deep-well rotor (*see* **Note 12**).

10. A PCR analysis on the bacmid samples is carried out to verify the presence of the gene of interest in the recombinant bacmid. Use the pUC/M13 primers and Novagen KOD Hot Start DNA Polymerase for amplification (*see* **Note 13**). Assemble the PCR reactions in 96-well plate(s). Prepare the following PCR master mix. (Note: It is designed for 110 reactions to account for loss during pipetting. Scale up or down depending on the number of samples).

10× Buffer for KOD Hot Start DNA Polymerase	550 μl
2 mM dNTPs	550 μl
25 mM $MgSO_4$	330 μl
pUC/M13 forward primer (30 μM)	55 μl
pUC/M13 reverse primer (30 μM)	55 μl
KOD Hot Start DNA Polymerase (1 U/μl)	110 μl
Sterile dH_2O	3.74 ml

Aliquot 49 μl of PCR mix into each well of a 96-well PCR plate (*see* **Note 14**). Use a multichannel pipette to add 1 μl (100 ng of recombinant bacmid DNA) bacmid DNA template into the corresponding well of the PCR plate.

Amplify the target genes with initial denaturation for 3 min at 93 °C, followed by 35 cycles of 94 °C for 45 s, 55 °C for 45 s, and 72 °C for 5 min. Final extension is at 72 °C for 7 min.

Use 5–10 μl from each reaction to analyze on 1% (w/v) agarose gel (containing ethidium bromide) by electrophoresis. Successful insertion of the gene into the bacmid will show a PCR product of approximately 2300 bp fragment of vector plus the size of your insert.

3.4 Transfection of Sf9 Cells with Bacmid

Transfection is defined as the process of introducing nucleic acids into eukaryotic cells by nonviral methods; in the present case the recombinant bacmid DNA is mixed with cationic lipid, cellfectin®II. The cationic head group of the lipid compound associates with negatively charged phosphates on the nucleic acid, thereby forming a unilamellar liposomal structure with a positive surface charge when in water [15, 16]. The positive surface charge of the liposomes mediates the interaction of the DNA and the cell membrane, allowing for fusion of the liposome/bacmid DNA with the negatively charged insect cell membrane. The cationic lipid-based transfection reagents have been used widely and are known for their consistently high transfection efficiencies.

1. Seed an appropriate number of 6-well plates with Sf9 cells, at 8×10^5 cells per well in a total volume of 2 ml. Allow at least 20 min for the cells to adhere (*see* **Note 15**).
2. Meantime, prepare the transfection mixture for 100 samples (96 bacmid samples and 4 controls (*see* **Note 14**)). In a sterile 15 ml Falcon tube dispense 11 ml of Grace's medium-unsupplemented (without antibiotics and serum; alternatively unsupplemeted growth medium like ESF 921 medium can also be used). Add 880 μl of Cellfectin II. Vortex briefly to mix.
3. In a sterile 96-well dispense 100 μl of Grace's medium-unsupplemented into each well and add 0.8–1 μg of bacmid to each well using a 12-channel pipette. Mix gently.
4. Add the diluted Cellfectin II in a 25 ml sterile pipetting reservoir. Dispense 108 μl of it into each well of the 96-well plate containing diluted bacmids. Mix gently and incubate for 15 min at room temperature.
5. Add ~210 μl of bacmid-Cellfectin mixture dropwise to the cells in each corresponding well of a 6-well plate. Incubate the cells 27 °C for 3–5 h.
6. Remove the transfection media and replace with 2 ml of growth media. Incubate cells at 27 °C for 5 days or until you see signs of viral infection (*see* **Note 16**).

3.5 Baculovirus Titering and Amplification

The consistent and high-level expression of recombinant proteins in insect cells or transducing vertebrate cells requires amplification and titer determination of recombinant virions. Often the titer is determined by detecting morphological changes in infected cells using traditional end-point dilution assays and/or laborious and time-consuming plaque formation. There are several other protocols like (a) Q-PCR-based primers and probes for gp64 [17]; (b) measurement of cell diameter change [18]; (c) immunostaning methods using an antibody to a DNA-binding protein [19]; (d) flow cytometric titering methods: cell surface expression of the gp64 protein [20], eGFP fluorescence [21], and staining of the baculovirus DNA with SYBR Green I [22]; (e) virus counter using a two-color method which involved staining the viral genome and the protein coat for baculoviruses [23]. In the current study, we describe simple 96-well tittering protocols using virus counter.

1. *Harvest P0 virus*: Collect the ~2 ml medium containing baculovirus after 5 days and transfer to a sterile 96-well storage block. Centrifuge the block at 500 × *g* for 5 min and transfer supernatant to a fresh sterile block. Seal with a sterile adhesive

aluminum foil. In order to protect from light, wrap the block with aluminum foil and store at 4 °C.

2. *Virus counting*: Aspirate 150 μl of viral suspension into an Eppendorf tube. Prepare the combo dye working solution by reconstituting the combo dye concentrate with 5 μl of acetonitrile. Add 1 ml of the combo dye buffer to combo dye concentrate. Then add 75 μl of combo dye to 150 μl of viral suspension sample. Incubate for 30 min in the dark. Then analyze the virus titer with virus counter.
3. *P1 amplification*: Take four sterile 24-well deep-well blocks (labels A–G and 1–12), and add 4 ml of Sf9 cells with a density of 1.5×10^6 viable cells/ml. Infect the culture with an MOI of 0.1 by adding appropriate amount of P0 viral stock from each well. Cover each block with a sterile adhesive breathable rayon film. Place all the four blocks in a shaker incubator and shake them at 300 rpm at 27 °C for 3 days. Centrifuge the blocks at $500 \times g$ for 5 min. Seal and wrap the block with aluminum foil.
4. *P2 amplification*: Use the same procedure as mentioned for P1 amplification to generate a high-titer P2 baculoviral stock.

3.6 Small-Scale Expression and Purification

Infection of Sf9 cells with recombinant baculovirus

1. Take four sterile 24-well deep-well blocks (labels A–H and 1–12), and add 4 ml of Sf9 cells with a density of 2.0×10^6 viable cells/ml.
2. Infect the culture with an MOI of 3 by adding appropriate amount of P2 viral stock to each well.
3. Cover each block with a sterile adhesive breathable rayon film.
4. Place all the four blocks in a shaker incubator and shake them at 300 rpm at 27 °C for 2–4 days.
5. Centrifuge the blocks at $500 \times g$ for 10 min.
6. Discard the media, seal the block, and store pellets at −80 °C.

Cell lysis and solubilization

1. Prepare the resuspension buffer supplemented with 0.5 mM TCEP (*tris*(2-carboxyethyl)phosphine), 4 μl/25 ml benzonase, protease inhibitors).
2. Add 500 μl of resuspension buffer to each well containing pellets, pipette up and down to resuspend the pellet, and then transfer the samples to a fresh 96-well deep-well block.
3. Lyse the cells with two rounds of sonication using a robotic or handheld sonicator.
4. Add 100 μl of resuspension buffer containing 12 % DDM (final concentration about 2 %) (*see* **Note 17**). Allow lysates to clear for at least 1 h (or overnight) using a Glas-Col shaking platform at 4 °C.

Nickel affinity protein purification

1. Pour desired amount of nickel resin into a clean 50 ml Falcon tube. Centrifuge at 500×*g* for 2 min, pour off supernatant, add Milli-Q water to nickel resin, and centrifuge. Repeat the step one more time with lysis buffer. Add equal amount of lysis buffer to the pelleted nickel resin and mix.
2. Transfer lysate to a bottom-sealed 96-well Thomson filter plate.
3. Add 50 μl of nickel slurry to each well.
4. Place the filter plate on a Glas-Col shaking platform and set it for 600 rpm at 4 °C to bind overnight.
5. The following day, unseal filter plate and place it on a vacuum manifold to draw out the lysate.
6. Reseal filter plate and add 1 ml of wash buffer to each well and shake at 600 rpm for 30 min on a shaking platform at 4 °C.
7. Unseal filter plate and place it on a vacuum manifold to draw out the wash buffer.
8. Repeat wash with an additional 1 ml of wash buffer and allow to shake at 600 rpm for 30 min at 4 °C.
9. Vacuum wash buffer from filter plate and place a 96-well U-shape microplate underneath.
10. Centrifuge for 2 min at 500×*g* in a plate rotor to remove excess wash buffer and replace bottom seal on filter plate.
11. Apply 35 μl of elution buffer into each well and shake at 600 rpm for 30 min at 4 °C on a shaking platform.
12. Remove bottom seal and place a new 96-well U-shape microplate underneath.
13. Centrifuge for 2 min at 500×*g* in a plate rotor to collect the eluates.
14. Analyze the eluted samples by SDS-PAGE followed by Coomassie blue staining to visualize expressed proteins (*see* **Note 18**). If the construct is a membrane protein–GFP fusion, then some of the sample can be used to run a fluorescence-based size-exclusion chromatography (FSEC) to check for aggregation/monodispersity [24, 25].

4 Notes

1. Amplification of target DNA should be done using cDNA version of the gene. We routinely use I.M.A.G.E clones for human, mouse, rat, bovine, frog, and zebrafish orthologs (image-consortium.org). Synthetic genes are a valid but pricey option. For amplification reactions any standard thermostable polymerase can be used. In our lab we routinely use KOD Hot-Start Polymerase (Novagen-EMD).

2. Overhangs for the pFastBac-C vectors are as follows:
For forward primers: 5′ TTAAGAAGGAGATATACT 3′ (followed by start codon of target gene). For reverse primes: 5′ GATTGGAAGTAGAGGTTCTCTGC 3′ (followed by reverse complement of last codon; do not include stop codon).

3. In most cases a simple purification to remove nucleotides and enzymes will suffice. If amplification results in several bands visible on an agarose gel, a gel purification step of the correct amplified band is necessary.

4. We routinely use T1-resistant bacterial strains (wherever possible) in our laboratory to avoid potential phage contamination. Chemically competent bacteria can be prepared using standard procedures. Cloning strains other than DH10B can also be used.

5. Sucrose is included for negative selection of parental vector containing the SacB gene (sucrose will have no effect on vectors lacking SacB gene).

6. Four 24-well blocks recreate a standard 8 × 12 96-well plate. We routinely reuse 24-well blocks multiple times by scraping out old agar, running them through a standard glassware washer, and autoclaving them.

7. We recommend sequencing constructs for insert verification. We use Beckman Genomics single-pass sequencing service (beckmangenomics.com). Restriction enzyme analysis (if possible) or amplification of insert are viable, albeit more ambiguous, options.

8. The DNA concentration of plasmids isolated from high-throughput miniprep can be as high as 100 ng/μl. If significantly higher, then they should be diluted in the working range of 20–100 ng/μl.

9. A 48-h incubation time is necessary to discriminate white colonies from blue ones. White colonies indicate that the β-galactoside gene residing in the parental bacmid has been replaced by the gene of interest from the pFastBac vector.

10. Pick two (or preferably three) colonies for each construct, as not all white colonies will harbor bacmid DNA recombined with the gene of interest.

11. It is a good practice to make the glycerol stocks of the pFastBac constructs and also the corresponding recombinant bacmids. As the BEVS is a multi-step process, things can go wrong at any step; frozen stocks will help to curb some of the delays.

12. Alternatively, bacmid DNA can also be isolated using the standard alkaline lysis miniprep procedure. Do not use Qiagen miniprep DNA columns, as they will not bind the bacmid DNA.

13. The pUC/M13 primers must be custom synthesized (pUC/M13 forward: 5′-CCCAGTCACGACGTTGTAAAACG-3′; pUC/M13 reverse: 5′-AGCGGATAACAATTTCACACAGG-3′).
14. It is important to include controls with your 96-sample batch of bacmids. The control should be as follows: (1) a bacmid that has worked in the past, with good transfection efficiency and virus production; (2) a bacmid with moderate efficiency; (3) a bacmid with poor efficiency; and (4) cells alone with transfection reagent but no bacmid.
15. Transfections can also be done with cells in suspension rather than monolayer, but we prefer monolayer as we can visually monitor signs of infection under a microscope.
16. When placing 6-well transfection plates in the 27 °C incubator for several days, it is important to minimize evaporation of the media in the wells. This is best achieved by placing the plates in an enclosed plastic dish containing dampened paper towels.
17. Sonication followed by dropwise addition of DDM detergent robustly solubilizes over 90 % of cell pellets. In case the lysate does not clear, shake the block for longer times in the cold room.
18. After the membrane protein isolation and first purification step, users should carry out a detergent screen to find the one which best suits their purposes.

Acknowledgement

This work was supported by NIH grant U54 GM095315-04.

References

1. Overington JP, Al-Lazikani B, Hopkins AL (2006) How many drug targets are there? Nat Rev Drug Discov 5:993–996
2. Bai XC, McMullan G, Scheres SH (2015) How cryo-EM is revolutionizing structural biology. Trends Biochem Sci 40:49–57
3. Cheng Y (2015) Single-particle cryo-EM at crystallographic resolution. Cell 161:450–457
4. Felberbaum RS (2015) The baculovirus expression vector system: a commercial manufacturing platform for viral vaccines and gene therapy vectors. Biotechnol J 10:702–714
5. Kloppmann E, Punta M, Rost B (2012) Structural genomics plucks high-hanging membrane proteins. Curr Opin Struct Biol 22:326–332
6. Pieper U, Schlessinger A, Kloppmann E, Chang GA, Chou JJ, Dumont ME, Fox BG, Fromme P, Hendrickson WA, Malkowski MG et al (2013) Coordinating the impact of structural genomics on the human alpha-helical transmembrane proteome. Nat Struct Mol Biol 20:135–138
7. Punta M, Love J, Handelman S, Hunt JF, Shapiro L, Hendrickson WA, Rost B (2009) Structural genomics target selection for the New York consortium on membrane protein structure. J Struct Funct Genomics 10:255–268
8. Love J, Mancia F, Shapiro L, Punta M, Rost B, Girvin M, Wang DN, Zhou M, Hunt JF, Szyperski T et al (2010) The New York consortium on membrane protein structure (NYCOMPS): a high-throughput platform for structural genomics of integral membrane proteins. J Struct Funct Genomics 11:191–199

9. Mancia F, Love J (2011) High throughput platforms for structural genomics of integral membrane proteins. Curr Opin Struct Biol 21:517–522
10. Hecht M, Bromberg Y, Rost B (2015) Better prediction of functional effects for sequence variants. BMC Genomics 16(Suppl 8):S1
11. Aslanidis C, de Jong PJ (1990) Ligation-independent cloning of PCR products (LIC-PCR). Nucleic Acids Res 18:6069–6074
12. Strain-Damerell C, Mahajan P, Gileadi O, Burgess-Brown NA (2014) Medium-throughput production of recombinant human proteins: ligation-independent cloning. Methods Mol Biol 1091:55–72
13. Luckow VA, Lee SC, Barry GF, Olins PO (1993) Efficient generation of infectious recombinant baculoviruses by site-specific transposon-mediated insertion of foreign genes into a baculovirus genome propagated in Escherichia coli. J Virol 67:4566–4579
14. Hitchman RB, Possee RD, King LA (2009) Baculovirus expression systems for recombinant protein production in insect cells. Recent Pat Biotechnol 3:46–54
15. Felgner PL, Gadek TR, Holm M, Roman R, Chan HW, Wenz M, Northrop JP, Ringold GM, Danielsen M (1987) Lipofection: a highly efficient, lipid-mediated DNA-transfection procedure. Proc Natl Acad Sci U S A 84: 7413–7417
16. Chesnoy S, Huang L (2000) Structure and function of lipid-DNA complexes for gene delivery. Annu Rev Biophys Biomol Struct 29:27–47
17. Hitchman RB, Siaterli EA, Nixon CP, King LA (2007) Quantitative real-time PCR for rapid and accurate titration of recombinant baculovirus particles. Biotechnol Bioeng 96:810–814
18. Janakiraman V, Forrest WF, Seshagiri S (2006) Estimation of baculovirus titer based on viable cell size. Nat Protoc 1:2271–2276
19. Kwon MS, Dojima T, Toriyama M, Park EY (2002) Development of an antibody-based assay for determination of baculovirus titers in 10 hours. Biotechnol Prog 18:647–651
20. Mulvania TH, Hedin D (2004) A flow cytometric assay for rapid, accurate determination of baculovirus titers. BioProcessing J 3:47–53
21. Karkkainen HR, Lesch HP, Maatta AI, Toivanen PI, Mahonen AJ, Roschier MM, Airenne KJ, Laitinen OH, Yla-Herttuala S (2009) A 96-well format for a high-throughput baculovirus generation, fast titering and recombinant protein production in insect and mammalian cells. BMC Res Notes 2:63
22. Shen CF, Meghrous J, Kamen A (2002) Quantitation of baculovirus particles by flow cytometry. J Virol Methods 105:321–330
23. Stoffel CL, Kathy RF, Rowlen KL (2005) Design and characterization of a compact dual channel virus counter. Cytometry A 65:140–147
24. Kawate T, Gouaux E (2006) Fluorescence-detection size-exclusion chromatography for precrystallization screening of integral membrane proteins. Structure 14:673–681
25. Hattori M, Hibbs RE, Gouaux E (2012) A fluorescence-detection size-exclusion chromatography-based thermostability assay for membrane protein precrystallization screening. Structure 20:1293–1299

Chapter 13

Small-Scale Screening to Large-Scale Over-Expression of Human Membrane Proteins for Structural Studies

Sarika Chaudhary, Sukanya Saha, Sobrahani Thamminana, and Robert M. Stroud

Abstract

Membrane protein structural studies are frequently hampered by poor expression. The low natural abundance of these proteins implies a need for utilizing different heterologous expression systems. *E. coli* and yeast are commonly used expression systems due to rapid cell growth at high cell density, economical production, and ease of manipulation. Here we report a simplified, systematically developed robust strategy from small-scale screening to large-scale over-expression of human integral membrane proteins in the mammalian expression system for structural studies. This methodology streamlines small-scale screening of several different constructs utilizing fluorescence size-exclusion chromatography (FSEC) towards optimization of buffer, additives, and detergents for achieving stability and homogeneity. This is followed by the generation of stable clonal cell lines expressing desired constructs, and lastly large-scale expression for crystallization. These techniques are designed to rapidly advance the structural studies of eukaryotic integral membrane proteins including that of human membrane proteins.

Key words Membrane protein expression, Mammalian cell culture, FSEC, Large-scale expression

1 Introduction

Membrane proteins constitute an integral component of the cellular proteome and participate in several physiological processes. Their clinical importance is emphasized by the fact that the majority (~60%) of prescription drugs act on membrane proteins. Despite their physiological importance, the success rate of structure determination of human membrane protein structures is low [1, 2]. Membrane protein crystallization requires extensive optimization of many parameters that are not critical for soluble proteins. A major bottleneck is the over-expression of proteins in order to obtain sufficient quantity of pure, homogeneous, and stable (PHS) proteins for structural studies [3].

Isabelle Mus-Veteau (ed.), *Heterologous Expression of Membrane Proteins: Methods and Protocols*, Methods in Molecular Biology, vol. 1432, DOI 10.1007/978-1-4939-3637-3_13, © Springer Science+Business Media New York 2016

Although prokaryotic expression systems are the most commonly utilized vehicles, they have not been very successful for over-expression of mammalian membrane proteins as they fail to provide the necessary folding machinery or posttranslational modifications. On the other hand mammalian cells, particularly human cells, offer inherent advantages for expressing human membrane proteins due to the endogenous translocation machinery, post-translational modifications, and lipid environment that is most native to mammalian membrane proteins. In the past, mammalian expression systems were not extensively manipulated for structural studies primarily due to technical difficulties associated with large-scale cultures and limited utility of traditional constitutive promoters for toxic proteins. With recent advances in inducible promoters, new methods have been developed and employed for expressing such proteins [3–5].

It is a widely known fact in the structural biology field that one cannot predict from the beginning if a particular protein can be successfully crystallized, purified, or even expressed. Considering all the hurdles at every single step, we developed a simplified yet robust methodology that can be employed towards screening a multiple numbers of constructs without a need for milligram quantities of protein. Our strategy involves generation of several desired constructs for a particular gene. Such constructs may include N-terminal or C-terminal truncations alongside the full-length construct. Different N- or C-terminal tags are added to the construct to allow for purification. For small-scale screening of these constructs, we utilize FSEC [6], where the gene of interest is cloned into an EGFP vector (pACMV-tetO-EGFP). Based on FSEC results, one can roughly compare the expression level between different constructs and then might consider re-cloning the gene into an expression vector of choice without EGFP (pACMV-tetO) [7] or can continue with the same vector simply by cleaving the EGFP later on. FSEC can also be utilized for several other optimizations including of buffer, additives, and detergents for achieving stability and homogeneity (Fig. 1a). Once constructs are chosen, one can move towards generating a stable clonal cell to be utilized for further screening (Fig. 1b). Post-detergent screening, thermostability assays that provide a proxy for protein stability can be useful in the pursuit of ideal conditions by optimizing buffers, pH, salt concentrations, and ligands to facilitate crystallization of a membrane protein that is also more stabilized conformationally by the environment, ligands, lipids, detergents, and many other variables [8].

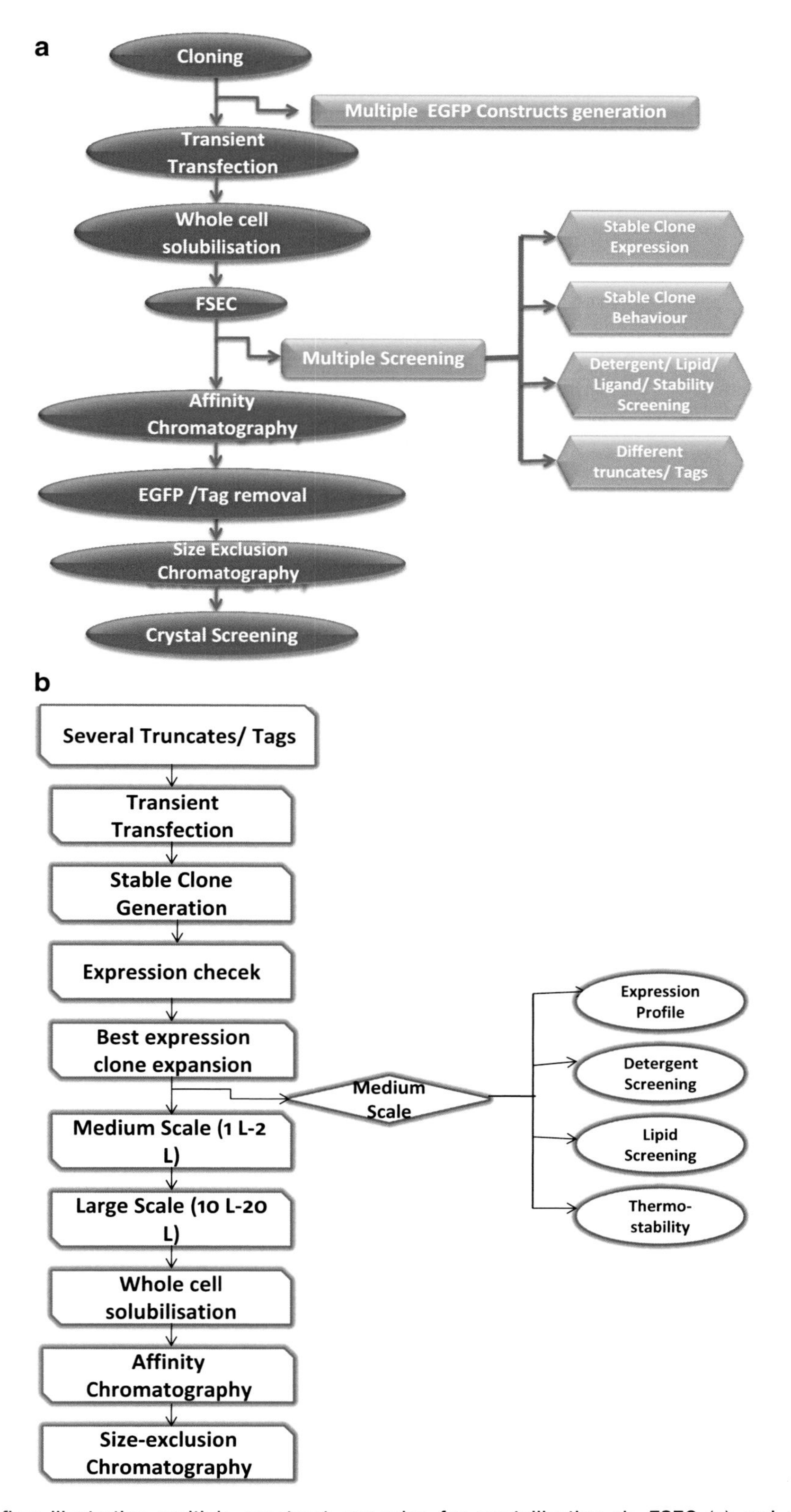

Fig. 1 Workflow illustrating multiple construct screening for crystallization via FSEC (**a**) and stable clone generation (**b**)

2 Materials

Prepare all solutions using distilled, deionized water and analytical grade reagents (unless specified otherwise). All reagents for cell culture work (media, antibiotics, chemicals, etc.) should be used of tissue culture grade. All prepared reagents should be stored at 4 °C (unless indicated otherwise). For cells and other waste disposal consult the material safety data sheets for all chemicals used and follow the safe procedures for handling and proper disposal of chemicals.

2.1 Cloning of the Target Gene

1. pACMV-tetO mammalian expression plasmid (*see* **Note 1**).
2. Cloning primers: Forward primers F1 and F2, and reverse primers R1 and R2, diluted to 25 μM in nuclease-free water. Store at −20 °C (*see* **Notes 2** and **3**).
3. Deoxyribonucleotide (dNTP) mix: 10 μM Stock diluted to 2 μM in nuclease-free water. Store at −20 °C.
4. PCR Purification Kit (Qiagen): Follow the manufacturer's instructions. Store at room temperature (*see* **Note 4**).
5. Gel Extraction Kit (Qiagen): Follow the manufacturer's instructions. Store at room temperature (*see* **Note 5**).
6. Phusion DNA polymerase (New England Biolabs) (*see* **Note 6**).
7. Restriction Endonuclease: XbaI and XhoI supplied with appropriate buffers (New England Biolabs).

2.2 Transfection and Adherent Cell Culture

1. HEK293S GnTI$^-$ cells (or another highly transfection efficient mammalian cell line).
2. Adherent DMEM (Dulbecco's modified Eagle's medium) medium: To 1 L DMEM high-glucose media add 1 % penicillin-streptomycin, and 10 % iron-supplemented bovine calf serum (BCS). We usually used the BCS provided by Hyclone (due to its premium quality).
3. Selection medium: 1 L DMEM high glucose, 1 % penicillin-streptomycin, 10 % FBS, 4 % Geneticin®, and 0.1 % blasticidin S HCl.
4. 5 mg/mL Blasticidin solution: Weigh 50 mg of blasticidin S HCl, dissolve in 10 mL of water. Filter-sterilize, aliquot, and store at −20 °C for 6–8 weeks.
5. 20 mg/mL Doxycycline hyclate: Weigh 100 mg of doxycycline solution, dissolve in 5 mL of autoclaved water. Filter-sterilize, aliquot, and store at −20 °C for 6–8 weeks.
6. 20 % (wt/vol) Glucose: Weigh 100 g of glucose in 500 mL of autoclaved water. Filter-sterilize and store at 4 °C for 2–3 months.

2.3 Expression of the Target Protein

1. 1× Solubilization buffer or lysis buffer: Mix 20 mM Tris–HCl (pH 7.4), 100 mM NaCl, and 10% (vol/vol) glycerol. Immediately before use, add 1 mM PMSF and one complete EDTA-free protease inhibitor cocktail tablet. Chill to 4 °C and discard any unused buffer.
2. Octyl-β-d-glucopyranoside (OG), *N*-dodecyl- β-d-maltopyranoside (DDM) or other detergents to test protein solubilization: Prepare appropriate concentration of respective detergents in ddH_2O. We order detergents from Anatrace.
3. 1 mM Phenylmethanesulfonyl fluoride (PMSF) in isopropanol. Prepare just before use.
4. Complete EDTA-free protease inhibitor cocktail tablets: Prepare the inhibitor cocktail as prescribed in the manufacturer's protocol.
5. 2× Solubilization buffer: Mix 40 mM Tris–HCl pH 7.4, 200 mM NaCl, and 20% v/v glycerol. Immediately before use, add 1 mM PMSF) and one complete EDTA-free protease inhibitor cocktail tablet. Chill the buffer to 4 °C and discard any unused buffer.

2.4 Fluorescence Detection Size-Exclusion Chromatography (FSEC)

1. Gel filtration column: Use either TSK-GEL G3000SW or Superdex200 10/300 GL, from GE Healthcare. Equilibrate the column with 1× solubilization buffer.

2.5 Suspension Culture

1. Suspension medium (for spinner flask): 1 L DMEM-high glucose without calcium salts, 1% penicillin-streptomycin, 10% iron-supplemented BCS, 1% of Pluronic, 0.3 g Primatone RL/UF, 3.7 g sodium bicarbonate.
2. Suspension medium (for WAVE bioreactor): Prepare 10 L DMEM-high glucose without calcium salts, to which add appropriate volume for getting a final concentration of 1% penicillin-streptomycin, 10% iron-supplemented BCS, 1% Pluronic, 3 g of Primatone RL/UF, and 37 g of sodium bicarbonate.
3. 500 mM Sodium butyrate: Weigh 27.5 g of sodium butyrate and dissolve in autoclaved water to a final volume of 500 mL. Filter-sterilize and store at room temperature (20–25 °C) for 6–8 weeks.
4. 10% (wt/vol) Primatone RL/UF: Weigh 3 g of Primatone RL/UF and dissolve in 30 mL of autoclaved water. Filter-sterilize and use immediately.
5. 10% (wt/vol) Pluronic: Weigh 50 mg Pluronic and dissolve in 500 mL of autoclaved water. Filter-sterilize and store at 4 °C for 6–8 weeks.

6. Spinner flask: Vigorously clean the spinner flask, by detaching the various components. For cleaning purpose, use 10% v/v glacial acetic acid (with water) and allow stirring overnight at room temperature. Following day, discard the glacial acetic acid and rinse flask thoroughly in order to remove all traces of acid. Perform two rounds of liquid autoclaves, for 30 min each filled with distilled water. Finally, perform a dry autoclave for another 30 min. Allow the flask to cool down before use.
7. WAVE bioreactor: Assemble the 20 L WAVE cellbag in a tissue culture room, following the manufacturer's protocols.

2.6 Affinity and Size-Exclusion Chromatography

1. SEC buffer: 20 mM Tris–HCl (pH 7.4), 100 mM NaCl, 10% (vol/vol) glycerol, and 40 mM OG (or 0.5 mM DDM). Store at 4 °C for up to 1 week.
2. FLAG resin: Equilibrate the resin with SEC buffer, following the manufacturer's protocol.
3. Tris(2-carboxyethyl)phosphine (TCEP).
4. Glass Econo-column: Use the column for loading the resin onto the column, followed by washing of the resin with SEC buffer and elution of the protein with SEC buffer, supplemented with FLAG peptide.
5. Superdex 200 10/300 GL: Equilibrate the column with three column volumes of SEC buffer.

3 Methods

Carry out all procedures at room temperature unless otherwise specified.

3.1 Expression Construct Design

Design the multiple constructs in a manner to introduce N- and C-terminal tags along with the protease cleavage site (Fig. 2). In an experiment, we normally utilize a FLAG tag with 3C cleavage site at the N-terminal and octa or deca-His tag with thrombin cleavage site at the C-terminal via a two-step PCR (Fig. 3). Introduce a five-amino-acid Gly/Ala repeat spacer sequence in between the N/C-terminus of the transgene and the thrombin and 3C site, to facilitate tag removal. Different tag sequences are given in Table 1.

1. PCR1: Perform first PCR step using F1 and R1 primers to insert 3C and thrombin at the N- and C-terminal of the transgene, respectively. Resolve PCR1 on agarose gel and carry out the agarose gel extraction to purify the first PCR product, to be used as a template for the second PCR.
2. PCR2: Use F2 and R2 primers for the second-step PCR to add the start codon-FLAG tag at the N-terminal and octa/deca-His-stop codon at the C-terminal along with the desired proteases.

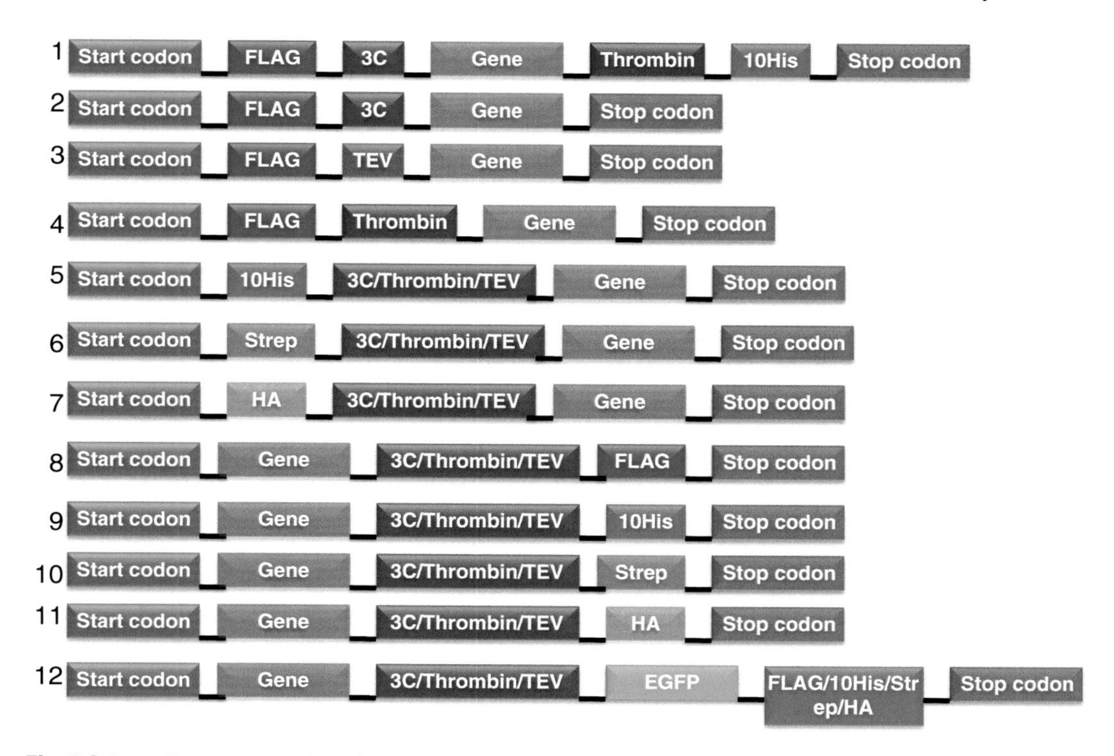

Fig. 2 Schematic representation of various possible constructs introducing the N-terminal and C-terminal tags along with the protease cleavage site. The affinity tags and protease cleavage sites can be introduced either at the N-terminal or C-terminal of the gene. Construct with EGFP tag can also be designed in for rapid screening through FSEC and EGFP can be cleaved along with tag post-affinity purification

3.2 EGFP-Tagged Expression Vector

In order to rapidly assess the suitability of human membrane protein expression constructs, yielding the pure and homogeneous proteins, transiently transfect the glycosylation-deficient GnT I-deficient HEK293S cells. For quick screening purpose, introduce enhanced green fluorescence protein (EGFP) tag at the C-terminal of pACMV-tetO vector. PCR amplify the EGFP sequence along with five-amino-acid (Gly/Ala) spacer.

1. Design oligo with the N-terminal consisting of spacer-thrombin-EGFP (N), and a C-terminal consisting of EGFP (C)-His_{8-10}.
2. Following standard protocols amplify PCR; digest the EGFP insert and pACMV-tetO vector.
3. Post-ligation, follow the manufacturer's protocol to transform the ligation products. Confirm the vector sequence.
4. Sub-clone the transgene of interest into pACMV-tetO-EGFP vector for rapid screening.

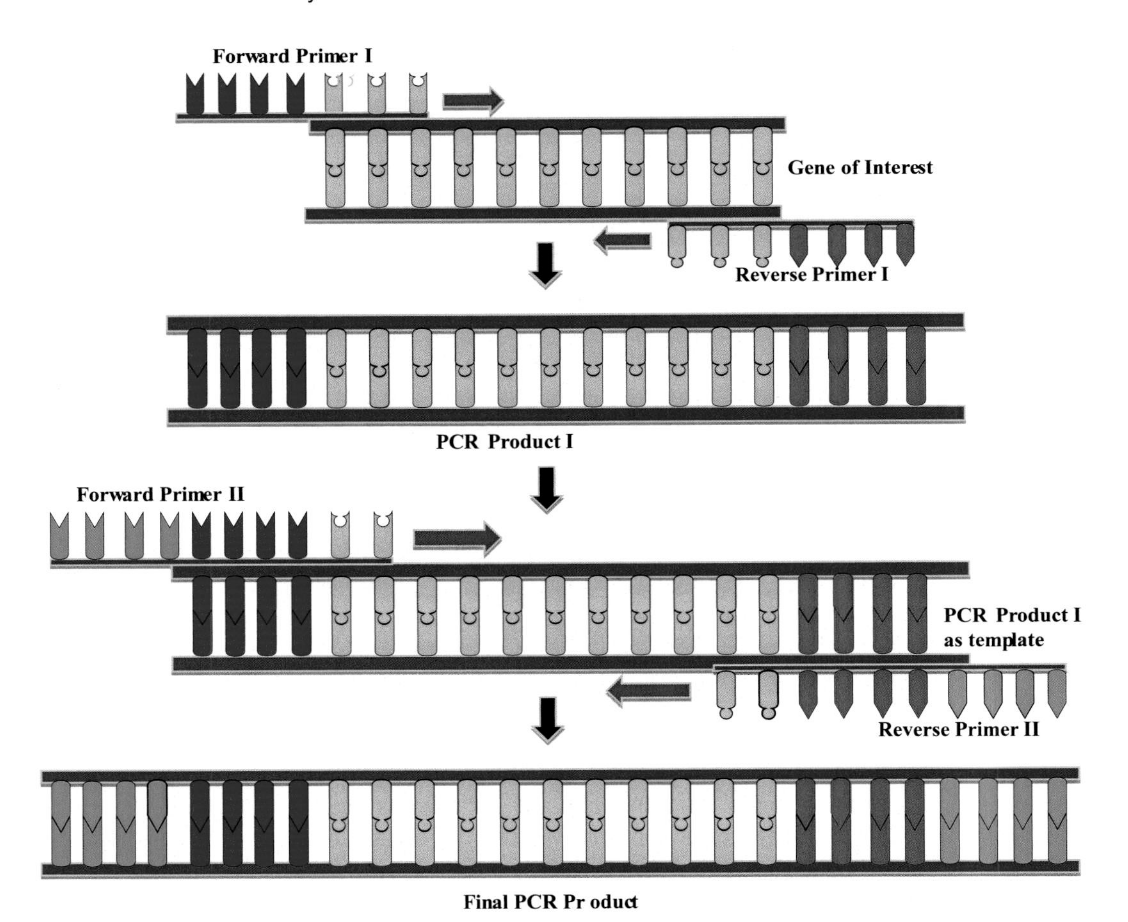

Fig. 3 Diagrammatic representation of a two-step PCR method utilized to introduce N-terminal and C-terminal affinity tags and protease cleavage site

Table 1
List of protein affinity tags and protease cleavage sites with their amino acid sequences, used for designing of several constructs

TAG (N)	Amino acid sequence
FLAG	Asp-Tyr-Lys-Asp-Asp-Asp-Asp-Lys
10His	His-His-His-His-His-His-His-His-His-His
Strep	Trp-Ser-His-Pro-Gly-Phe-Glu--Lys
HA	Tyr-Pro-Tyr-Asp-Val-Pro-Asp-Tyr-Ala
Protease	Amino acid sequence
Thrombin	Leu-Val-Pro-Arg-Gly-Ser
3C	Leu-Glu-Val-Leu-Phe-Gln-Gly-Pro
TEV	Glu-Asn-Leu-Tyr-Phe-Gln-Gly

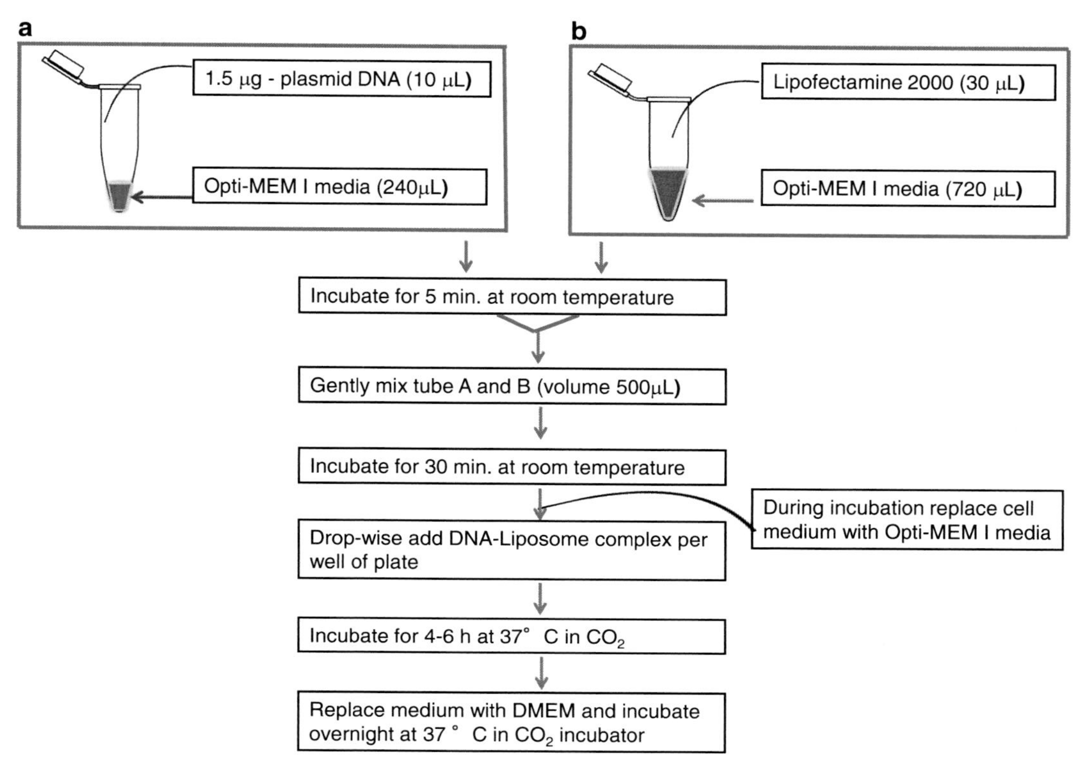

Fig. 4 Transient transfection

3.3 Transient Transfection and Induction of HEK293S GnTI⁻-Deficient Cells

Transiently transfect HEK293S GnTI⁻ cells with pACMV-tetO plasmid containing the gene of interest (Fig. 4). Over-expression of human membrane proteins is favorable while using an inducible cell line as high cell density can be achieved before induction, thereby attenuating the toxicity.

1. Seed three 10 cm plates (~40% confluence) with HEK293S cells (DMEM + 10% FBS) (*see* **Notes 7** and **8**).
2. Next day, visualize cells under the microscope to confirm cell confluence (~70–80%).
3. Gently mix 10 μg of plasmid DNA (Eppendorf tube A) in 1.5 mL of Opti-MEM I (*see* **Note 9**).
4. Simultaneously, following the manufacturer's instructions, dilute 60 μL of Lipofectamine 2000 in 1.5 mL of Opti-MEM I medium (Eppendorf tube B).
5. Incubate both reactions (Eppendorf tube A and tube B) at room temperature for 10 min.
6. Mix diluted DNA samples (from tube A) to the diluted Lipofectamine 2000 (tube B).
7. Incubate at room temperature for 30 min.

8. In the meantime, aspirate off media and add 7 mL of Opti-MEM I.
9. Gently add DNA-liposome complex to the plate containing Opti-MEM I media and mix by gently swirling the plate.
10. Incubate at 37 °C in a CO_2 incubator.
11. Post-4–6 h aspirate off media and add fresh adherent medium (DMEM + 10% FBS).
12. Incubate at 37 °C overnight.
13. Post-24 h, use 2 μg/mL doxycycline to induce the cells.
14. 32–36 h post-induction, wash cell monolayer on plate with PBS (twice) and trypsinize cells.
15. Harvest the cells and perform the whole-cell membrane solubilization.

3.4 Expression Assessment

1. Resuspend cells in 1 mL of 2× solubilization buffer by pipetting up and down gently.
2. For solubilization, based on total number of detergents to be screened, split resuspended cells into equal-volume aliquots and transfer to 1.5 mL ultracentrifuge tube.
3. Add 2× detergent solution (final 1× concentration).
4. Routinely, we screen expression levels in four different detergents per sample: β-octylglucoside (β-OG) (400 mM), β-dodecylmaltoside (β-DDM; 40 mM), lauryldimethylamine-oxide (LDAO; 200 mM), and fos-choline 14 (FC-14; 40 mM) (*see* **Notes 10** and **11**).
5. Solubilization time might vary with different detergents. For screening purpose solubilize by 2-h stirring at 4 °C.
6. To compare the extent of solubilization, save 12 μL as pre spin aliquot per detergent solubilization.
7. Remove insolubilized material by centrifugation in TLA45 ultracentrifuge rotor at high speed (100,000 ×*g*) for 1 h.
8. After transferring supernatant to 1.5 mL microfuge tube, take out 12 μL as post-spin sample.
9. Perform western blot with pre-spin and post-spin samples to analyze detergent solubilization. We follow standard western blotting protocols to compare pre-spin and post-spin samples.

3.5 Fluorescence Detection Size-Exclusion (FSEC) Chromatography

1. Wash a gel filtration column (TSK-GEL G3000SW or Superdex200 10/300 GL) connected to Shimadzu chromatography system with 2× column volume with water to remove the 20% ethanol used to store column.
2. Equilibrate the column with 1× solubilization buffer.

Table 2
Various antibiotics used for the purpose of screening for optimal antibiotic dose. Selection concentrations of these antibiotics have also been listed above

Selection marker	Formula	FW (g/mol)	Selection conc. (μg/mL)
Geneticin	$C_{20}H_{40}N_4O_{10}.2H_2SO_4$	496.6•196.1	600–800
Zeocin	$C_{55}H_{83}N_{19}O_{21}S_2Cu$	1137.41	200–400
Blasticidin S	$C_{17}H_{26}N_8O_5$ HCl	458.9	5–10
Hygromycin B	$C_{20}H_{37}N_3O_{13}$	527.5	100–200

3. Use RF-10AXL Shimadzu fluorescence detector (λ_{ex} = 488 nm and λ_{em} = 509 nm) with initially set to a sensitivity of "medium" and a gain of "16×." Use SPD-10A UV-Vis Shimadzu detector to detect absorbance at 280 nm (A280).
4. Following the manufacturer's instructions, inject ~100 μL of 0.45 μm filtered solubilized sample. Perform each experiment in triplicates.
5. Compare the expression levels and profile among the different constructs over FSEC chromatogram (*see* **Note 12**).

3.6 Antibiotic Kill Curve

1. One day prior to performing antibiotic selection, seed cells in 6-well plates containing 0.5 mL complete growth medium per well. At the time of antibiotic selection, cell density should be nearly ~60–80 % ($0.8–3.0 \times 10^5$ cells/mL).
2. Add increasing amounts of the G418 to duplicate wells of cells. Maintain one well as a control and do not add any antibiotic. We add 0, 100, 200, 400, 800, and 1000 μg/mL selection antibiotic in duplicate wells. Plasmids with different antibiotic markers can also be utilized. An optimal concentration range for different selection markers is given in Table 2.
3. Examine cultures for 1 week for toxicity signs while replacing media (containing selection antibiotic) on alternate days (*see* **Note 13**).
4. The antibiotic concentration at which the cells are dead post-1 week of antibiotic selection is considered as an optimal dose. Once the optimal antibiotic dose is decided, seed cells for stable transfection.

3.7 Stable Clone Generation

Based on FSEC analysis, one can select the construct(s) providing the best expression and profile on chromatogram among all others. Alternatively, one can also go back and reclone the best expressing construct into any other vector of choice without EGFP. After selecting the best construct, proceed towards stable clonal cell generation (Fig. 5) (*see* **Note 14**).

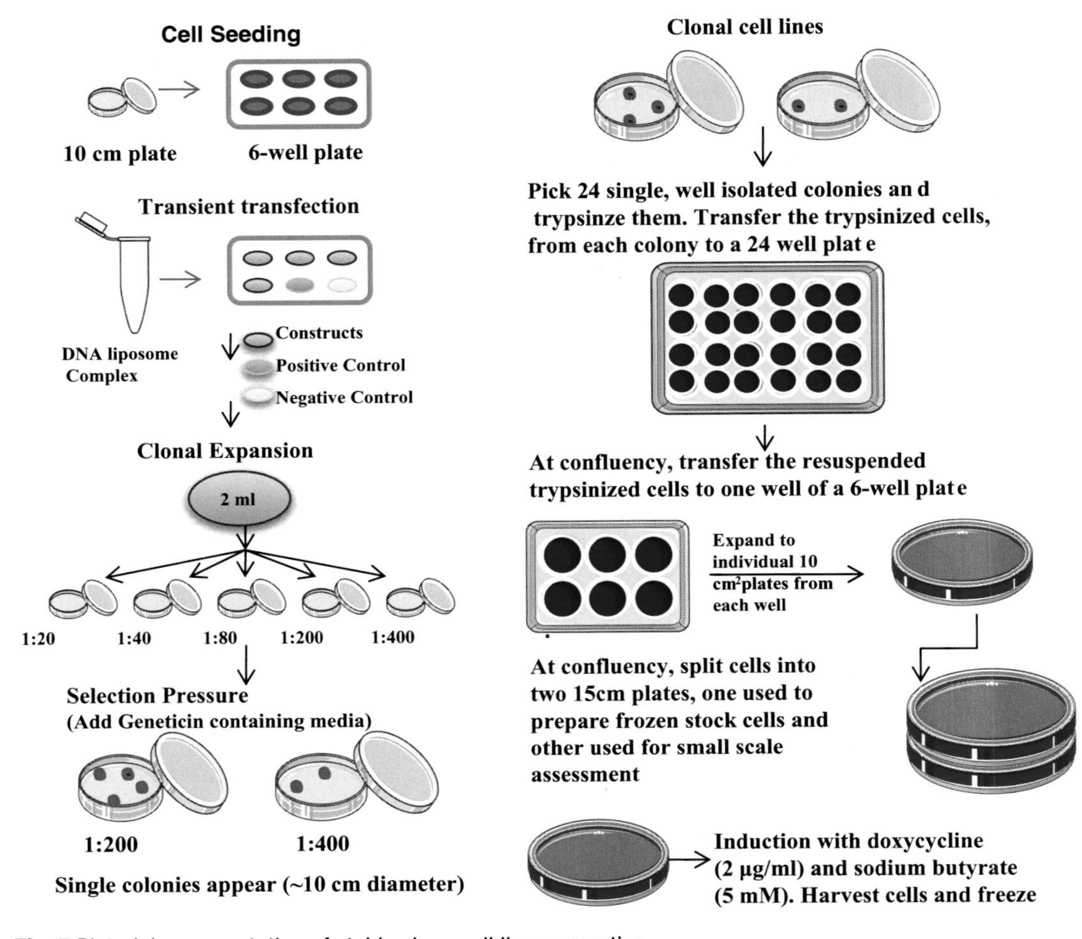

Fig. 5 Pictorial representation of stable clone cell line generation

1. One day prior to transfection, plate cells in a 6-well tissue culture plate so that the cell density is nearly ~70–80% at the time of transfection.
2. Post-18–24 h of seeding, perform transfection following the protocol as explained under Fig. 4.
3. On day three, aspirate off overnight DMEM, gently wash with PBS, and add 500 μL trypsin. Detach cells from the plate surface by tapping it against the hood surface. Neutralize trypsin by adding 1.5 mL DMEM (*see* **Note 8**).
4. Prepare five 10 cm plates for different dilutions (1:400, 1:200, 1:80, 1:40, and 1:20).
5. Transfer 50 μL (1:400 dilution), 100 μL (1:200), 250 μL (1:80), 500 μL (1:40), and 1 mL (1:20) to the plates and add adherent DMEM media to bring the final volume to 10 mL accordingly. Incubate overnight at 37 °C in a CO_2 incubator.

6. Next day (on day 4), replace media with 10 mL of selection media containing the selection marker quantity calculated from antibiotic kill curve per 10 cm plate (*see* **Note 14**). Incubate overnight at 37 °C in CO_2 incubator.
7. On day 5, replace overnight media with 10 mL of selection media containing the selection marker quantity selected based upon antibiotic kill curve per 10 cm plate (*see* **Note 14**). Incubate overnight at 37 °C in CO_2 incubator.
8. Once the single and isolated colonies appear, mark them. Out of all the colonies from five plates, we choose only 24 individual colonies.
9. Aspirate off media from the plate and gently wash the cells with D-PBS.
10. Based on colony diameter, using a sterile forceps, pick a sterile cloning cylinder, and gently press it over the autoclaved grease for applying grease layer at the bottom of cylinder.
11. Move the greased cylinder over a colony of cells, firmly press it, add trypsin to the cloning cylinder depending upon cylinder's size, and gently pipette up and down to detach cells.
12. Transfer the detached cells to one well of a 24-well culture plate and make up volume to 1 mL with adherent DMEM medium. If required, rinse the area covered by cloning cylinder with media to collect the cells left after first transfer. Incubate cells overnight at 37 °C in CO_2 incubator.
13. Start selection post-24 h by replacing the DMEM media with selection media. Till cells reach ~80–90% confluence stage, aspirate off old media and provide cells with fresh selection media on every alternate day.
14. First expand the colonies to 6-well plate after trypsinization with 100 μL of trypsin and post-attainment of cell confluence, expand each to 10 cm plate containing 10 mL media (*see* **Note 8**).
15. At 80–90% confluence, expand cells from one 10 cm to two 15 cm plates (plate A and plate B) containing 15 mL media (*see* **Note 8**).
16. At this step, plate A is used to freeze cells for future use while plate B is used to compare expression level among all 24 clonal cells (*see* **Note 15**).

3.8 Solubilization of Membrane Proteins

Membrane protein purification from yeast and insect cells requires membrane preparation while in our experience with HEK 293S cells whole-cell solubilization provides comparatively higher yield of pure protein and eliminates the prerequisite of membrane preparation. Based on detergent used for solubilization, cell pellets are solubilized for 1–2 h at 4 °C in solubilization buffer. Once the best expressing clone is identified (via either western blotting or FSEC), expand the particular clone from 10 cm plate to 1 L suspension culture.

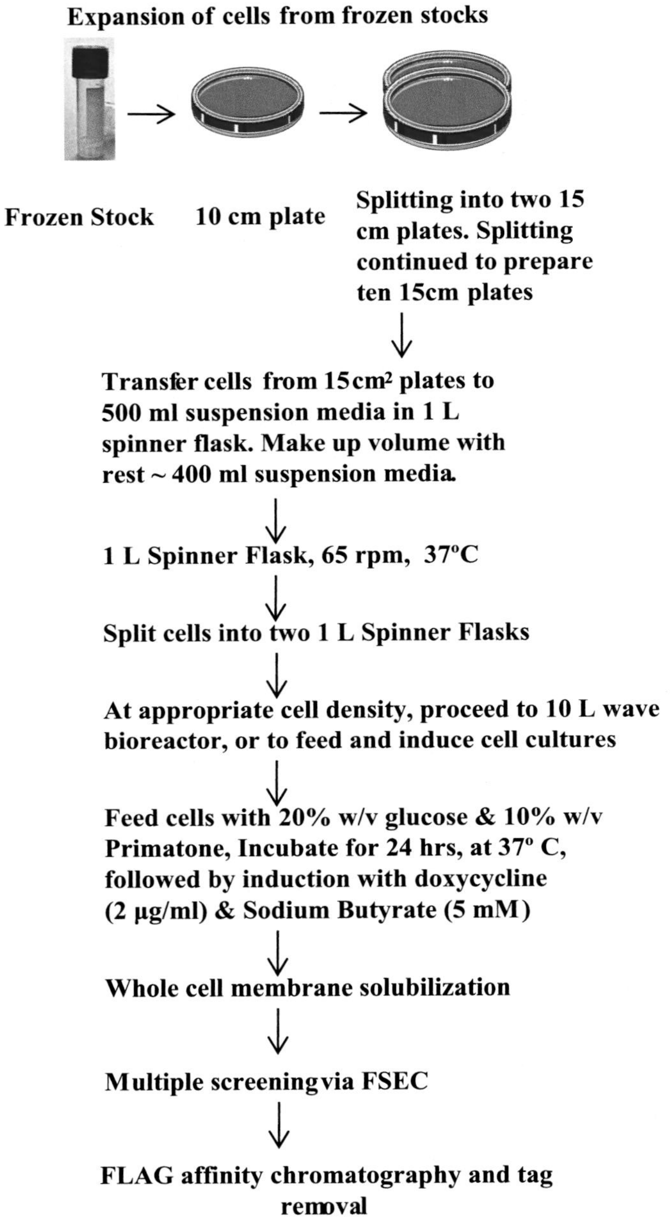

Expansion of cells from frozen stocks

Frozen Stock → 10 cm plate → Splitting into two 15 cm plates. Splitting continued to prepare ten 15cm plates

↓

Transfer cells from 15 cm² plates to 500 ml suspension media in 1 L spinner flask. Make up volume with rest ~ 400 ml suspension media.

↓

1 L Spinner Flask, 65 rpm, 37°C

↓

Split cells into two 1 L Spinner Flasks

↓

At appropriate cell density, proceed to 10 L wave bioreactor, or to feed and induce cell cultures

↓

Feed cells with 20% w/v glucose & 10% w/v Primatone, Incubate for 24 hrs, at 37° C, followed by induction with doxycycline (2 µg/ml) & Sodium Butyrate (5 mM)

↓

Whole cell membrane solubilization

↓

Multiple screening via FSEC

↓

FLAG affinity chromatography and tag removal

Fig. 6 Expansion of frozen stocks to medium- and large-scale suspension culture

3.9 Suspension Culture

Following standard protocols, thaw frozen cell vials of best expression clone and for expansion move from adherent towards suspension cultures (Fig. 6). For 1 L suspension culture, use twelve 15 cm plates at 70–50% confluence.

1. Pre-warm 1 L suspension medium at 37 °C water bath.
2. Inside the hood, aseptically add about 800 mL pre-warmed suspension medium to the spinner flask through side arm.

3. Out of 12 plates aspirate off DMEM from four 15 cm plates at a time, gently wash with DPBS, and trypsinize cells. Add 8 mL DMEM suspension to resuspend cells very well.
4. Carefully, transfer the cells to the spinner flask via side arm and follow the same procedure for transferring cells from all twelve 15 cm plates to the same spinner flask.
5. Transfer the leftover suspension media to the flask and close the side arm lid. Post-moving the flask to 37 °C incubator, loosen both side arm lids for gas exchange.
6. Take cell count on a daily basis and once the suspension cell density reaches at ~70–80 % confluence, split cells by transferring half of the cells to the new spinner flask containing 500 mL of freshly prepared pre-warmed suspension medium. Incubate the spinner flask at 37 °C incubator.
7. Determine the total number of cells and percent viability using a hemocytometer and trypan blue exclusion on a daily basis.

For suspension culture using large-scale WAVE Bioreactor Cellbags, we commonly use 20 L wave bag for 10 L culture volume. For bringing the final volume to 10 L, pre-warm 7 L of suspension medium at 37 °C water bath, and thaw 700 mL of bovine calf serum and antibiotics accordingly.

1. Take the WAVE cellbag inside the hood and using a sterile razor remove the plastic wrapping and tighten all the inlets and outlets.
2. Aseptically, transfer suspension media to cellbag and following the manufacturer's instructions secure the cellbag on the holder tray of a rocking unit, close off the outlet air filters, and with 10 % CO_2/air inflate the cellbag bioreactor. Rock at 15 rpm for 20–30 min till the bag is completely inflated. Once bag is completely inflated, open the outlet air filters and rock for another 1–2 h for the complete equilibration of pH and temperature.
3. Following the manufacturer's protocol, clamp the inlet and outlet air filters, carry cellbag to the biosafety cabinet, and inoculate with cells by aseptically pouring all the cells.
4. Bring bag back to bioreactor, secure it on holder tray, and inflate it following **step 2**. Once bag is completely inflated unclamp the outlet air filters and rock at 22 rpm.
5. On a daily basis, monitor the cell density and at ~1.0×10^6 cell count feed the cells with 20 % w/v glucose and 10 % w/v Primatone RL/UF.
6. Post-24 h of feeding, induce cells with 2 μg/mL doxycycline and 5 mM sodium butyrate.
7. Harvest cells post-36 h (or optimized time) for centrifugation for 10 min at $5000 \times g$ at 4 °C (*see* **Note 16**).

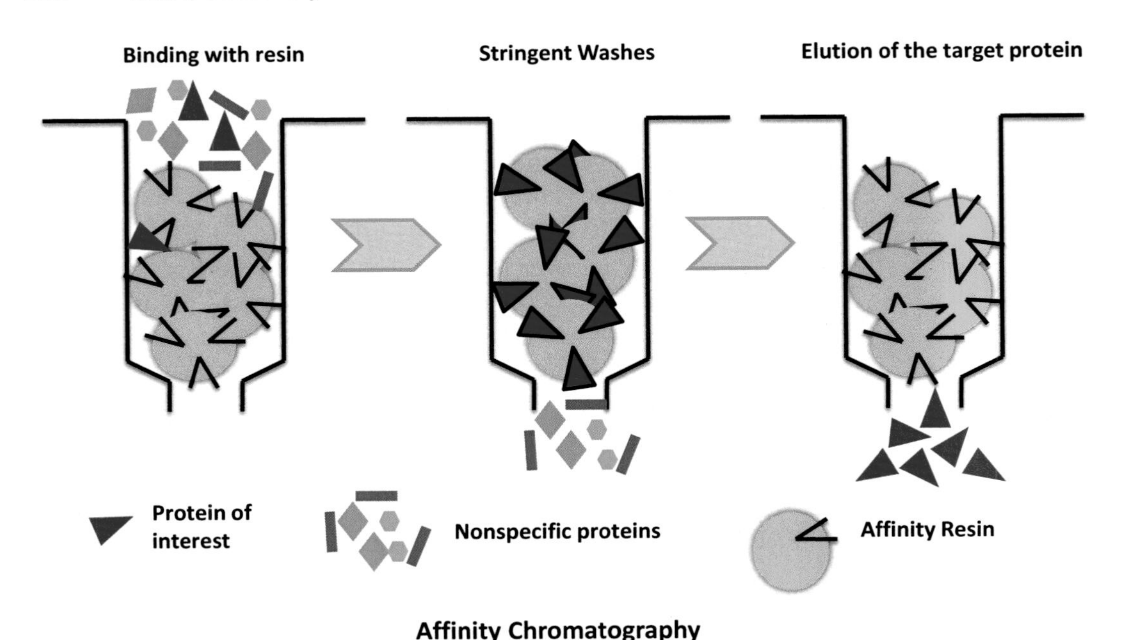

Fig. 7 Principle behind affinity chromatography has been shown in the above figure. The first step defines the binding of protein with the resin, followed by the removal of nonspecific proteins with the help of stringent washes. Finally, the target resin-bound protein is eluted out

3.10 Affinity Chromatography

Post-whole-cell lysis and protein solubilization, based on the tag choice affinity chromatography is performed (Fig. 7) (FLAG affinity purification discussed below) (*see* **Note 17**). All procedures should be performed on ice or at 4 °C.

1. Pre-equilibrate FLAG resin with solubilization buffer following the manufacturer's protocols and incubate for binding at 4 °C for 2–3 h.
2. Post-binding transfer the resin with supernatant to Econo-column and collect the flow-through fractions.
3. Wash the column with SEC buffer (10 column volume).
4. Elute target protein with 1 mL of 100 μg/mL FLAG peptide (in SEC buffer). Collect 10 elutes of 1 mL each and add 3 mM final concentration of TCEP, a reducing agent (or as per the manufacturer's instructions) right away.
5. For separation on SDS-PAGE, load maximal amount of sample onto an SDS-PAGE gel and run for desired amount of time for adequate separation.
6. We routinely separate the recovered material by SDS-PAGE and pool overexpressing elution fractions.
7. At this step we cleave off the tag(s) (including EGFP tag in case of EGFP constructs) with appropriate proteases, using the manufacturer's recommended concentrations or proceed for overnight dialysis in SEC buffer.

3.11 SEC Chromatography

1. One can choose a column for size-exclusion chromatography based on protein size or resolution requirement. We routinely use Superdex 200 10/300 GL size-exclusion chromatography column.
2. Pre-equilibrate the column with two column volumes of degassed SEC buffer, inject the filtered sample, and collect the pool fractions of desired peak.
3. Based on protein molecular weight, use a centrifugal filtration device to concentrate a sample to the desired protein concentration, and proceed to crystallization trials.

4 Notes

1. Gene of interest can be cloned in between Kpn1 and XhoI restriction sites of pACMV-tetO vector.
2. Remember to include a stop codon immediate downstream of the fusion tag on the reverse primer to prevent read-through.
3. Primers can be synthesized or ordered from any vendor.
4. Any PCR purification kit can be utilized though we have used the one from Qiagen.
5. Any gel extraction kit can be used though we have used the one from Qiagen. One should take proper care during single-band excision from gel.
6. For PCR step, any polymerase can be utilized though we had good results using Phusion® high-fidelity DNA polymerase.
7. For each individual transfection include positive and negative controls. For a positive control typically a pACMV-tetO construct containing a well-expressed transgene is suitable, whereas a transfection without a plasmid serves as a negative control.
8. For a 6-well tissue culture plate add 2 mL DMEM medium per well and 500 μL of trypsin, for a 10 cm plate we add 10 mL DMEM medium and 1 mL of trypsin, and for a 15 cm plate we add 25 mL DMEM medium and 2 mL of trypsin.
9. For transfection purpose we use the plasmid eluted with pre-warmed ddH_2O instead of elution buffer.
10. One can analyze the expression of different constructs solubilized in a couple of detergents to remove the possibility of expression being hindered due to the insolubility in a particular detergent itself.
11. From our experience, for crystallography purpose most of the membrane proteins are solubilized in DDM and OG. FC14, being a harsh detergent, solubilizes the membrane proteins but might leave the protein inactive.

12. FSEC normally is used as a quality measure but can also be used as a quantity measure with or without combination of western blotting.
13. For obtaining the stable transfection, cells go under selection media resistance/pressure for 4–6 weeks. At the beginning of first week we notice a lot of cell death while changing the media on every alternate day; during rest of the selection pressure media is changed twice to thrice per week until resistant foci can be identified. Cell death decreases post-2–3 weeks and single colonies start appearing. Pick several foci and expand cells in 24-well plate.
14. Since our technique relies upon random integration of the gene of interest into the genome, the expression levels obtained are strongly dependent on where the transgene integrates. Selection of clonal cells is then required to identify clones with high expression that are stable under prolonged culture.
15. On a routine basis, we perform western blotting by following standard protocols to find the best expression clones among all the 24 stable clones. We perform all the screening at the small scale by solubilizing whole cells, without preparing membranes as explained under Expression Assessment section. For detergent screening, we again take the advantage of EGFP fusion tag and employ FPLC to check membrane protein homogeneity in a particular detergent as explained under fluorescence detection size-exclusion chromatography section.
16. Post-induction harvest time should be optimized for each protein. In our experience, 36-h incubation post-induction gave the best results.
17. We normally prefer N-terminal FLAG and/or C-terminal HIS tag. Here, we explored the FLAG purification.

References

1. Andréll J, Tate CG (2013) Overexpression of membrane proteins in mammalian cells for structural studies. Mol Membr Biol 30(1):52–63
2. Tate CG (2001) Overexpression of mammalian integral membrane proteins for structural studies. FEBS Lett 504:94–8
3. Chaudhary S, Pak JE, Gruswitz F, Sharma V, Stroud RM (2012) Overexpressing human membrane proteins in stably transfected and clonal human embryonic kidney 293S cells. Nat Protoc 7:453–466
4. Chaudhary S, Pak JE, Pedersen BP, Bang LJ, Zhang LB, Ngaw SM, Green RG, Sharma V, Stroud RM (2011) Efficient expression screening of human membrane proteins in transiently transfected Human Embryonic Kidney 293S cells. Methods 55:273–280
5. Gruswitz F, Chaudhary S, Ho JD, Schlessinger A, Pezeshki B, Ho CM, Sali A, Westhoff CM, Stroud RM (2010) Function of human Rh based on structure of RhCG at 2.1 A. Proc Natl Acad Sci U S A 25:9638–9643
6. Kawate T, Gouaux E (2006) Fluorescence-detection size-exclusion chromatography for precrystallization screening of integral membrane proteins. Structure 14(4):673–81

7. Reeves PJ, Callewaert N, Contreras R, Khorana HG (2002) Structure and function in rhodopsin: high- level expression of rhodopsin with restricted and homogeneous N-glycosylation by a tetracycline- inducible N-acetylglucosaminyltransferase I-negative HEK293S stable mammalian cell line. Proc Natl Acad Sci U S A 99:13419–24

8. Tomasiak TM, Pedersen BP, Chaudhary S, Rodriguez A, Colmanares YR, Roe-Zurz Z, Thamminana S, Tessema M, Stroud RM (2014) General qPCR and Plate Reader Methods for Rapid Optimization of Membrane Protein Purification and Crystallization Using Thermostability Assays. Curr Protoc Protein Sci 77: 29.11.1–29.11.14

Chapter 14

Purification of Human and Mammalian Membrane Proteins Expressed in *Xenopus laevis* Frog Oocytes for Structural Studies

Rajendra Boggavarapu, Stephan Hirschi, Daniel Harder, Marcel Meury, Zöhre Ucurum, Marc J. Bergeron, and Dimitrios Fotiadis

Abstract

This protocol describes the isolation of recombinant human and mammalian membrane proteins expressed in *Xenopus laevis* frog oocytes for structural studies. The cDNA-derived cRNA of the desired genes is injected into several hundreds of oocytes, which are incubated for several days to allow protein expression. Recombinant proteins are then purified via affinity chromatography. The novelty of this method comes from the design of a plasmid that produces multi-tagged proteins and, most importantly, the development of a protocol for efficiently discarding lipids, phospholipids, and lipoproteins from the oocyte egg yolk, which represent the major contaminants in protein purifications. Thus, the high protein purity and good yield obtained from this method allows protein structure determination by transmission electron microscopy of single detergent-solubilized protein particles and of 2D crystals of membrane protein embedded in lipid bilayers. Additionally, a radiotracer assay for functional analysis of the expressed target proteins in oocytes is described. Overall, this method is a valuable option for structural studies of mammalian and particularly human proteins, for which other expression systems often fail.

Key words 2D crystallization, Human, Mammalian, Membrane protein, Protein structure, Single particle analysis

1 Introduction

Membrane proteins represent 20–30 % of all genes encoded in most genomes [1]. They are of critical importance in human health and disease, and represent about 60 % of the approved drug targets [2]. However, the structural information available on membrane proteins is limited in comparison with soluble proteins [3]. In particular, there is very little structural information on human and mammalian membrane proteins (*see* Membrane Protein Structure Database; http://blanco.biomol.uci.edu/mpstruc/). The majority of membrane protein structures were determined by X-ray crystallography. A prerequisite for this well-established and powerful

Isabelle Mus-Veteau (ed.), *Heterologous Expression of Membrane Proteins: Methods and Protocols,* Methods in Molecular Biology, vol. 1432, DOI 10.1007/978-1-4939-3637-3_14, © Springer Science+Business Media New York 2016

method is the availability of highly ordered three-dimensional (3D) crystals, which diffract to high resolution. In general, obtaining 3D crystals of a membrane protein is a real challenge [3, 4] because milligram amounts of protein are required for their growth. Furthermore, high-quality recombinant protein, e.g., functional, stable and homogenous protein, is an additional prerequisite for successful crystallization [5]. Nowadays, *Escherichia coli* is mainly used as an expression system for the homologous and heterologous overexpression of prokaryotic membrane proteins [6, 7]. However, this system has limited success with eukaryotic proteins. Yeast, e.g., *Saccharomyces cerevisiae* [8] and *Pichia pastoris* [9, 10], baculovirus-insect cell system [11], and mammalian cells, e.g., HEK293 and BHK-21 cells [12] are often used for the expression of eukaryotic membrane proteins. In spite of different expression systems being available for heterologous expression of membrane proteins, there is no general or simple expression system for mammalian membrane proteins. Every expression system has its advantages and disadvantages, and the adequate system for the corresponding target protein has to be found [13, 14]. Therefore, there is an urgent need for expression systems that are able to reliably and consistently provide milligram amounts of functional human and mammalian membrane proteins for structural studies.

Structural biology is undergoing a revolution since the introduction of direct electron detection devices (DDD) in cryo-transmission electron microscopy (cryo-TEM) at the end of 2013 [15, 16]. The DDD technology resulted in dramatic improvements in the quality of electron micrographs. Therefore, cryo-TEM of single particles is currently gaining great momentum because microgram amounts of solubilized, homogenous, and stable membrane protein preparations suffice to determine 3D structures by cryo-TEM and single particle analysis (SPA). One of the current highlights represents the structure of the mammalian transient receptor potential (TRP) channel TRPV1, which was solved at 3.4 Å resolution, thus breaking the side chain resolution barrier for membrane proteins without crystallization [17]. In comparison to cryo-TEM, negative stain TEM of purified protein is a fast, easy, and straightforward method [18–20]. Important information about the low-resolution structure, oligomeric state, shape, dimensions, and supramolecular organization of protein and protein complexes can be obtained [19, 21]. The majority of structure determination methods are not optimal for studying the membrane protein structure in its native environment, the lipid bilayer. To do this, 2D crystallization is the method of choice [22]. From 2D crystals the membrane protein structure is determined by TEM and electron crystallography [22, 23]. In addition to TEM, atomic force microscopy of 2D crystals provides subnanometer resolution information on the topography of membrane proteins in lipid bilayers [24, 25].

Oocytes from the wild South-African clawed frog *Xenopus laevis* are being used successfully as expression system since 1971 [26]. In the last couple of decades *Xenopus laevis* oocytes have been successfully used for the functional expression of membrane proteins [27, 28]. In this method, in vitro synthesized cRNA is injected into oocytes, whereas in a few cases cDNA has been injected. In the year 2011, we established a method to purify human and mammalian membrane proteins heterologously expressed in *Xenopus laevis* oocytes [29]. Our notion for the establishment of such a method at that time was to use an expression system, which had been used very successfully for the functional expression of membrane proteins. By using frog oocytes, we expressed and purified five mammalian membrane transport proteins, i.e., human aquaporin-1 (hAQP1), human glutamate transporter (hEAAC1), human peptide transporter-1 (hPEPT1), human sodium-coupled glucose transporter (hSGLT1), and mouse potassium chloride symporter-4 (msKCC4). As proof of concept for the use of such recombinant protein in structural biology, we used hAQP1 for negative stain TEM and SPA. Our work showed similar oligomeric state, shape, and dimensions of hAQP1 particles to previous reports of hAQP1 isolated from human erythrocytes [30]. Furthermore, 2D crystals of hAQP1 were produced successfully [29]. We also took advantage of this established method to express and purify human vitamin C transporter-1 (SVCT1). In this study, we explored the low-resolution structure and oligomeric states of hSVCT1. This was the first report on the purification of a human member from the vitamin C transporter family [31]. Apart from our research group, other researchers have started using this method recently [32, 33]. The above mentioned examples establish *Xenopus laevis* oocytes as a valid and successful expression system for mammalian and particularly human membrane proteins. The main objective of this protocol is to provide detailed information on the expression and purification of mammalian membrane proteins. Possible problems and pitfalls are explained as well. High-quality membrane proteins produced by the here presented method together with new generation state-of-the-art TEMs and DDD cameras have great potential to produce high resolution human and mammalian membrane protein structures in the near future.

The presented method includes the following steps: (1) cRNA synthesis of the desired gene, (2) preparation and defolliculation of *Xenopus laevis* oocytes, (3) cRNA injection into oocytes, (4) functional analysis of the target protein by radiotracer uptake assay using oocytes (optional), (5) isolation of yolk free oocyte membranes, (6) protein purification from isolated oocyte membranes, (7) biochemical analysis of purified proteins, (8) 2D crystallization of purified proteins, (9) negative staining and TEM of detergent-solubilized proteins and 2D protein crystals. Examples of expected results are shown in Figs. 1, 2, 3, and 4.

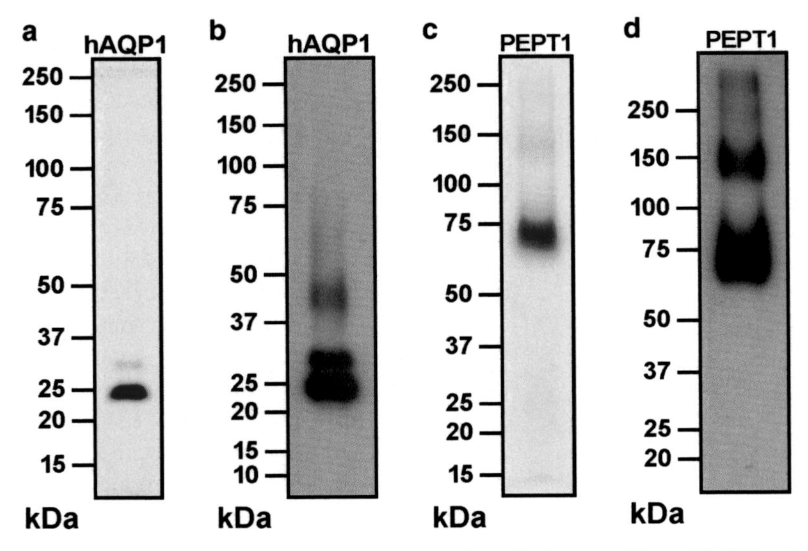

Fig. 1 SDS-PAGE and Western blot analyses of purified, recombinant hAQP1 and hPEPT1 expressed in *Xenopus laevis* oocytes. Silver-stained SDS/polyacrylamide gels (**a** and **c**) and Western blots using anti-HA antibody (**b** and **d**) from representative purifications. (**a** and **b**) The recombinant, nonglycosylated human AQP1 monomer runs at ~25 kDa similar to native AQP1 isolated from human erythrocytes [40]. Additional higher molecular mass bands are discerned corresponding to glycosylated and dimeric AQP1 forms. (**c** and **d**) Monomeric (~70 kDa) and dimeric (~140 kDa) recombinant human PEPT1. All bands on silver-stained SDS/polyacrylamide gels can be assigned to bands observed on corresponding Western blots. Figure was adapted from Fig. 3 in Bergeron et al. [29]

2 Materials

2.1 cRNA Synthesis

1. cDNA of interest in pMJB08 or an alternative *Xenopus laevis* expression vector (including T7 RNA polymerase promoter site and affinity tags (His-tag for this protocol) or epitopes for purification and detection).
2. Restriction enzyme for linearization of the plasmid.
3. mMessage mMachine T7 transcription kit (Ambion).
4. Recombinant ribonuclease inhibitor (RNase OUT, 40 U/μL).

2.2 Preparation and Defolliculation of Xenopus laevis Oocytes for Protein Expression

1. *Xenopus laevis* female.
2. Anesthetic: 10 g/L Ethyl 3-aminobenzoate solution (methanesulfonate salt).
3. Surgical instruments and autoclaved tooth picks.
4. Petri dishes (plastic).
5. Penicillin-Streptomycin (Pen-Strep) solution stabilized with 10,000 U penicillin and 10 mg/mL streptomycin, sterile-filtered, suitable for cell culture.

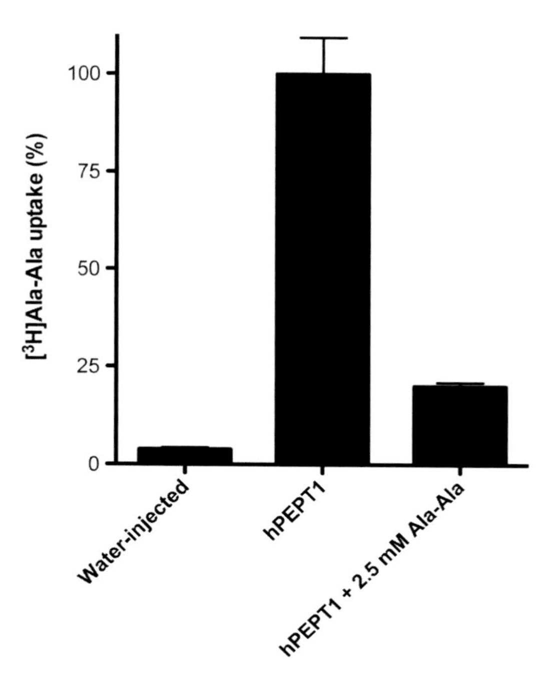

Fig. 2 Function of human PEPT1 expressed in *Xenopus laevis* oocytes. Oocytes injected with hPEPT1 cRNA mediate uptake of [^{3}H]Ala-Ala. Controls are water-injected oocytes and competitive inhibition of [^{3}H]Ala-Ala uptake in hPEPT1 cRNA injected oocytes with 2.5 mM Ala-Ala. Data represents mean ± SEM from 12 oocytes (one representative experiment from three independent experiments is shown)

6. Modified Barth's Medium (MBM): 10 mM HEPES-NaOH pH 7.4, 88 mM NaCl, 1 mM KCl, 0.75 mM $CaCl_2$, 2.4 mM $NaHCO_3$, 0.66 mM $NaNO_3$, 0.82 mM $MgSO_4$ supplemented with 10 μL/mL of Pen-Strep.
7. Calcium-free MBM (MBM without $CaCl_2$, *see* **step 6**) supplemented with 10 μL/mL of Pen-Strep.
8. Tissue culture dishes (Falcon tissue culture dishes, polystyrene, sterile, 100 × 20 mm).
9. 100× collagenase from *Clostridium histolyticum* stock solution: 300 mg/mL in calcium-free MBM.
10. Orbital shaker.
11. 50 mL Falcon tubes.
12. Stereomicroscope (e.g., stereoscopic zoom microscope SMZ745T, Nikon).
13. 18 °C incubator.

2.3 cRNA Injection for Uptake Assays or Protein Purification

1. Defolliculated *Xenopus laevis* oocytes (*see* Subheading 3.2).
2. cRNA of your gene of interest (including affinity tags and epitopes for purification and detection) (*see* Subheading 3.1).

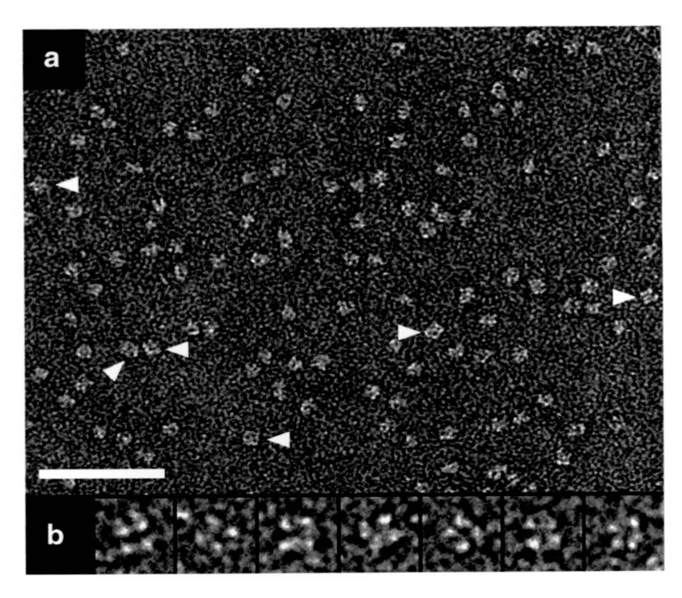

Fig. 3 Negative stain TEM of purified recombinant human AQP1. The homogeneity of the purified hAQP1 protein is reflected in the overview electron micrograph (**a**). Numerous particles exhibit a square shape (*arrowheads*), which is typical for AQP1 top views [30]. The gallery in (**b**) displays well-preserved top views of hAQP1. In the raw images four densities are clearly visible. The scale bar represents 50 nm (**a**) and the frame sizes of the magnified particles are 13.8 nm (**b**). Figure was adapted from Fig. 4 in Bergeron et al. [29]

3. Glass 3.5″ microcapillaries (Drummond Scientific Company).
4. Electrode puller (e.g., Zeitz DMZ-universal puller, AutoMate Scientific).
5. Platinum loop for moving of the oocytes and support grid for oocytes while injecting.
6. Injector (e.g., Nanoject II, Drummond Scientific Company).
7. Stereomicroscope (e.g., stereoscopic zoom microscope SMZ745T, Nikon).
8. Light source.

2.4 Functional Analysis by Radiotracer Uptake Assay Using Oocytes

1. *Xenopus laevis* oocytes expressing the protein of interest and water-injected oocytes as control (*see* Subheading 3.3).
2. Eppendorf tubes (1.5 and 2 mL).
3. Radiolabeled substrates (preferentially ^{3}H or ^{14}C labeled for easy handling).
4. Shaker.
5. 96-well white opaque microplate (e.g., Optiplate 96, PerkinElmer).
6. 5 % (w/v) sodium dodecyl sulfate (SDS) solution.
7. Scintillation cocktail (e.g., MicroScint 40, PerkinElmer).

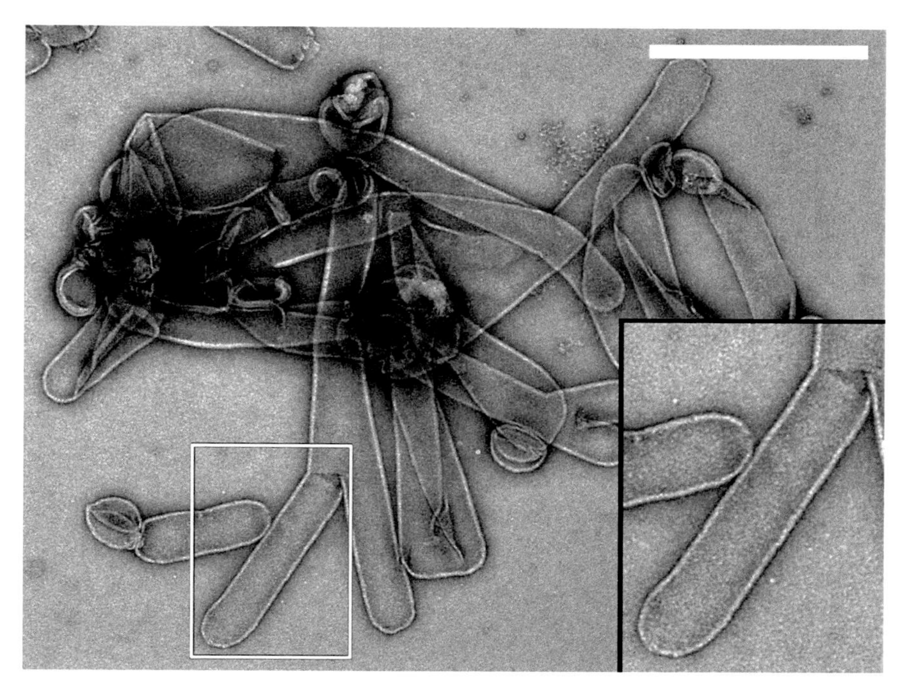

Fig. 4 Negatively stained tubular crystals of recombinant human AQP1. The area marked by the *white box* was magnified and is displayed as *inset*: tubular hAQP1 crystals with a typical width of ~110 nm are seen. The scale bar represent 500 nm and the frame size of the inset is 503 × 362 nm. Figure was adapted from Fig. 5 in Bergeron et al. [29]

8. Clear adhesive seal for 96-well microplate (e.g., TopSeal 96-well microplate seal, PerkinElmer).
9. Microplate scintillation counter.

2.5 Isolation of Yolk-Free Oocyte Membranes

1. *Xenopus laevis* oocytes expressing the protein of interest (*see* Subheading 3.3).
2. Protease inhibitor cocktail (e.g., SIGMA*FAST* protease inhibitor cocktail tablets, EDTA-free).
3. Oocyte homogenization buffer: 20 mM Tris–HCl pH 8.0 supplemented with one tablet of protease inhibitor cocktail per 50 mL.
4. 90 mL homogenizer.
5. 50 mL Falcon tubes.
6. Centrifuge with swinging bucket rotor.
7. Ultracentrifuge tubes (94 mL).
8. Ultracentrifuge.
9. Short and long neck glass Pasteur pipette.
10. Vacuum pump.

11. Membrane wash buffer: 20 mM Tris–HCl pH 8.0, 1 M NaCl supplemented with one tablet of protease inhibitor cocktail per 50 mL.
12. Resuspension buffer: 20 mM Tris–HCl pH 8.0, 150 mM NaCl.
13. Liquid nitrogen for flash freezing.

2.6 Protein Purification from Isolated Oocyte Membranes

1. Yolk free oocyte membranes (*see* Subheading 3.5).
2. Solubilization buffer: 20 mM Tris–HCl pH 8.0, 150 mM NaCl, 2 % (v/v) Triton X-100 (*see* **Note 1**).
3. Rotatory shaker.
4. Ultracentrifuge tubes (13.5 mL).
5. Ultracentrifuge.
6. Cobalt resin for metal affinity chromatography (e.g., HisPur cobalt resin, Thermo Fisher Scientific).
7. Equilibration buffer: 20 mM Tris–HCl pH 8.0, 150 mM NaCl, 5 mM imidazole, 0.05 % (v/v) Triton X-100 (*see* **Note 1**).
8. Eppendorf tubes.
9. Microcentrifuge.
10. Binding buffer: 20 mM Tris–HCl pH 8.0, 150 mM NaCl, 10 mM imidazole, 0.05 % (v/v) Triton X-100 (*see* **Note 1**).
11. 50 mL Falcon tube.
12. Gravity flow column (e.g., Wizard Midicolumn, Promega).
13. Wash buffer 1: 20 mM Tris–HCl pH 8.0, 150 mM NaCl, 10 mM imidazole, 0.05 % (v/v) Triton X-100 (*see* **Note 1**).
14. Wash buffer 2: 20 mM Tris–HCl pH 8.0, 150 mM NaCl, 20 mM imidazole, 0.05 % (v/v) Triton X-100 (*see* **Note 1**).
15. Cleavage buffer: 20 mM Tris–HCl pH 8.0, 150 mM NaCl, 0.025 % (v/v) Triton X-100 (*see* **Note 1**).
16. Laboratory film.
17. His-tagged human rhinovirus 3C (HRV3C) protease (e.g., Turbo3C HRV3C protease, recombinant, BioVision).
18. Concentrator with appropriate molecular weight cut-off (*see* **Note 2**, e.g., Amicon Ultra-0.5 mL centrifugal filter tubes 100 K, Millipore).

2.7 Biochemical Analysis of Purified Proteins

1. Purified protein (*see* Subheading 3.6).
2. Bicinchoninic acid assay kit (e.g., Pierce BCA protein assay kit, Thermo Fisher Scientific).
3. Reagents for silver staining of polyacrylamide gels.
4. Protein electrophoresis system.

5. Gel blotting system and general Western blot equipment.
6. Primary antibody: Mouse anti-HA monoclonal IgG antibody (Sigma).
7. Primary antibody: Mouse anti-penta His IgG antibody (Qiagen).
8. Secondary antibody: Goat anti-mouse IgG (H + L)-HRP conjugate (Bio-Rad).

2.8 2D Crystallization of Purified Proteins

1. Purified protein (*see* Subheading 3.6).
2. Detergent-free dialysis buffer.
3. Lipids solubilized in detergent (Avanti Polar Lipids, Inc.).
4. Eppendorf tubes.
5. Dialysis buttons (Hampton Research).
6. Dialysis membrane (e.g., Spectra/Por Biotech cellulose acetate membrane 100 kDa MWCO, Spectrum Labs).

2.9 Negative Staining and TEM of Detergent-Solubilized Proteins and 2D Protein Crystals

1. Purified protein (*see* Subheading 3.6) or 2D protein crystals (*see* Subheading 3.8).
2. Parlodion carbon-coated copper TEM grids.
3. Small pointed forceps.
4. Glow discharge system.
5. 0.75 % (w/v) uranyl formate or a different negative stain.
6. Transmission electron microscope (e.g., Philips CM12) with an image acquisition system (e.g., CCD or DDD camera).

3 Methods

3.1 cRNA Synthesis

1. Linearize plasmid DNA containing the cDNA of the desired gene without cutting the gene. Choose an appropriate enzyme according to your gene.
2. Follow the manufacturer's protocol of the mMessage mMachine T7 transcription kit for cRNA synthesis from your desired cDNA. Supplement additionally 2 μL of RNase inhibitor per reaction.

3.2 Preparation and Defolliculation of Xenopus laevis Oocytes for Protein Expression

Note that all animal experiments have to be in accordance with the animal welfare law of your home country and approved by the local veterinary authority.

1. Anesthetize the frog in a bath of ethyl 3-aminobenzoate solution until no movement is observed.
2. Place the frog on its back on top of crushed ice with some paper towels in between—this is to prevent the skin from drying and to keep the temperature of the body low. Note that

this procedure may differ from one country to another. Refer to veterinary protocol of your home country.

3. Surgically remove the ovary from one side of the frog (*see* ref. 34 for a detailed description of the surgery). Separate and open the ovarian follicles into small pieces using autoclaved toothpicks or fine scissors. Place these small parts of ovary in a Petri dish containing 25 mL of calcium-free MBM (*see* **Note 3**).
4. Prepare two Falcon tissue culture dishes (*see* **Note 4**) with 25 mL of 3 mg/mL collagenase solution each (500 μL of 100× collagenase stock in 49.5 mL of calcium-free MBM).
5. Split the oocytes from one ovary into two parts and put them into the two culture dishes containing collagenase solution.
6. Shake the two dishes gently on an orbital shaker at 50–60 rpm for 1.5 h at room temperature.
7. Place the oocytes from both dishes into a 50 mL Falcon tube with some collagenase solution from the dishes (~20 mL total volume).
8. Invert the tube gently several times to detach the oocytes from follicles.
9. Let the oocytes settle and remove the collagenase solution. Add calcium-free MBM up to 20 mL. Repeat this washing step 2–3 times (*see* **Note 5**).
10. Transfer the oocytes from the Falcon tube to a Petri dish and check them visually under the stereomicroscope: (a) usually 99 % of the oocytes are free (i.e., not accumulated) and (b) around 80 % of the oocytes have lost their collagen membrane (defolliculation).
11. After checking the defolliculation state, put the oocytes back into two new Petri dishes and add 25 mL of 3 mg/mL collagenase solution to each.
12. Depending on the efficiency of the defolliculation, shake the dishes gently on an orbital shaker for another period of time, e.g., from 30 min to 1.5 h (*see* **Note 6**).
13. Transfer the oocytes into a 50 mL Falcon tube and invert the tube gently several times by hand.
14. Discard as much of collagenase solution as possible.
15. Wash the oocytes with calcium-free MBM tempered at 18 °C (*see* **Note 7**) three times in a 50 mL Falcon tube, i.e., remove supernatant and fill up to 40–50 mL with calcium-free MBM.
16. Wash the oocytes again with MBM (containing calcium) in a 50 mL Falcon tube (*see* **step 15**). Wash them until the liquid in the Falcon tube looks transparent and the oocytes appear as clean as possible (usually after 2–3 washing steps).

17. Fill up to 25 mL total volume with MBM and separate the oocytes into three or four Petri dishes.
18. Store the oocytes at 18 °C in an incubator.
19. On the next day (best) or at least 4 h after treatment, select morphologically intact (undamaged), spherical stage V–VI oocytes (healthy) for cRNA injection (*see* Subheading 3.3) (*see* **Note 8**). A typical yield of oocytes from one ovary is about 7000 injectable oocytes. Proceed directly with injection or store until next day (suggested).

3.3 cRNA Injection for Uptake Assays or Protein Purification

1. Sort the oocytes selected in Subheading 3.2 prior to injection and remove damaged oocytes. Dilute the cRNA (Subheading 3.1) to 400 μg/mL and inject 25 nL of cRNA per oocyte (10 ng of cRNA, *see* **Note 9**).
2. Inject oocytes using glass microneedles with a diameter of less than 20 μm (prepared with an electrode puller).
3. For uptake experiments inject a minimum of 50 oocytes per condition (e.g., for control experiments and to test different compounds), and for protein purification 800–5600 (*see* **Note 10**).
4. Oocytes injected with the same volume of water should be used as control in uptake experiments.
5. Incubate the oocytes for 2–3 days (*see* **Note 11**).
6. Sort the oocytes every day after injection, remove the dead oocytes, and renew the MBM.
7. Proceed with radiotracer uptake assays (*see* Subheading 3.4) or membrane isolation (*see* Subheading 3.5).

3.4 Functional Analysis by Radiotracer Uptake Assay Using Oocytes

1. On the day of the uptake experiment sort the oocytes expressing the transporter and water-injected oocytes, and select only healthy looking oocytes for the uptake experiment.
2. Put 15 oocytes in a 2 mL Eppendorf tube, remove liquid carefully with a pipette (*see* **Note 12**).
3. Add 200 μL of MBM containing 0.5–1 μCi of the radioactive substrate. The molarity can be adjusted with cold substrate to keep the desired concentration of the substrate, thus preventing substrate depletion (*see* **Note 13**).
4. The typical uptake solution contains the radiotracer, cold substrate, and MBM with or without competitor.
5. Incubate the reaction at room temperature for 1–60 min and shake gently from time to time. The duration must be optimized for each individual protein (*see* **Note 14**).
6. After the incubation time is completed, aspirate the liquid (save it for scintillation counting in **step 9**) and wash with

1 mL of cold MBM. Aspirate the liquid again and repeat the washing step 4 times (*see* **Note 15**).

7. Transfer single oocyte (at least 12) into individual wells of a 96-well plate (*see* **Note 16**).
8. Add 50 μL of 5% SDS solution to each oocyte. Seal the wells with TopSeal and put the plate on a shaker at 900 rpm until the oocytes are completely lysed and homogenized (*see* **Note 17**).
9. Add 5 μl of the supernatant of the uptake reactions (**step 6**) as control to empty wells.
10. Add 150 μL of scintillation cocktail to each well, reseal, and place the plate on a shaker at 500 rpm to mix.
11. Immediately measure the plate with a scintillation counter, measuring each well for 2 min (Packard TopCount; *see* **Note 18**). Figure 1 shows results from an uptake experiment using hPEPT1 and [^{3}H]Ala-Ala.

3.5 Isolation of Yolk Free Oocyte Membranes

1. Select healthy oocytes 2–3 days after cRNA injection (*see* **Note 11**). This protocol gives the volumes for a preparation from ~5600 oocytes and can be scaled down accordingly.
2. Pool ~5600 oocytes and homogenize them in 60 mL of oocyte homogenization buffer (*see* **Note 19**) using the homogenizer.
3. Equally distribute the homogenized oocytes into four 50 mL Falcon tubes. Wash the homogenizer with 25 mL of the same buffer and also transfer equally into the Falcon tubes (finally ~20 mL per Falcon tube, *see* **Note 19**).
4. Centrifuge the homogenized oocytes at 1000 × *g* for 15 min.
5. Transfer the supernatants carefully into one 94 mL ultracentrifuge tube using a short neck glass Pasteur pipette (*see* **Note 20**).
6. Centrifuge at 100,000 × *g* and 4 °C for 1 h.
7. Remove the supernatant using a vacuum pump with a long neck glass Pasteur pipette or by simply decanting (*see* **Note 21**). Before proceeding to the next step, make sure that there is no yellow colored material or any sticky substance on the wall of the centrifuge tube.
8. Add 60 mL of membrane wash buffer to the membrane pellet, detach it by scratching with a 1 mL pipette tip, and transfer all to a 90 mL homogenizer.
9. Wash the centrifuge tube with additional 25 mL of buffer and transfer it to the homogenizer.
10. Homogenize the membranes and transfer them to a 94 mL ultracentrifuge tube, wash the homogenizer with 5 mL of buffer, and add it to the ultracentrifuge tube.
11. Pellet down the membranes by ultracentrifugation at 100,000 × *g* and 4 °C for 1 h.

12. Aspirate the supernatant with a vacuum pump or by decanting, and repeat **steps 8–11**.
13. Discard the supernatant and resuspend the membranes in 5 mL resuspension buffer per ~5600 oocytes used.
14. After resuspending the membranes either flash freeze them in batches corresponding to 800 oocytes in liquid nitrogen and store at −80 °C, or proceed to purification without freezing.

3.6 Protein Purification from Isolated Oocyte Membranes

Keep a small aliquot of all steps for SDS-PAGE analysis. Make several parallel purifications for increased protein amounts.

1. Gently resuspend (to avoid foam formation) the membranes from 800 oocytes in a total of 8 mL solubilization buffer.
2. Transfer the resuspended membranes into a 15 mL Falcon tube. Solubilize membranes at 4 °C for 1 h under gentle rotation.
3. Centrifuge the solubilized membranes at 100,000 × *g* and 4 °C for 1 h.
4. In the meantime equilibrate 500 μL cobalt resin (bed volume) for membranes from 800 oocytes with equilibration buffer by centrifugation (1000 × *g*, 1 min) in an Eppendorf tube and washing at least twice.
5. Dilute the supernatant 1:1 with binding buffer and add the supernatant to the pre-equilibrated cobalt resin in a 50 mL Falcon tube. Incubate on a rotatory shaker for 2 h at 4 °C.
6. Load the gravity flow column with your sample.
7. Wash the column with at least five column volumes of wash buffer 1.
8. Wash column with at least five column volumes of wash buffer 2.
9. Wash the column with at least five column volumes of cleavage buffer.
10. Remove all the liquid from the column (i.e., centrifuge at 1000 × *g* and 4 °C for 1 min) and cut-off the Wizard Midicolumn tip containing the resin.
11. Seal the tip with laboratory film at the bottom and add 250 μL of cleavage buffer and 50 μl of HRV3C protease (250 U). Seal the top as well and keep it on a rotatory shaker overnight at 4 °C.
12. Elute the protein by centrifugation into an Eppendorf tube at 3000 × *g* and 4 °C for 2 min.
13. If the concentration of the eluted protein is too low for subsequent biochemical analysis, concentrate the protein solution in a concentrator with an appropriate molecular weight cut-off (*see* **Note 2**) to the desired concentration.

3.7 Biochemical Analysis of Purified Proteins

1. Determine the protein concentration using the BCA assay following the manufacturer's instructions.
2. Assess the purity of the protein by SDS-PAGE and silver staining (*see* Fig. 2 for successfully purified membrane proteins: also consider bands corresponding to oligomers and glycosylated proteins).
3. Perform a Western blot by using anti-HA antibody (or a suitable antibody of your choice) to assess the presence of your recombinant target protein (*see* Fig. 2 and consider comment in **step 2**).
4. Once the purity of the isolated protein has been confirmed, it can be used for structural studies, e.g., SPA of purified protein and 2D crystallization (Subheading 3.8). Both, purified protein and 2D crystals are visualized by negative stain and/or cryo-TEM (Subheading 3.9).

3.8 2D Crystallization of Purified Proteins

2D crystallization of membrane proteins is a method by itself and beyond the scope of this chapter. However, a brief description is provided here. For introduction and detailed information on 2D crystallization, the following publications are recommended [22, 35–37].

1. Prepare the required detergent-free buffer(s) for dialysis in advance (*see* **Note 22**).
2. Prepare the crystallization mixture at different lipid-to-protein ratios (LPRs)—a good range to test for 2D crystallization is between 0.1 and 1.5 (w/w). Mix in an Eppendorf tube the appropriate amount of protein, preferably at about 1 mg/mL final concentration, and the detergent-solubilized lipid. The choice of the lipid needs to be optimized by screening experiments (*see* **Note 23**).
3. Transfer the protein/lipid/detergent mixture into a dialysis button. Place a dialysis membrane of appropriate size (*see* **Note 2**) on top and close the button properly (*see* **Note 24**).
4. Place the dialysis buttons in the buffer prepared in **step 1**.
5. Incubate the dialysis buffers in an incubator maintained at 24 °C (or the temperature of choice) or on the laboratory bench at room temperature (*see* **Note 25**) until the detergent has been completely removed from the buttons (*see* **Note 26**).
6. After appropriate incubation time harvest the samples transferring them into Eppendorf tubes (*see* **Note 27**).
7. Analyze the reconstitution samples using negative stain TEM (Subheading 3.9).

3.9 Negative Staining and TEM of Detergent-Solubilized Proteins and 2D Protein Crystals

Handle the copper grids with small pointed forceps.

1. Place three drops (~200 μL) of ultrapure water and two drops of 5–20 μL of uranyl formate (or another stain of choice) on a piece of laboratory film.
2. Render the Parlodion carbon-coated copper grids hydrophilic by glow discharge at low pressure in air.
3. Place 4–5 μL of purified protein or 2D crystals to adsorb on a grid and wait for around 15–30 s for detergent-solubilized protein and 1 min for 2D crystals.
4. Remove the protein solution by blotting on a filter paper.
5. Wash the grid by gently moving it on the surface of the water drops, while blotting on filter paper between each drop.
6. Prestain the grid in the first drop of uranyl formate and blot immediately. Then stain again in the second drop of uranyl formate for 10 s and blot. Let the grids dry for a few minutes and store them in a dry place.
7. Analyze samples in TEM and pay special attention to commonly occurring artifacts when working with purified protein at low concentration (*see* **Note 28**). For examples of outcomes *see* Fig. 3 (single particle analysis of purified hAQP1 protein), Fig. 4 (2D crystals of hAQP1), and Fig. 5 (typical artifact in negative stain TEM).

Fig. 5 Electron micrograph of negatively stained parlodion carbon-coated copper grids incubated with protein-free, detergent containing buffer solution reveals the presence of staining artifacts, i.e., particles with diameters ranging from 10 to 17 nm. A selection of the two most prominent populations, i.e., smaller and larger particles of about 10 nm and 14–17 nm, is shown in the galleries (**b** and **c**). Scale bar in (**a**) is 50 nm and frame size in (**b** and **c**) is 35 nm

4 Notes

1. The choice of the detergent for solubilization and purification can vary from protein to protein and is of major importance for the stability and function of the purified membrane protein [38, 39]. Generally, non-ionic detergents with low critical micelle concentration (CMC) are mild and convenient. For solubilization, detergent concentrations in great excess of the CMC are used [5]. The detergent concentration is reduced in the washing and elution steps of the purification to a several-fold of the CMC, e.g., for *n*-dodecyl-β-D-maltopyranoside (DDM, CMC = 0.0087 %) 1.5–2 % are commonly used for solubilization and 0.04 % for washing and elution [5].
2. There are concentrators and dialysis membranes with different molecular weight cut-offs (e.g., 100 or 50 kDa). With membrane proteins a significant amount of mass is added to their molecular weight by the presence of endogenous lipids and the detergent micelle [5]. Therefore, a higher molecular weight cut-off can usually be used (e.g., a 30 kDa protein will most likely not go through a 50 kDa cut-off membrane).
3. Cut the ovarian follicles as small as possible to maximize the accessibility of the collagenase solution. When you take out both ovaries, try to keep approximately one fourth of the total oocytes per plate. This will facilitate handling and separation of follicles.
4. Use only Falcon tissue culture dishes for the collagenase treatment because the follicle membrane sticks nicely to this type of plate. Take several plates depending on the number of oocytes to be treated.
5. Make sure that the liquid in the Falcon tube looks as transparent as possible and the oocytes are clearly visible.
6. Check after the first 1.5 h collagenase treatment and decide how long to continue the second treatment. Sometimes it may need longer treatment than 1.5 h, and sometimes only ~30 min after the first treatment is sufficient.
7. Never add cold buffer to the oocytes after collagenase treatment. Make sure that your calcium-free or calcium-containing MBM is at 18 °C and supplemented with Pen-Strep.
8. This step is crucial for the rest of the experiment; therefore make sure that all the selected oocytes for injection are of good quality, i.e., morphologically intact (undamaged), spherical stage V–VI oocytes and similar size.
9. The amount of cRNA to be injected must be optimized in advance to maximize the protein expression level, but 10 ng per oocyte is a good starting point for injection. The volume of cRNA to inject may vary depending on the stock concentration,

e.g., 20–50 nL. Preferably use a 400 μg/mL cRNA solution. Before starting the injection process check the injector settings.

10. Keep in mind that survival and quality of the injected oocytes might vary with different target proteins, oocyte batches, and experiments. Therefore, it is advisable to always inject additional oocytes, e.g., 30 % more. A minimum of 800 oocytes are required for the here presented purification protocol.
11. The optimal incubation time may vary from protein to protein and should be evaluated in advance, e.g., by Western blot analysis using an appropriate antibody.
12. Place the pipette tip opening at the bottom/wall of the tube to prevent sucking in and destroying oocytes.
13. For peptide transporters, 50 μM final concentration of the dipeptide Ala-Ala was used. Since the radiotracer was used at 2 μM, 48 μM of cold substrate Ala-Ala was added to maintain the final substrate concentration.
14. When the uptake experiment is performed for the first time it is advisable to test different time points, e.g., 1, 5, 10, 15, 30, and 60 min to identify the desired linear initial rate region. Uptake velocity can be increased and decreased by performing the experiment at higher and lower temperatures, respectively.
15. You may release the medium with some pressure to swirl the oocytes around for better washing efficiency.
16. Keep a small amount of the last washing solution to make the oocytes float and easy to transfer with a cut pipette tip or plastic pipette with large neck opening. Transfer oocytes with as little liquid as possible.
17. Make sure that the oocytes are completely lysed. Depending on the batch of oocytes, this may take between 30 min to 2 h.
18. Postponing the scintillation measurement may result in quenching effects due to color development from the pigments of the oocytes.
19. The volume used for homogenizing oocytes will vary depending on the number of oocytes processed. Adapt volume correspondingly, e.g., 60 mL per 5600 oocytes.
20. Avoid aspirating the small egg yolk layer (yellow colored film) at the top of the supernatant.
21. In the 94 mL ultracentrifuge tube most of the yolk material will float and stick to the lid of the tube and upper tube surface. Therefore, carefully decant without mixing. Hold the tube vertically to ensure that the yolk material sticking to the wall will not settle down and contaminate the membranes. Remove the sticky egg yolk material with soft paper towels.

22. A large amount of detergent-free buffer should be prepared: at least 1 L per four dialysis buttons. For AQP1, the following dialysis buffer was used successfully: 20 mM MES-NaOH pH 6, 200 mM NaCl, 10 mM $MgCl_2$, 2 mM DTT, 10% glycerol, and 0.01% NaN_3 [29, 40]. For other membrane proteins, optimal dialysis buffer conditions for successful crystallization have to be screened and identified [22, 35–37].
23. Examples of successful 2D crystallization include the use of dimyristoyl phosphatidylcholine (DMPC), dioleoyl phosphatidylcholine (DOPC), *E. coli* polar lipids extract, and soy bean and egg phosphatidylcholine [22]. For AQP1 *E. coli* polar lipids were successfully used [29, 40].
24. Avoid the formation of air bubbles in the dialysis button as the reduced contact surface may alter the dialysis kinetics significantly.
25. The temperature must be optimized for the corresponding lipid, e.g., consider transition temperature of the used lipid.
26. Time of incubation will vary in function of the detergent used for protein purification and solubilization of lipids. The higher the CMC of the detergent, the faster it will be removed via dialysis. E.g., if both, protein and lipid are in *n*-decyl-β-D-maltopyranoside (DM) the dialysis will take around 2 weeks, with DDM around 4 weeks. To determine the minimal dialysis time, redundant samples can be set up and analyzed by TEM at different time points.
27. Use a pipette to harvest the sample by poking through the dialysis membrane without opening the dialysis button. Slight resuspension might be necessary by pipetting up and down few times.
28. When preparing electron microscopy grids of purified protein at low concentration, always also prepare grids of protein-free sample, i.e., same buffer without protein, for comparison. On grids incubated with protein-free sample and negatively stained, particles in the range of 10–17 nm might be observed (Fig. 5) and misinterpreted as protein molecules. For a recent example, *see* ref. 32, where such particles were misinterpreted as monomeric and oligomeric TRPM4 channel proteins.

Acknowledgements

Financial support from the University of Bern, the Swiss National Science Foundation (grant 31003A_162581), the Bern University Research Foundation, and the National Centre of Competence in Research (NCCR) TransCure and NCCR Molecular Systems Engineering to DF is gratefully acknowledged.

References

1. Krogh A, Larsson B, von Heijne G, Sonnhammer EL (2001) Predicting transmembrane protein topology with a hidden Markov model: application to complete genomes. J Mol Biol 305:567–580
2. Yildirim MA, Goh KI, Cusick ME, Barabasi AL, Vidal M (2007) Drug-target network. Nat Biotechnol 25:1119–1126
3. Carpenter EP, Beis K, Cameron AD, Iwata S (2008) Overcoming the challenges of membrane protein crystallography. Curr Opin Struct Biol 18:581–586
4. Bill RM, Henderson PJ, Iwata S, Kunji ER, Michel H, Neutze R et al (2011) Overcoming barriers to membrane protein structure determination. Nat Biotechnol 29:335–340
5. Ilgü H, Jeckelmann JM, Gachet MS, Boggavarapu R, Ucurum Z, Gertsch J et al (2014) Variation of the detergent-binding capacity and phospholipid content of membrane proteins when purified in different detergents. Biophys J 106:1660–1670
6. Hockney RC (1994) Recent developments in heterologous protein production in *Escherichia coli*. Trends Biotechnol 12:456–463
7. Zoonens M, Miroux B (2010) Expression of membrane proteins at the *Escherichia coli* membrane for structural studies. Methods Mol Biol 601:49–66
8. Emmerstorfer A, Wriessnegger T, Hirz M, Pichler H (2014) Overexpression of membrane proteins from higher eukaryotes in yeasts. Appl Microbiol Biotechnol 98:7671–7698
9. Byrne B (2015) *Pichia pastoris* as an expression host for membrane protein structural biology. Curr Opin Struct Biol 32C:9–17
10. Looser V, Bruhlmann B, Bumbak F, Stenger C, Costa M, Camattari A et al (2015) Cultivation strategies to enhance productivity of *Pichia pastoris*: A review. Biotechnol Adv 33(6 Pt 2):1177–1193
11. Massotte D (2003) G protein-coupled receptor overexpression with the baculovirus-insect cell system: a tool for structural and functional studies. Biochim Biophys Acta 1610:77–89
12. Andrell J, Tate CG (2013) Overexpression of membrane proteins in mammalian cells for structural studies. Mol Membr Biol 30:52–63
13. Eifler N, Duckely M, Sumanovski LT, Egan TM, Oksche A, Konopka JB et al (2007) Functional expression of mammalian receptors and membrane channels in different cells. J Struct Biol 159:179–193
14. Bernaudat F, Frelet-Barrand A, Pochon N, Dementin S, Hivin P, Boutigny S et al (2011) Heterologous expression of membrane proteins: choosing the appropriate host. PLoS One 6:e29191
15. Grigorieff N (2013) Direct detection pays off for electron cryo-microscopy. eLife 2:e00573
16. Bai XC, McMullan G, Scheres SH (2015) How cryo-EM is revolutionizing structural biology. Trends Biochem Sci 40:49–57
17. Liao M, Cao E, Julius D, Cheng Y (2013) Structure of the TRPV1 ion channel determined by electron cryo-microscopy. Nature 504:107–112
18. Bremer A, Häner M, Aebi U (1998) In: Celis JE (ed) Cell biology: a laboratory handbook, vol 3, 2nd edn. Academic, San Diego, pp 277–284
19. Meury M, Harder D, Ucurum Z, Boggavarapu R, Jeckelmann J-M, Fotiadis D (2011) Structure determination of channel and transport proteins by high-resolution microscopy techniques. Biol Chem 392:143–150
20. Ohi M, Li Y, Cheng Y, Walz T (2004) Negative staining and image classification - powerful tools in modern electron microscopy. Biol Proced Online 6:23–34
21. Rosell A, Meury M, Alvarez-Marimon E, Costa M, Perez-Cano L, Zorzano A et al (2014) Structural bases for the interaction and stabilization of the human amino acid transporter LAT2 with its ancillary protein 4F2hc. Proc Natl Acad Sci U S A 111:2966–2971
22. Fotiadis D, Engel A (2015) Two-dimensional crystallization of membrane proteins and structural assessment. eLS:1–10.
23. Hite RK, Raunser S, Walz T (2007) Revival of electron crystallography. Curr Opin Struct Biol 17:389–395
24. Fotiadis D (2012) Atomic force microscopy for the study of membrane proteins. Curr Opin Biotechnol 23:510–515
25. Müller DJ, Engel A (2007) Atomic force microscopy and spectroscopy of native membrane proteins. Nat Protoc 2:2191–2197
26. Gurdon JB, Lane CD, Woodland HR, Marbaix G (1971) Use of frog eggs and oocytes for the study of messenger RNA and its translation in living cells. Nature 233:177–182
27. Sigel E, Minier F (2005) The Xenopus oocyte: system for the study of functional expression and modulation of proteins. Mol Nutr Food Res 49:228–234

28. Wagner CA, Friedrich B, Setiawan I, Lang F, Broer S (2000) The use of *Xenopus laevis* oocytes for the functional characterization of heterologously expressed membrane proteins. Cell Physiol Biochem 10:1–12
29. Bergeron MJ, Boggavarapu R, Meury M, Ucurum Z, Caron L, Isenring P et al (2011) Frog oocytes to unveil the structure and supramolecular organization of human transport proteins. PLoS One 6:e21901
30. Walz T, Smith BL, Agre P, Engel A (1994) The three-dimensional structure of human erythrocyte aquaporin CHIP. EMBO J 13: 2985–2993
31. Boggavarapu R, Jeckelmann JM, Harder D, Schneider P, Ucurum Z, Hediger M et al (2013) Expression, purification and low-resolution structure of human vitamin C transporter SVCT1 (SLC23A1). PLoS One 8: e76427
32. Clemencon B, Fine M, Lüscher B, Baumann MU, Surbek DV, Abriel H et al (2014) Expression, purification, and projection structure by single particle electron microscopy of functional human TRPM4 heterologously expressed in *Xenopus laevis* oocytes. Protein Expr Purif 95:169–176
33. Clemencon B, Lüscher BP, Fine M, Baumann MU, Surbek DV, Bonny O et al (2014) Expression, purification, and structural insights for the human uric acid transporter, GLUT9, using the *Xenopus laevis* oocytes system. PLoS One 9:e108852
34. Markovich D (2008) Expression cloning and radiotracer uptakes in *Xenopus laevis* oocytes. Nat Protoc 3:1975–1980
35. Hasler L, Heymann JB, Engel A, Kistler J, Walz T (1998) 2D crystallization of membrane proteins: rationales and examples. J Struct Biol 121:162–171
36. Jap BK, Zulauf M, Scheybani T, Hefti A, Baumeister W, Aebi U et al (1992) 2D crystallization: from art to science. Ultramicroscopy 46:45–84
37. Kühlbrandt W (1992) Two-dimensional crystallization of membrane proteins. Q Rev Biophys 25:1–49
38. Privé GG (2007) Detergents for the stabilization and crystallization of membrane proteins. Methods 41:388–397
39. Sonoda Y, Newstead S, Hu NJ, Alguel Y, Nji E, Beis K et al (2011) Benchmarking membrane protein detergent stability for improving throughput of high-resolution X-ray structures. Structure 19:17–25
40. Fotiadis D, Suda K, Tittmann P, Jeno P, Philippsen A, Müller DJ et al (2002) Identification and structure of a putative Ca^{2+}-binding domain at the C terminus of AQP1. J Mol Biol 318:1381–1394

Chapter 15

Membrane Protein Solubilization and Composition of Protein Detergent Complexes

Katia Duquesne, Valérie Prima, and James N. Sturgis

Abstract

Membrane proteins are typically expressed in heterologous systems with a view to in vitro characterization. A critical step in the preparation of membrane proteins after expression in any system is the solubilization of the protein in aqueous solution, typically using detergents and lipids, to obtain the protein in a form suitable for purification, structural or functional analysis. This process is particularly difficult as the objective is to prepare the protein in an unnatural environment, a protein detergent complex, separating it from its natural lipid partners while causing the minimum destabilization or modification of the structure. Although the process is difficult, and relatively hard to master, an increasing number of membrane proteins have been successfully isolated after expression in a wide variety of systems. In this chapter we give a general protocol for preparing protein detergent complexes that is aimed at guiding the reader through the different critical steps. In the second part of the chapter we illustrate how to analyze the composition of protein detergent complexes; this analysis is important as it has been found that compositional variation often causes irreproducible results.

Key words FTIR, Density gradient, Lipid analysis

1 Introduction

The objective behind heterologous production of membrane proteins is frequently the purification of the membrane protein for in vitro structural or functional analysis. This step is often limiting for studies on membrane proteins with solubilization and purification giving poor yields and irreproducible samples. Here we describe an approach to optimizing the preparation of protein detergent complexes (PDC) for subsequent analysis. Three different steps are discussed: preparing and purifying membrane fragments; solubilizing membranes with detergents; and analyzing the composition of PDC.

Membranes are a complex mixture of many different proteins and lipids. It is usually well worthwhile preparing specific membrane fractions, thus reducing the heterogeneity and diversity of the sample. Unfortunately membranes are very variable and purification

Isabelle Mus-Veteau (ed.), *Heterologous Expression of Membrane Proteins: Methods and Protocols,* Methods in Molecular Biology, vol. 1432, DOI 10.1007/978-1-4939-3637-3_15, © Springer Science+Business Media New York 2016

methods are often context dependent; however most depend on physical disruption and centrifugation. The method we develop below concerns purifying the different membranes from the bacterium *Roseobacter* (*Rsb.*) *denitrificans*, and illustrates the preparation of three different membrane fractions. This method, like the majority of membrane purification protocols, relies on differential and density gradient centrifugation. However the details of the centrifugation protocol are quite variable.

Extracting membrane proteins from their natural environment and putting them in aqueous solution is typically a critical step in studies of membrane proteins. All too often all activity is lost, or extraction is far from quantitative. Another regular problem is difficulty during the scale-up of extraction from initial small-scale trials to a more preparative scale. Here we have concentrated on a simple protocol to optimize solubilization, and also to understand how to scale-up the solubilization.

The reproducibility of measurements on membrane proteins is frequently critically dependant on the reproducibility of the PDC formed during solubilization. In particular the amount of bound lipid seems often to play an important role [1]. Furthermore, the reproducibility of the extraction step depends strongly on the quality of the membranes. So in a third part of the protocol we introduce analytical methods that allow a reasonable characterization of the chemical composition of PDC.

2 Materials

2.1 Preparation of Bacterial Cytoplasmic Membranes

1. Bacterial Cells.
2. Centrifuge and Rotors.
3. Ultracentrifuge and Rotors.
4. Cell disruptor; various types can be used but this protocol is optimized for a French press (*see* **Note 1**).
5. Isolation buffer, e.g., 10 mM Tris–HCl, pH 8.0; 250 mM sucrose. Many other buffers can also be used in place of Tris with little or no modification.
6. Enzymes and inhibitors:
   (a) Deoxyribonuclease I (DNase I).
   (b) Ribonuclease (RNase).
   (c) Anti-protease cocktail.
   (d) Lysozyme.
7. Sucrose solutions for density gradient in 10 mM Tris–HCl, pH 8.0; 5 mM EDTA:
   (a) 70% sucrose.
   (b) 35% sucrose.
   (c) 15% sucrose.

2.2 Solubilization of Membranes

1. Purified membranes.
2. Ultracentrifuge suitable for small volumes.
3. Spectrophotometer for absorption and turbidity measurements.
4. Buffer: for example, 10 mM Tris–HCl, pH 8.0.
5. Concentrated detergent solution: for example 1 M *n*-octyl-β-D-glucopyranoside in buffer.

Most of the material required is discussed in the methods and notes sections. However two points merit discussion here: the choice of buffers for solubilizing membranes; and turbidity measurements. Buffers for solubilization should be chosen bearing in mind the pH range where the protein of interest is stable and active, and the various assays that are envisioned. For example, phosphate buffers must be avoided if phospholipids will be measured as phosphate. However most buffers good for soluble proteins are also good for membrane proteins. Turbidity and light scattering measurements are easily made with an absorption spectrophotometer. When making such measurements you should bear in mind the wavelength dependence of light scattering; the $1/\lambda^4$ dependence means higher readings are obtained at lower wavelengths. It is also important to realize that as absorption spectrophotometers are not optimized for turbidity measurements the readings cannot generally be simply transferred from one machine to another.

2.3 Analysis of Protein Detergent Complexes

The analysis method that we propose here is able to give an estimation of the composition of a PDC solution estimating the concentrations of protein, detergent, lipid, and buffer in the sample. The method is based on FTIR spectroscopy, which though not universally available is more and more widespread. As an added bonus the method can provide information on the secondary structure of the protein [2]. This method requires:

1. FTIR spectrophotometer with ATR attachment.
2. Standard solutions or spectra (buffer, detergent, lipids).

As an alternative to using infrared spectroscopy we propose using chemical tests to estimate protein, detergent, and lipid concentrations. These tests have been adapted for a micro-plate reader and glycosidic detergents require:

1. Micro-plate reader.
2. Bradford protein assay reagent diluted as per manufacturers instructions.
3. 10 % Phenol (for glycosidic detergents).
4. Concentrated sulfuric acid (for glycosidic detergents).
5. Ferrous sulfate ammonium molybdate reagent (for phospholipids as PO_4 (*see* **Note 2**)).

3 Methods

3.1 Preparation of Bacterial Cytoplasmic Membranes

The protocol is loosely based on that published online by the Hancock laboratory for the purification of bacterial outer membranes [3], and that used for the preparation of bacterial intracytoplasmic membranes [4]. The procedure is based on a differential centrifugation to separate membrane fragments largely on size, followed by density gradient centrifugation, to separate fragments on density. The bacterial outer membrane is stronger than the inner membrane; thus fragments tend to be larger after mechanical disruption. Furthermore, the outer membrane and intracytoplasmic membranes are more protein rich and is thus denser than the cytoplasmic membrane.

1. The washed bacteria are resuspended by gentle pipetting in isolation buffer to a final concentration of 250 UOD. (Typically bacterial cell density is measured by the solution turbidity at 600 nm or some wavelength free from absorption.) The isolation buffer is supplemented with: 100 μg/mL DNaseI; 20 μg/mL RNase; complete protease inhibitors (as per manufacturers instructions).
2. Add lysozyme to a final concentration of 1 mg/mL (to digest peptidoglycan cell wall).
3. Break the cells by two passages through a French pressure cell at 8300 kPa (1200 psi).
4. Remove unbroken cells and large debris by centrifugation at 13,000 × *g* for 5 min; this gives the "crude extract" (*see* Fig. 1).
5. Collect the large fragments of outer membrane by differential centrifugation, for example 72,000 × *g* for 5 min (40,000 rpm in a TLA55 rotor).
6. Prepare sucrose density gradients for purification of membrane fragments:
   (a) Layer the supernatant onto sucrose gradients made, for a SW40 swinging bucket rotor, with 2 mL of 70 % sucrose, 9 mL of a continuous 70–15 % sucrose gradient.
   (b) Resuspend pelleted crude outer membranes (OM) in a few mL of buffer.
   (c) Layer the resuspended pellet onto sucrose gradients made, for a SW40 swinging bucket rotor, with 2 mL of 70 % sucrose, 9 mL of a continuous 70–35 % sucrose gradient.
7. Centrifuge overnight (16 h at 39,000 rpm on a SW40 rotor).
8. Fractionate the gradients and identify membrane fractions.
9. To collect purified membranes pool selected fractions.
10. Dilute membrane fractions threefold with sucrose-free buffer and collect by ultracentrifugation, for example 40,000 rpm for 60 min in a 70Ti rotor (120,000 × *g*).

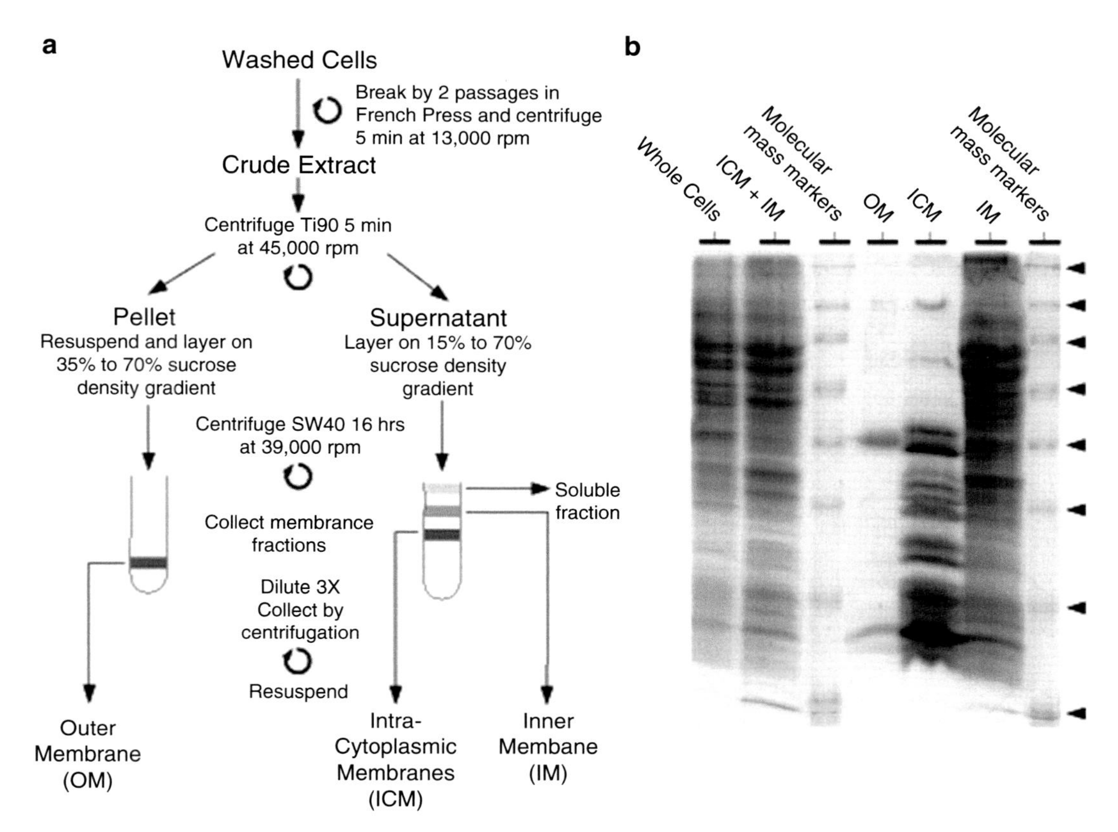

Fig. 1 Illustration of the protocol for the purification of *Rsb. denitrificans* membranes. *Left panel*: flow diagram showing the different steps in the protocol for preparing: outer membrane (OM), intracytoplasmic membrane (ICM), and inner (cytoplasmic) membrane (IM) fragments. *Right panel*: 12.5 % SDS-PAGE analysis of the fractions. Samples were heated to 95 °C for 2 min in loading buffer (4 % SDS, *see* **Note 3**). Markers are at relative molecular masses of 250,000, 98,000, 64,000, 50,000 36,000, 30,000 16,000, and 6000

11. Resuspend purified membranes (outer, inner, and intracytoplasmic) in about 1 mL of buffer.

The object of the membrane purification protocol is to prepare a sample that is homogeneous and reproducible. If you are lucky your protein of interest might even survive being frozen in the membranes. It is a good idea to control the quality of the membranes, for example by measuring the protein concentration (wt/wt) and then verifying the gel profile. In Fig. 1 the gel on the right illustrates the quality of the purification, with little cross contamination between the different fractions and relatively high yield. The vast majority of membrane purification protocols, like this one, rely on differential centrifugation and sucrose density gradients though the precise conditions vary with the biological sample, for example between OM from different bacteria [5].

3.2 Solubilization of Membranes

The development of a membrane solubilization protocol is a difficult and repetitive process. Often, it is necessary to restart the work because some aspect of the final preparation is inadequate for the desired use or some critical aspect of the starting material changes. Nevertheless, the route we trace is relatively straightforward and allows the rapid establishment of reliable, reproducible routine conditions for solubilization. The process is inherently multivariant, with many parameters playing an important role. It is usual to select a protein and membrane first, usually determined by the project, and then chose a detergent that is promising for solubilization and finally to try to optimize solubilization conditions by modifying the concentrations, temperature, and solubilization time. We treat these different aspects here and try to provide some insight into the optimization.

3.2.1 Basic Solubilization Procedure

The basic protocol that needs to be optimized is remarkably simple to write in four steps.

1. Incubate a known amount of membrane sample (mg) in a known volume (mL) of solution with a known final concentration of your selected detergent (mM or mg/mL) at a given temperature (°C) for a given time (minutes). As can be immediately appreciated, there are a large number of variables that need to be optimized, and these are treated below in slightly more detail. It cannot be overemphasized enough that the amount of membrane, the volume, and the final concentration of detergent are three independent variables (*see* **Note 4**). Typical initial values could be 2 mg of membrane in 1 mL of solution with 50 mM octyl-glucoside detergent at 20 °C for 30 min (*see* **Note 5**).
2. Separate the solubilized protein from the unsolubilized detergent-resistant membranes. This step most usually involves an ultracentrifugation necessary to sediment the remaining membrane fragments. For this purpose, a small benchtop ultracentrifuge (type airfuge or TL100) is particularly useful (*see* **Note 6**). Conditions we frequently use are centrifugation at 40,000 rpm for 30 min in a TLA45 rotor (120,000 × g at 20 °C) (*see* **Notes** 7 and **8**).
3. Separate the supernatant of solubilized material from the pellet, and if necessary for analysis resuspend the pellet. It is usually easier to resuspend the pellet in about the same volume of buffer as it was separated from (*see* **Note 9**).
4. Analyze the solubilized and unsolubilized fractions to determine the yield, purity, activity, and stability of the protein of interest. When considering the purity of membrane protein fractions, it is important to look beyond other proteins and consider the presence of other nonprotein contaminants (e.g., lipids). Several variations on this general scheme can be found

in the literature; perhaps the most interesting involves repeated extractions. Often, it is possible to extract certain types of impurities selectively and by using two successive solubilizations obtain considerable purification [6].

As can be appreciated readily the first step contains a multitude of different parameters all potentially important. In the sections below we try to demystify the decisions that go into choosing these parameters.

3.2.2 Choice of Detergent

The number of different detergents available is considerable and these have very different properties and prices. It is not always possible to test more than a dozen or two for the extraction and formation of PDC. The choice of detergent is primarily dictated by the sensitivity of the protein of interest to the detergent. It is thus necessary initially to determine in which detergents the protein maintains its activity (*see* **Note 10**). Unfortunately, selection of a detergent for a protein of interest, while the most important aspect of detergent selection is the hardest to predict; indeed there is no guarantee that there exists an appropriate detergent for your favorite protein.

The choice of detergent also depends to some extent on how you will use the PDC. Some detergents lend themselves to certain analytical methods, for example analytical ultracentrifugation and C8E5 [7, 8], while other like Triton X-100 are contraindicated. In Table 1 we have collected some of the critical information on several detergents. We have selected those that are the most commonly used, and a few others to illustrate particular points. Different laboratories often have different cultures in testing detergents and different prejudices. Some laboratories swear by Triton X-100 while others refuse to use it. The wide variety of detergents available, with extremely varied hydrophobic and hydrophilic parts, coupled with the rarity of systematic studies [12], make this choice of detergent a particularly daunting task for the novice.

As a general rule, when developing isolation protocols it is necessary to try 5–30 different detergents, in the search for one in which your protein of interest is stable for a few hours. I would suggest the following plan of attack.

1. Use a variety of different chemistries. Include in the initial screen for example:
   (a) Several different head group chemistries: glycosidic, poly-oxyethylene and ionic;
   (b) Aliphatic and polycyclic detergents;
   (c) Detergents with varied CMC.
2. Avoid detergents that are incompatible with anticipated experiments. It is very disheartening, after having established a protocol to need to start again because a change in detergent is necessary for a particular test.
3. After initial tests variants on promising chemistries can be tried.

Table 1
Table of properties for a selection of detergents

Detergent name	CMC (mM)[a,c]	*N* agg[b,c]	Notes
Glycosidic group			
n-dodecyl-β-D-maltoside	0.17	78–149	Reputed mild, used in crystallography
n-decyl-β-D-maltoside	1.8	69	Reputed mild, used in crystallography
n-octyl-β-D-glucoside	18.0	78	Reputed mild, used in crystallography
Poly-oxy-ethylene group			
C8E5	7.1	41	Reputed mild, used in crystallography
C10E5	0.81	73	
C12E8	0.09	90–180	
Trixon X-100	0.23	75–165	Polydisperse, strong UV absorption
Tween 20	0.059		
Brij-58	0.08	70	Polydisperse
Polycyclic group			
Cholate	9.5	2–3	
Deoxycholate	6.0	22	
CHAPS	8.0	10	Reputed mild
BigChap	2.9	10	Reputed mild
Digitonin	<0.5	60	Reputed mild
Ionic group			
SDS[d]	2.6	62–101	Anionic, strong and denaturing
Fos-choline-12	1.5	50–60	Used for NMR
Lauryl-DAO[e]	1.0	76	Zwitterionic, used in crystallography
ANZERGENT 3 12	2.8	55–87	
DHPC[f]	10.0	19	Reputed mild, phospholipid, makes bicelles

[a]The critical micellar concentration (CMC)
[b]The aggregation number N_{agg}
[c]Values are mostly from the Anatrace and Sigma literature [9, 10], in which values for many other detergents can also be found
[d]*SDS* sodium dodecyl sulfate
[e]*DAO* dimethylamine-*N*-oxide
[f]The values for di-hexanoyl phosphatidyl choline (DHPC) are from ref. 11

Initial tests can often be relatively quick, determining if the protein is relatively stable in the detergent and is at least partially solubilized. It must be stressed that detergent selection should still be considered an art rather than a science and that there are no hard and fast rules. This reflects our incomplete understanding of protein–detergent and protein–lipid interactions (*see* **Notes 11** and **12**).

There are however a number of indications to aid in selection. A lower CMC makes it cheaper to use a detergent for purifications but harder to get rid off the detergent or exchange it. During purifications, to maintain PDC in solution, it is necessary to include detergent in all buffers at a concentration at or slightly above the CMC. For some detergents such as octyl-glucoside this can become very expensive. However, low CMC reflects stronger hydrophobic interactions between the detergent molecules but also between the detergent molecules and the hydrophobic parts of the protein. These strong interactions are harder to undo if you wish to exchange detergent. Furthermore a low CMC makes detergent removal by dialysis impractical.

A lower N_{agg} tends to be better for purification. Lower N_{agg} means that the detergent has less tendency to aggregate on the protein when forming a PDC. This leads to easier purification as there is relatively more protein in each PDC. The number of detergent molecules bound to proteins in PDC can be very high (see below).

Why are detergents mild or strong? This classification is largely empirical and reflects experience on their use in protein purification and lipid dissolution. Mild detergents dissolve lipids less well, which is probably useful for protein isolation. There is some evidence that protein inactivation during solubilization and isolation can be caused by the removal of essential lipid molecules [1].

Polar groups can play an important role in stabilizing proteins or destabilizing them. This often depends on specific interactions. However poly-oxy-ethylene and glycosidic detergents may have a general stabilizing effect, much like the role of polyols in cryo-protection.

Detergents are often available at different levels of purity. Frequently lower purity can adversely affect reproducibility, through lot to lot variation. This is often discussed for the α anomer contamination of β-D-maltosides, which drastically change micelle properties [13]. For this reason it is best to use the purest detergent you can afford.

3.2.3 Optimization

Time and temperature both have a profound effect on protein solubilization. Low temperature and short times can offer kinetic selection of rapidly solubilized components. Low temperature can also modify the lipid phase diagram [14], and change how lipids and detergents mix [15, 16]. In extreme cases this can result with certain combinations of time temperature and detergent in detergent-resistant membranes. For example cholesterol is very poorly solubilized by octyl-glucoside at low temperature.

In general, the range of reasonable temperatures is relatively limited, as we do not wish to expose the sample to elevated temperatures with the inherent risk of denaturation or proteolysis. As a general rule solubilization is most reproducible away from any lipid phase transition temperatures and typical temperatures to test are: 0, 4–10, 20, and 37 °C. Time should also be varied during optimization. The equilibration time for solubilization can be varied and times between a few minutes to overnight are regularly used. It is also possible to vary the speed of detergent addition.

3.2.4 Detergent Titration, Reproducibility, and Scale-Up

In order to scale-up solubilization from initial tests to a preparative scale it is necessary to understand better how the detergent interacts with the different components of your sample.

The best way to obtain this information is from a detergent titration as described here and illustrated in Fig. 2. Analysis of such titrations is treated in more detail in various places for example [17].

1. Prepare a set of samples containing different concentrations of membrane, between about 1 and 10 mg/mL.
2. Thermostat the samples, and the spectrophotometer, at the desired temperature.
3. Measure sample turbidity in the spectrophotometer at an appropriate wavelength.
4. Add detergent, allow to equilibrate, and repeat measurement (**step 3**).
5. Continue until samples are clear (fully solubilized).
6. Prepare graphs like Fig. 2a showing how turbidity varies with detergent concentration in each sample.
7. Determine detergent concentrations for critical points (changes in slope, 50 % reduction etc.) on the curves.
8. Prepare a graph like Fig. 2b showing how the positions of critical points depend on membrane concentration.
9. Determine slopes and intercepts of the different lines.

This experiment provides a wealth of information on the solubilization process, and importantly allows the determination of conditions necessary for reproducibility with different membrane concentrations or amounts. In Fig. 2 from panel **a** it is immediately clear that the shape of the curve depends on the membrane concentration, with the scattering declining much more rapidly for the more dilute sample. Less clearly visible is that the scattering declines to a minimum before rising to a later maximum and then eventually declining. In panel **b** three specific points on the curves have been replotted as a function of the membrane concentration, including some extra points. This shows how the positions of these points are linearly related to the membrane concentration. Thus on each line the objects in solution have the same composition, the slope giving the lipid:detergent ratio and the intercept the "free" detergent concentration. Also shown in panel **b** is the CMC of the

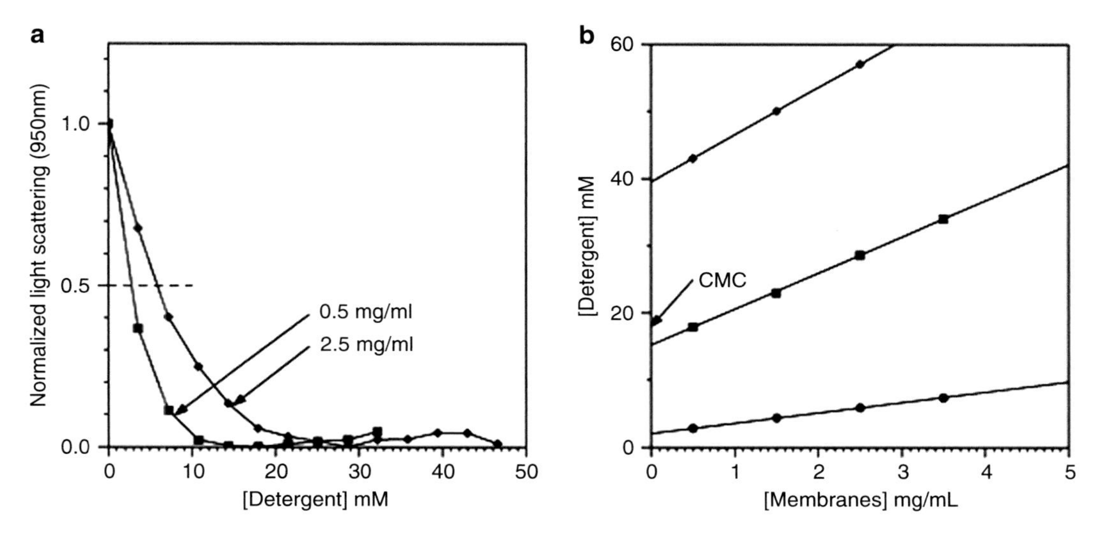

Fig. 2 Detergent titration of membrane solubilization. (**a**) Bacterial intracytoplasmic membranes were solubilized with *n*-octyl-β-D-glucopyranoside by slowly titrating the detergent from 0 to 50 mM and following solubilization by changes in turbidity, measured at 950 nm (a wavelength at which the sample has minimal absorbance). The curves have been normalized to run from an initial value of 1 to a minimum of 0. Curves are shown for sample concentrations of 0.5 and 2.5 mg/mL. (**b**) Selected points on the curves in *panel* (**a**), here 50 and 100 % clarification and the later maximum, are replotted as a function of membrane concentration, and fit to linear regression lines (2.06 + 1.54 × [Mb])mM; (15.22 + 5.37 × [Mb])mM; and (15.22 + 5.37 × [Mb])mM respectively

detergent used to emphasize that this concentration (a property of the detergent in solution without membranes) is not particularly helpful as a reference.

This figure is primordial for understanding how to scale-up solubilization as the objective while scaling up is to remain on the same line of constant composition. Thus the detergent concentration necessary depends on the membrane concentration in a non-trivial way.

3.3 Analysis of Protein Detergent Complexes

3.3.1 FTIR Analysis of PDC

Infrared spectroscopy is becoming increasingly easy to use with modern FTIR spectrophotometers and the convenience of ATR accessories. It is a method ideally suited to chemical analyses as almost all molecules have clear infrared absorption signatures. However as all molecules give signatures one must be careful about all the different components in the solution. The method here is derived from that of da Costa et al. [18].

1. Obtain ATR-FTIR spectra of reference samples (as shown in Fig. 3):
   (a) buffer, in our example phosphate buffer;

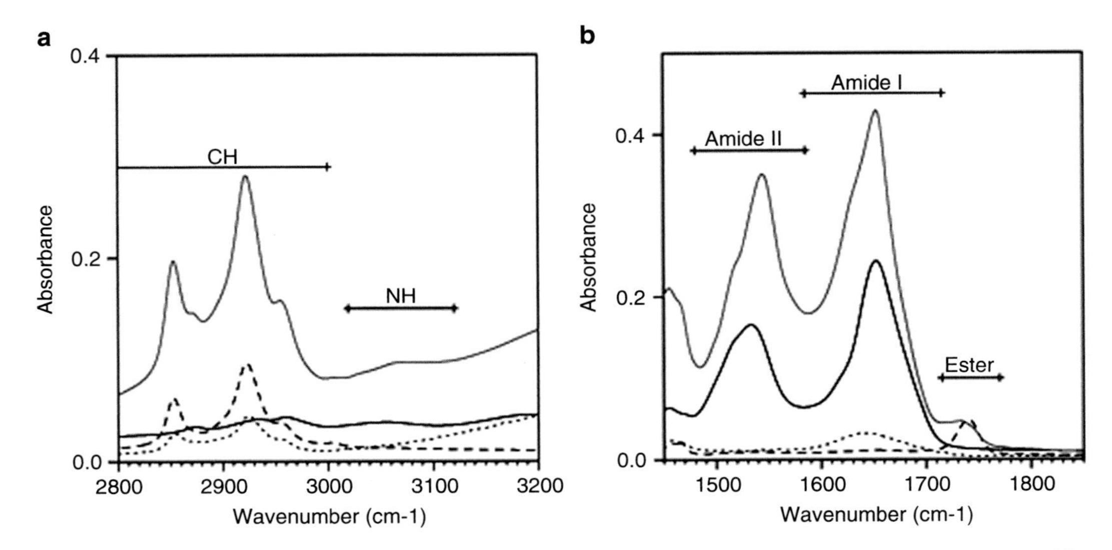

Fig. 3 ATR-FTIR spectra of different standards and PDC. (**a**) High frequency region showing XH stretching modes. (**b**) Mid frequency region showing specific group modes. *Bars* indicate different spectral bands and integration regions. In each panel the four spectra shown are: PDC (*solid gray line*), standards (*black lines*), protein (lysozyme) (*solid*), *Escherichia (E.) coli* lipids (*dashed*), and *n*-octyl-β-D-glucopyranoside (*dotted*). Spectra were obtained in Phosphate buffer (10 mM pH 7.5) containing 20 mM NaCl. Spectra of standard samples were obtained at concentrations between 0.5 and 20 mg/ml. Spectra have had solvent contributions subtracted separately in the two regions shown

(b) detergent, in our example *n*-octyl-β-D-glucopyranoside;

(c) lipids, in our example *E. coli* polar lipids;

(d) protein, for this illustration we have used lysozyme

and subtract the solvent (buffer) contribution from the different spectra.

2. Identify absorption bands that vary strongly between the different samples (as shown in Fig. 3).
3. Calculate the integrated intensities of all the spectra in all the samples normalized to their concentrations (*see* Table 2 and **Note 13**).
4. Use the values from the standards to calculate the deconvolution matrix coefficients (*see* **Note 14**).
5. Calculate the composition of the PDC using these coefficients and the integrated intensities from the PDC spectrum.

The CH stretching region shown in Fig. 3a has strong contributions from both detergent and lipid, as is to be expected, while the protein shows much smaller contributions in this region in line v its chemical composition. The Amide A (NH stretching) v can also be clearly seen in the protein (and PDC) spectra. In Fig. 3b several group modes can be seen. Most prominently the broad and intense Amide I and Amide II bands of the protein around 1650 and 1560 cm^{-1} respectively. There is little

Table 2
Integrated intensities of spectral bands used for deconvolution

	Integrated intensity[a, b]		
	νCH	νEster	Amide I/II
Sample (mL/mg)	2800–3000 cm^{-1}	1770–1715 cm^{-1}	1715–1480 cm^{-1}
Protein	0.0457	−0.0056	0.9241
Lipids	0.2148	0.0433	−0.0133
Detergent	0.1093	−0.0050	0.0703
PDC	0.6973	0.0113	1.7247

[a]A linear baseline correction was used between the integration limits
[b]Integrations are normalized for concentration (1 mg/mL)

contribution in this region from lipids or detergent, though a small OH bending contribution in the detergent spectrum can be seen. Also near 1730 cm^{-1} is the prominent ester carbonyl stretching mode of the fatty acyl chains in the lipid sample. This absorption is also visible in the PDC spectrum as a shoulder.

Unfortunately for the type of analysis proposed here the exact positions and widths of the different peaks depend on the molecular environment. Therefore to reliably use the infrared absorption to determine the composition it is preferable to use integrated intensities rather than specific wavenumbers. The values we obtain for the different bands are shown in Table 2.

The normalized integrals in Table 2 can be used to deconvolute the spectrum of the PDC (*see* **Note 14**). In the example shown this deconvolution gives the concentrations of protein, lipid, and detergent in the sample as respectively 1.58 mg/mL (25%), 0.92 mg/mL (14%), and 3.91 mg/mL (61%). This composition illustrates that PDC often contain considerably less protein than other molecules (here just 25%), and that there is often considerable lipid bound to the PDC. The errors in the composition are probably dominated by systematic errors (the average molecular mass of a phospholipid per ester group, or protein per amide); some of these errors can be reduced by using more IR absorption bands for the deconvolution.

3.3.2 Chemical Analyses

The spectroscopic method above is fast and easy, once you have developed the skills necessary to obtain the spectra and subtract baselines. However FTIR spectrophotometers are not standard lab equipment. The same analysis or a similar one can be performed using simple chemical tests. Glycosidic detergents can be conveniently assayed using a small-scale phenol-sulfuric acid assay [19,

20]; phospholipids can be readily determined by small-scale versions of the iron-molybdate phosphate assay [21]; and protein concentrations can be estimated using the Bradford method [22]. These chemical measurements are however considerably more time-consuming and take some time to perfect; in particular the phosphate and sugar assays are particularly prone to problems due to contaminants giving high background readings.

1. Bradford protein assay [22] adapted for micro-plate reader: in a 96-well plate place samples and standard, 40 μL per well, containing up to 2 μg of protein; add 160 μL of diluted Bradford reagent and mix well; measure absorbance at 595 nm after 10 min incubation.
2. Iron-molybdate phosphate assay [21] adapted for micro-plate reader: prepare fresh reagent (*see* **Note 2**); in the 96-well plate place samples and standard, 60 μL per well, containing up to 2 nmol of phosphate; add 40 μL of fresh Iron-Molybdate reagent and mix well; measure absorbance at 600 nm after 30 min incubation.
3. Phenol-Sulfuric acid sugar assay [19], adapted for micro-plate reader (the pipetting should be done under a fume hood): in a 96-well plate place samples and standard, 20 μL per well, containing up to 3 μg of sugar; add 10 μL of 10% phenol to each tube, and mix; carefully add 100 μL of sulfuric acid to each tube, and mix (the sample will get hot); measure absorbance at 490 nm after 30 min incubation.

Use the standard curves to calculate the amount of sugar, phosphate, and protein in each sample. The sugar is a measure for glycosidic detergents, the phosphate from phosphoglycerol lipids and the protein of your PDC. Using these methods the composition of PDC can be measured with similar precision to that obtained from FTIR spectroscopy, at the cost of considerably more sample (especially if numerous replicates are necessary). For the same PDC as examined by FTIR using these chemical assay we obtained a composition of protein:lipid:detergent of 21 ± 3%, 22 ± 3%, and 57 ± 3% respectively. The errors here are probably dominated by experimental difficulties in the selection of standards, the complete reaction of sugars or phospholipids, and baseline subtraction.

4 Notes

1. Various cell disruption process are available, these include: sheering as in a French press, bead-beater or cell disruptor; ultrasonification; or explosion as with a Yeda press. For membrane protein preparation most people prefer shearing disruption

or explosion rather than ultrasonification. This is both because of the quality of the membrane fragments obtained (important for membrane purification) and the amount of energy put into the sample and thus potentially modifying it.

2. The iron-molybdate reagent should be prepared fresh immediately before use. This solution is made from:
   (a) 50 mg $FeSO_4{\cdot}H_2O$ in 700 μL H_2O;
   (b) 100 μL 10% $(NH_4)_6$ $Mo_7O_{24}{\cdot}4H_2O$ in 10 N H_2SO_4 (this solution can be prepared in advance);
   (c) Adjust volume to 1.0 mL

   This preparation should be done in a fume hood taken the necessary precautions for manipulating concentrated acid.
3. To cook or not to cook? Membrane proteins often behave (very) badly on SDS gels, and one of the known difficulties is solubility in SDS! The situation is simply summarized:
   (a) if membrane proteins are heated even in SDS they can form insoluble aggregates that do not enter the gel.
   (b) if membrane proteins are not heated they often retain some (all) of their native structure and remain associated in higher level oligomers and complexes.

   The consequence is that some proteins apparently disappear on heating while others appear.
4. The concentration of the detergent needed depends on the lipid and protein concentrations (*see* Fig. 2). For this reason it is important to consider separately the membrane concentration, the detergent concentration, and the volume.
5. It is usually best to add detergent slowly, if possible while mixing; this avoids as much as possible large local changes in concentration.
6. Benchtop ultracentrifuges are very useful for membrane protein studies; however they are able to generate very high centrifugal forces, sufficient to precipitate proteins and PDC. So care must be taken not to sediment proteins and to bear in mind the potential inhomogeneity of the supernatant.
7. Do not dilute samples containing detergent to make up the volume necessary for ultracentrifugation. Use an appropriate tube and volume from the start. Unnecessary changes in volumes and concentrations modify the parameters you are trying to optimize.
8. Do not unnecessarily change the sample temperature. These can radically change the nature of the solution leading to phase separations and changes in solubility. Try to work at one fixed temperature.

9. The detergent-resistant, unsolubilized membranes usually contain a large amount of detergent and thus behave quite differently from untreated membranes. Furthermore on addition of detergent-free buffer, detergent and possibly some proteins will leave the membranes for the aqueous phase. Try as far as possible to avoid foaming during resuspension.
10. Initial detergent testing should probably concentrate on relatively dilute solutions some way above the CMC. A starting point for stability tests could be the CMC of the detergent plus an equal weight of detergent to total membrane.
11. It is sometimes beneficial for the stability of proteins to add extra lipids during solubilization.
12. Several surfactants and assemblies have been recently described, while most do not allow solubilization they offer promise in stabilizing membrane proteins: these include amphipols [23], fluorinated surfactants [24], and protein nanodiscs [25]. In addition protein solubilization in polymer nanodiscs has also been reported [26, 27].
13. Many modern FTIR spectrophotometers have spectral deconvolution programs integrated in their software. All offer the possibility of solvent subtraction.
14. Calculation of coefficients for deconvolution and spectral deconvolution. Using the integrated absorptions for the standard samples it is relatively simple to calculate the coefficients necessary to determine the contribution of each of these types of molecule to a test spectrum. First the various integrals are inserted in a matrix as shown below:

$$M = \begin{bmatrix} I_{\text{Protein}}^{3000-2800} & I_{\text{Lipid}}^{3000-2800} & I_{\text{Detergent}}^{3000-2800} \\ I_{\text{Protein}}^{1770-1715} & I_{\text{Lipid}}^{1770-1715} & I_{\text{Detergent}}^{1770-1715} \\ I_{\text{Protein}}^{1715-1480} & I_{\text{Lipid}}^{1715-1480} & I_{\text{Detergent}}^{1715-1480} \end{bmatrix}$$

Here $I_{\text{Protein}}^{3000-2800}$ is the integrated absorption of the protein sample between 3000 and 2800 cm^{-1}. This matrix is then inverted (this can be done with a spreadsheet program for example giving the matrix M^{-1}), and the resulting coefficients are then used to deconvolute a sample spectrum. The deconvolution is then done as follows:

$$\begin{bmatrix} \text{Protein} \\ \text{Lipid} \\ \text{Detergent} \end{bmatrix} = M^{-1} \times \begin{bmatrix} I_{\text{Sample}}^{3000-2800} \\ I_{\text{Sample}}^{1770-1715} \\ I_{\text{Sample}}^{1715-1480} \end{bmatrix}$$

References

1. Palsdottir H, Hunte C (2004) Lipids in membrane protein structures. Biochim Biophys Acta 1666(1–2):2–18. doi:10.1016/j.bbamem.2004.06.012
2. Goormaghtigh E, Raussens V, Ruysschaert JM (1999) Attenuated total reflection infrared spectroscopy of proteins and lipids in biological membranes. Biochim Biophys Acta 1422(2):105–185. doi:10.1016/S0304-4157(99)00004-0
3. Hancock REW (1999) Hancock laboratory methods (visited July 2015). http://www.cmdr.ubc.ca/bobh/methods.htm
4. Niederman RA, Gibson KD (1971) The separation of chromatophores from the cell envelope in *Rhodopseudomonas spheroides.* Prep Biochem 1(2):141–150
5. Jarosławski S, Duquesne K, Sturgis JN, Scheuring S (2009) High-resolution architecture of the outer membrane of the Gram-negative bacteria *Roseobacter denitrificans.* Mol Microbiol 74(5):1211–1222. doi:10.1111/j.1365-2958.2009.06926.x
6. Deisenhofer J, Epp O, Miki K, Huber R, Michel H (1984) X-ray structure analysis of a membrane protein complex. Electron density map at 3 A resolution and a model of the chromophores of the photosynthetic reaction center from *Rhodopseudomonas viridis.* J Mol Biol 180(2):385–398
7. Ebel C (2011) Sedimentation velocity to characterize surfactants and solubilized membrane proteins. Methods 54(1):56–66. doi:10.1016/j.ymeth.2010.11.003
8. Fleming KG (2008) Determination of membrane protein molecular weight using sedimentation equilibrium analytical ultracentrifugation. Curr Protoc Protein Sci Chapter 7:Unit 7.12.1–7.12.13. doi:10.1002/0471140864.ps0712s53
9. Anatrace (2015) Anatrace company web site (visited July 2015). http://www.anatrace.com
10. Sigma Aldrich (2015) Sigma Aldrich company web site (visited July 2015). http://www.sigmaaldrich.com
11. Lin TL, Chen SH, Roberts MF (1987) Thermodynamic analyses of the structure and growth of asymmetric linear short-chain lecithin micelles based on small-angle neutron scattering data. J Am Chem Soc 109(8):2321–2328. doi:10.1021/ja00242a013
12. Le Maire M, Champeil P, Møller JV (2000) Interaction of membrane proteins and lipids with solubilizing detergents. Biochim Biophys Acta 1508(1–2):86–111
13. Focher B, Savelli G, Torri G, Vecchio G, McKenzie D, Nicoli D, Bunton C (1989) Micelles of 1-alkyl glucoside and maltoside: anomeric effects on structure and induced chirality. Chem Phys Lett 158(6):491–494
14. Clarke JA, Heron AJ, Seddon JM, Law RV (2006) The diversity of the liquid ordered (Lo) phase of phosphatidylcholine/cholesterol membranes: a variable temperature multinuclear solid-state NMR and x-ray diffraction study. Biophys J 90(7):2383–2393. doi:10.1529/biophysj.104.056499
15. Keller S, Heerklotz H, Jahnke N, Blume A (2006) Thermodynamics of lipid membrane solubilization by sodium dodecyl sulfate. Biophys J 90(12):4509–4521. doi:10.1529/biophysj.105.077867
16. Sot J, Manni MM, Viguera AR, Castaneda V, Cano A, Alonso C, Gil D, Valle M, Alonso A, Goni FM (2014) High-melting lipid mixtures and the origin of detergent-resistant membranes studied with temperature-solubilization diagrams. Biophys J 107(12):2828–2837. doi:10.1016/j.bpj.2014.10.063
17. Lichtenberg D, Ahyayauch H, Goni FM (2013) The mechanism of detergent solubilization of lipid bilayers. Biophys J 105(2):289–299. doi:10.1016/j.bpj.2013.06.007
18. DaCosta CJB, Baenziger JE (2003) A rapid method for assessing lipid:protein and detergent:protein ratios in membrane-protein crystallization. Acta Crystallogr D Biol Crystallogr 59(1):77–83. doi:10.1107/S0907444902019236
19. DuBois M, Gilles KA, Hamilton JK, Rebers PA, Smith F (1956) Colorimetric method for determination of sugars and related substances. Anal Chem 28(3):350–356. doi:10.1021/ac60111a017
20. Duquesne K, Sturgis JN (2010) Membrane protein solubilization. Methods Mol Biol (Clifton, NJ) 601:205–217. doi:10.1007/978-1-60761-344-213
21. Taussky HH, Shorr E (1953) A microcalorimetric method for the determination of inorganic phosphorous. J Biol Chem 202:675–685

22. Bradford MM (1976) A rapid and sensitive method for the quantitation of microgram quantities of protein utilizing the principle of protein-dye binding. Anal Biochem 72:248–254. doi:10.1016/0003-2697(76)90527-3
23. Popot JL (2014) Amphipols: where from? Where to? J Membr Biol 247(9-10):755–757
24. Breyton C, Gabel F, Abla M, Pierre Y, Lebaupain F, Durand G, Jl P, Ebel C, Pucci B (2009) Micellar and biochemical properties of (hemi) fluorinated surfactants are controlled by the size of the polar head. Biophys J 97(4):1077–1086. doi:10.1016/j.bpj.2009.05.053
25. Bayburt TH, Grinkova YV, Sligar SG (2002) Self-assembly of discoidal phospholipid bilayer nanoparticles with membrane scaffold proteins. Nano Lett 2(8):853–856. doi:10.1021/nl025623k
26. Knowles TJ, Finka R, Smith C, Lin YP, Dafforn T, Overduin M (2009) Membrane proteins solubilized intact in lipid containing nanoparticles bounded by styrene maleic acid copolymer. J Am Chem Soc 131(22):7484–7485. doi:10.1021/ja810046q
27. Scheidelaar S, Koorengevel MC, Pardo JD, Meeldijk JD, Breukink E, Killian JA (2015) Molecular model for the solubilization of membranes into nanodisks by styrene maleic acid copolymers. Biophys J 108(2):279–290. doi:10.1016/j.bpj.2014.11.3464

Chapter 16

Detergent-Free Membrane Protein Purification

Alice J. Rothnie

Abstract

Membrane proteins are localized within a lipid bilayer; in order to purify them for functional and structural studies the first step must involve solubilizing or extracting the protein from these lipids. To date this has been achieved using detergents which disrupt the bilayer and bind to the protein in the transmembrane region. However finding conditions for optimal extraction, without destabilizing protein structure, is time consuming and expensive. Here we present a recently-developed method using a styrene-maleic acid (SMA) co-polymer instead of detergents. The SMA co-polymer extracts membrane proteins in a small disc of lipid bilayer which can be used for affinity chromatography purification, thus enabling the purification of membrane proteins while maintaining their native lipid bilayer environment.

Key words Membrane proteins, Solubilization, Purification, SMALP, Polymer, Nanodisc, Detergent

1 Introduction

Transmembrane proteins carry out a wide range of vital roles, including controlling what enters and leaves a cell and mediating intracellular communication. Consequently they are the target of a large number of prescribed drugs. However understanding exactly what they look like and how they work is more difficult than for soluble proteins due to their location within a membrane bilayer. To purify membrane proteins they must be extracted or solubilized from the lipid bilayer. To date this has been achieved using detergents/surfactants, which destabilize the membrane and interact with the membrane protein, creating a micellar structure around the regions that would normally be in the membrane. This approach has proven successful for many proteins, including all those for which an X-ray crystal structure has been obtained so far. However the use of detergents is not without difficulties, such as (1) striking the right balance between efficient extraction from the membrane without also denaturing the protein; (2) stripping away annular lipids from the protein which are crucial for function; and (3) loss of lateral pressure provided by the membrane which affects

Isabelle Mus-Veteau (ed.), *Heterologous Expression of Membrane Proteins: Methods and Protocols,* Methods in Molecular Biology, vol. 1432, DOI 10.1007/978-1-4939-3637-3_16,

both structure and function. Exactly which detergent will work best for a given protein cannot be easily predicted and therefore tends to involve a lengthy and expensive trial-and-error process, further complicated by the fact that the best detergents for extraction are not always the best for downstream applications. Recently we, and others, have demonstrated a new approach to membrane protein extraction/solubilization without the use of detergents, instead using a styrene-maleic acid (SMA) co-polymer [1–9]. The SMA co-polymer inserts into a biological membrane and forms small discs of bilayer encircled by the polymer (Fig. 1b) [10], which we term SMA lipid particles (SMALPs), but are also called lipodisqs [8, 9] or native nanodiscs [5]. SMALPs are small, soluble and stable, and proteins within SMALPs can easily be purified using affinity chromatography. SMALP-encapsulated proteins have been shown to be more thermostable than detergent-solubilized proteins [1, 3, 5]. The small-sized SMALPs do not significantly scatter light like proteoliposomes, and the polymer does not provide a large interfering signal, making them useful for various spectroscopic and biophysical techniques [1–4, 9, 11]. As both sides of the membrane are freely available they are ideal for membrane protein-binding assays [1, 3]. They have also been successfully utilized for structural studies using either negative stain or

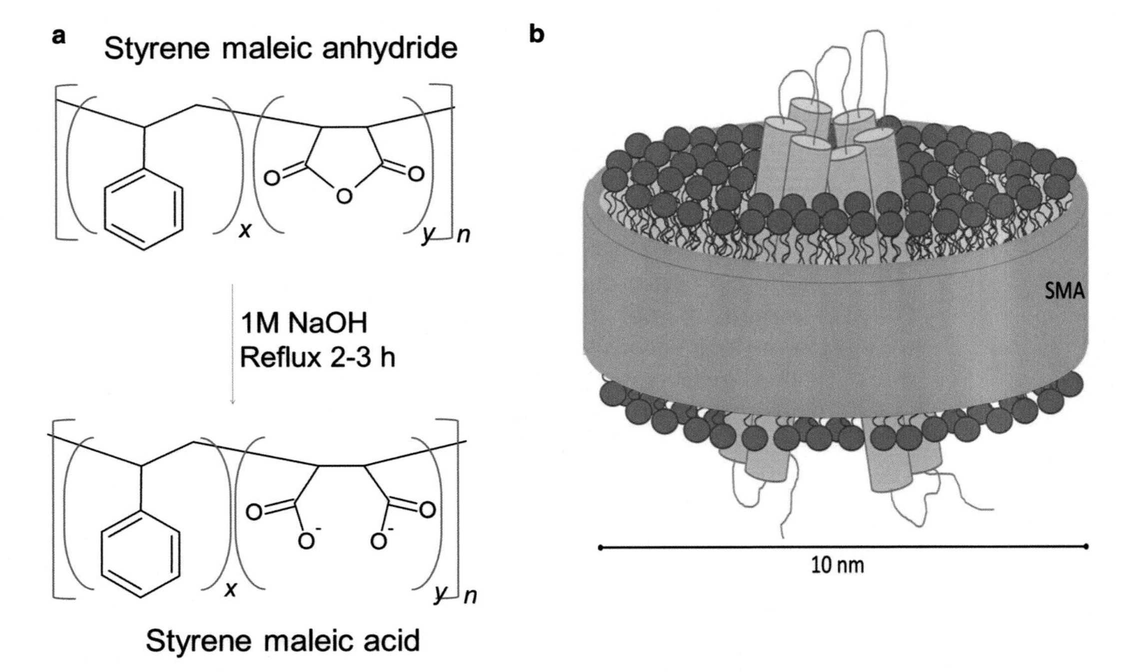

Fig. 1 SMA co-polymer and SMALP structures. (**a**) The SMA2000 polymer (Cray Valley) is a styrene-maleic anhydride co-polymer that must be hydrolyzed to form a styrene-maleic acid co-polymer. (**b**) The SMA co-polymer encircles a disc of lipid bilayer, effectively solubilizing the transmembrane protein while maintaining its lipid environment

cryo-electron microscopy [1, 7]. It is therefore possible to extract, purify and study the structure and function of a membrane protein while retaining its natural bilayer environment.

2 Materials

2.1 SMA Polymer Preparation

1. SMA2000 (styrene-maleic anhydride) powder: This polymer has a 2:1 ratio of styrene:maleic anhydride and a molecular weight of 7.5 kDa (Cray Valley).
2. 1 M NaOH solution.
3. Concentrated HCl (SG 1.18).
4. 0.6 M NaOH solution.
5. Distilled water.

2.2 Membrane Protein Extraction and Purification

1. Membrane preparations from cells expressing the target protein (*see* **Note 1**).
2. Buffer 1: 20 mM Tris–HCl pH 8, 150 mM NaCl, 10% (v/v) glycerol (*see* **Note 2**).
3. Ni^{2+}-NTA (Ni^{2+}-nitrilotriacetate) agarose resin, and an empty gravity flow column.
4. 2 M Imidazole.
5. Standard SDS-PAGE and Western blotting equipment and reagents.

3 Methods

3.1 SMA Co-polymer Preparation

SMA2000 is a styrene-maleic anhydride co-polymer. To be active for membrane solubilization it must be hydrolyzed to form styrene-maleic acid (Fig. 1a).

1. Dissolve 25 g SMA2000 powder in 250 ml 1 M NaOH overnight at room temperature using a magnetic stirrer and a round-bottomed flask (*see* **Note 3**).
2. In a fume hood put the round-bottomed flask containing the dissolved SMA2000 on a heating mantle and attach a condenser. Bring the solution to a boil and then reflux the polymer solution for 2–3 h. Allow to cool.
3. Divide the polymer solution between four 250 ml centrifuge tubes. In a fume hood gradually add concentrated HCl to each one, mixing well, to precipitate the polymer. Approximately 1 ml HCl per 6 ml polymer solution is required. Then add 100 ml distilled water to each tube and mix well.
4. Centrifuge at 10,000 × *g* for 10 min at room temperature and carefully pour off supernatant.

5. Add 150 ml distilled water to each tube and resuspend the polymer by shaking.
6. Centrifuge at 10,000 × *g* for 10 min at room temperature and carefully pour off supernatant.
7. Repeat **steps 5** and **6** four times.
8. Dissolve the polymer by adding 60 ml 0.6 M NaOH to each tube and either shaking or stirring for several hours.
9. Check the pH and adjust to pH 8.
10. Freeze-dry the SMA co-polymer.
11. Store at room temperature.

3.2 Membrane Protein Extraction and Purification

1. Resuspend the membrane preparation in buffer 1 at a concentration of 30 mg/ml wet weight of membrane pellet (*see* **Note 4**).
2. Add SMA co-polymer powder (from Subheading 3.1, **step 11**) to the resuspended membranes to give a final concentration of 2.5 % (w/v) (*see* **Notes 5** and **6**).
3. Incubate at room temperature for 1 h, shaking (*see* **Notes 7** and **8**).
4. Centrifuge at 100,000 × *g* for 20 min at 4 °C and harvest the supernatant containing the solubilized protein.
5. Measure solubilization efficiency by running a Western blot of solubilized sample against insoluble (resuspend the pellet in buffer 1 supplemented with 2 % (w/v) SDS) (Fig. 2a).
6. Mix the solubilized protein with Ni^{2+}-NTA resin (pre-washed in buffer 1), at a ratio of 100 μl resin/ml solubilized protein, and mix gently overnight at 4 °C (*see* **Notes 9** and **10**).
7. Pour into an empty gravity-flow column and wash the resin five times with 10 bed volumes (bv) of buffer 1 supplemented with 20 mM imidazole (*see* **Note 11**).
8. Wash resin twice with 10 bv of buffer 1 supplemented with 40 mM imidazole.
9. Wash once with 1 bv of buffer 1 supplemented with 60 mM imidazole.
10. Elute six times with ½ bv of buffer 1 supplemented with 200 mM imidazole.
11. Run samples from each step on SDS-PAGE and stain (either silver stain or Coomassie, depending on the abundance of the protein) (Fig. 2b).
12. Pool elution fractions containing the protein of interest, and remove any remaining free SMA by gel filtration and/or concentrate the sample using centrifugal concentrators (*see* **Note 12**).
13. Store sample for short term at 4 °C, or long term at −70 °C.

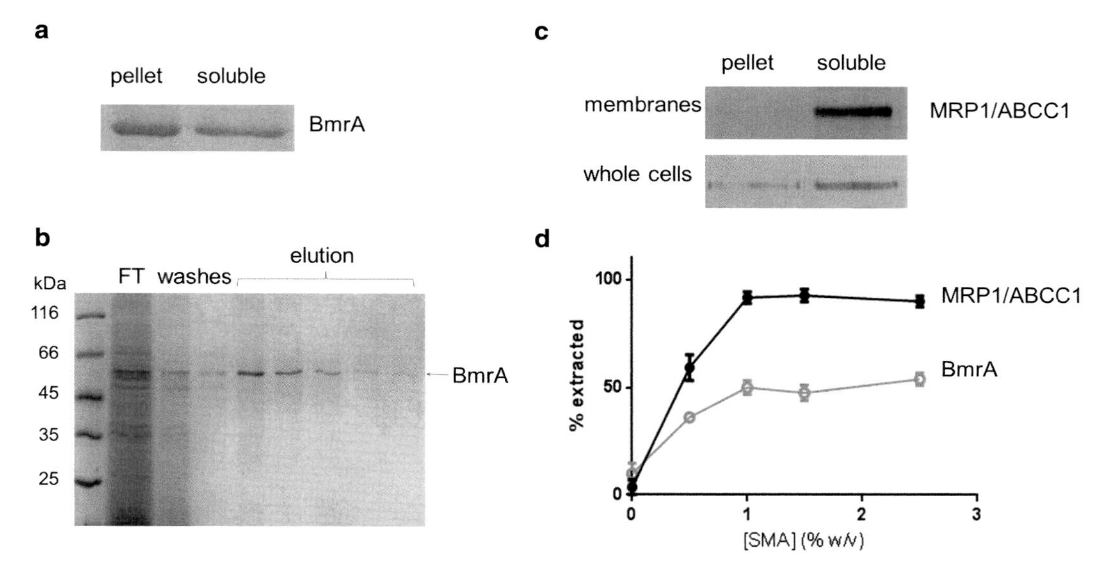

Fig. 2 Extraction and purification using SMA co-polymer. (**a**) Membrane preparations from C41 (DE3) E. coli overexpressing the ABC transporter BmrA were solubilized with 2.5 %(w/v) SMA for 1 h at room temperature, and then centrifuged at 100,000 g for 20 min. Samples of soluble and insoluble material were assayed by Western blotting using an anti-his primary antibody. (**b**) SMA-solubilized BmrA was purified using Ni^{2+}-NTA affinity chromatography. Samples of unbound protein, washes, and eluted protein were run on SDS-PAGE and stained with InstantBlue (Expedeon). (**c**) Human ABC transporter MRP1/ABCC1 overexpressed in H69AR cancer cells was extracted either from membrane preparations or whole cells. Soluble and insoluble fractions were analyzed by Western blot, using QCRL-1 as a primary antibody. (**d**) BmrA (open circles, grey) and MRP1 (closed circles, black) were solubilized with varying concentrations of SMA, and the % extracted (solubilized) analyzed by Western blot as in A & C, data are mean $\pm$ sem, $n \geq 3$

4 Notes

1. Membrane preparations are ideal, but whole cells can also be used (Fig. 2c) [5, 6], which will require addition of DNase. Membranes from all common expression systems (bacteria, insect cells, yeast, and mammalian cells [1] (Fig. 2)) work effectively. This protocol details the method for polyhistidine-tagged proteins, but other affinity chromatography methods can be used [1, 6].
2. The composition of this buffer is generally quite flexible, but the pH is important; a pH of 8 is ideal, certainly no lower than 7.5. Also divalent cations should be avoided. A low pH or divalent cations will prevent efficient extraction with the SMA co-polymer, or if added once the SMALP is formed, may cause it to precipitate.
3. Weigh out the SMA2000 in a fume hood. If some of the SMA2000 does not dissolve this does not matter. Always wear gloves when handling SMA since it can penetrate skin.

4. Many methods using detergents use a specific concentration of total membrane protein; however we use wet pellet weight of the membrane not protein concentration as it is the lipids the SMA co-polymer interacts with. To measure wet pellet weight, weigh an empty ultracentrifuge tube, then add your membrane, spin at 100,000 × *g* to pellet your membranes, carefully remove all of the supernatant, and weigh the tube again. The difference in weight from the empty tube gives you the wet pellet weight.
5. We simply add powder to the membrane suspension, because it is more convenient for long-term storage of the polymer, but alternatively a concentrated stock solution in buffer 1 can be prepared and mixed with the membrane suspension.
6. Although we use 2.5 % (w/v) SMA co-polymer as a standard, successful solubilization can be achieved with lower concentrations (Fig. 2d).
7. Although the solution will noticeably appear clearer almost instantly, we have found that it often takes longer to achieve a good extraction of your membrane protein of interest. While 1 h at room temperature appears sufficient for most proteins we have tested [1], some proteins may require longer [3, 7].
8. The temperature is important because of the phase transition of the lipids. Since the SMA co-polymer interacts with the lipids rather than the protein it is important the temperature is above the phase-transition temperature so that the lipids are in the liquid phase [12]. While this may worry many membrane protein researchers who are used to maintaining everything at 4 °C during detergent solubilization, we have not found this to be a problem, presumably because the SMALP maintains stability of membrane proteins much better than detergents [1, 3]. In fact, for proteins that prove difficult to solubilize increasing the temperature to 37 °C for solubilization has been used [1, 3].
9. Binding of polyhistidine-tagged SMALP-encapsulated proteins to the Ni^{2+}-NTA resin is sometimes problematic. Possible reasons include interactions between the polyhistidine-tag and the co-polymer, steric hindrance from the co-polymer, or column spoiling by excess free SMA co-polymer. We have found that a dodeca-histidine tag is much better than a hexa-histidine tag for efficient binding. Sometimes increasing the concentration of NaCl in the buffer can improve binding, or decreasing the concentration of SMA co-polymer in the sample may help. We have also found that resins from different suppliers can affect binding efficiency in a protein-dependent manner; for example Ni^{2+}-NTA agarose (Qiagen) is better for some proteins tested, whereas HisPur Ni^{2+}-NTA (ThermoFisher) is better for others.
10. Rather than using Ni^{2+}-NTA resin and a gravity flow column, it is possible to use a HisTrap FF column (GE Healthcare) and the Akta system. In this case, load the column very slowly,

wash with buffer 1 until the A_{280} returns to baseline, and then elute using an imidazole gradient from 20 to 200 mM.

11. One of the biggest advantages of this method is that once formed the SMALPs are stable, and, unlike with detergent, it is not necessary to supplement the purification buffers or assay buffers with further SMA co-polymer.
12. Another big advantage of this method is that the protein can be concentrated easily using centrifugal concentrators. This is unlike detergent-solubilized proteins where problems are caused by simultaneously concentrating the detergent, protein is lost by sticking to the membranes of the concentrator, and the protein is prone to aggregation.

References

1. Gulati S, Jamshad M, Knowles TJ, Morrison KA, Downing R, Cant N, Collins R, Koenderink JB, Ford RC, Overduin M, Kerr ID, Dafforn TR, Rothnie AJ (2014) Detergent-free purification of ABC (ATP-binding-cassette) transporters. Biochem J 461:269–278
2. Knowles TJ, Finka R, Smith C, Lin YP, Dafforn T, Overduin M (2009) Membrane proteins solubilized intact in lipid containing nanoparticles bounded by styrene maleic acid copolymer. J Am Chem Soc 131:7484–7485
3. Jamshad M, Charlton J, Lin Y P, Routledge S J, Bawa Z, Knowles T J, Overduin M, Dekker N, Dafforn T R, Bill R M, Poyner D R, Wheatley M (2015) G-protein coupled receptor solubilization and purification for biophysical analysis and functional studies, in the total absence of detergent. Biosci Rep 35(2). pii: e00188
4. Swainsbury DJ, Scheidelaar S, van Grondelle R, Killian JA, Jones MR (2014) Bacterial reaction centers purified with styrene maleic acid copolymer retain native membrane functional properties and display enhanced stability. Angew Chem Int Ed Engl 53:11803–11807
5. Dorr JM, Koorengevel MC, Schafer M, Prokofyev AV, Scheidelaar S, van der Cruijsen EA, Dafforn TR, Baldus M, Killian JA (2014) Detergent-free isolation, characterization, and functional reconstitution of a tetrameric K^+ channel: the power of native nanodiscs. Proc Natl Acad Sci U S A 111:18607–18612
6. Paulin S, Jamshad M, Dafforn TR, Garcia-Lara J, Foster SJ, Galley NF, Roper DI, Rosado H, Taylor PW (2014) Surfactant-free purification of membrane protein complexes from bacteria: application to the staphylococcal penicillin-binding protein complex PBP2/PBP2a. Nanotechnology 25
7. Postis V, Rawson S, Mitchell JK, Lee SC, Parslow RA, Dafforn TR, Baldwin SA, Muench SP (2015) The use of SMALPs as a novel membrane protein scaffold for structure study by negative stain electron microscopy. Biochim Biophys Acta 1848:496–501
8. Long AR, O'Brien CC, Malhotra K, Schwall CT, Albert AD, Watts A, Alder NN (2013) A detergent-free strategy for the reconstitution of active enzyme complexes from native biological membranes into nanoscale discs. BMC Biotechnol 13:41
9. Orwick-Rydmark M, Lovett JE, Graziadei A, Lindholm L, Hicks MR, Watts A (2012) Detergent-free incorporation of a seven-transmembrane receptor protein into nanosized bilayer Lipodisq particles for functional and biophysical studies. Nano Lett 12:4687–4692
10. Jamshad M, Grimard V, Idini I, Knowles TJ, Dowle MR, Schofield N, Sridhar P, Lin YP, Finka R, Wheatley M, Thomas ORT, Palmer RE, Overduin M, Govaerts C, Ruysschaert JM, Edler KJ, Dafforn TR (2015) Structural analysis of a nanoparticle containing a lipid bilayer used for detergent-free extraction of membrane proteins. Nano Res 8:774–789
11. Sahu ID, McCarrick RM, Troxel KR, Zhang R, Smith HJ, Dunagan MM, Swartz MS, Rajan PV, Kroncke BM, Sanders CR, Lorigan GA (2013) DEER EPR measurements for membrane protein structures via bifunctional spin labels and lipodisq nanoparticles. Biochemistry 52:6627–6632
12. Scheidelaar S, Koorengevel MC, Pardo JD, Meeldijk JD, Breukink E, Killian JA (2015) Molecular model for the solubilization of membranes into nanodisks by styrene maleic Acid copolymers. Biophys J 108:279–290

Chapter 17

Conformational Dynamics and Interactions of Membrane Proteins by Hydrogen/Deuterium Mass Spectrometry

Eric Forest and Petr Man

Abstract

Hydrogen/deuterium exchange associated with mass spectrometry has been recently used to characterize the dynamics and the interactions of membrane proteins. Here we describe experimental workflow enabling localization of the regions involved in conformational changes or interactions.

Key words Hydrogen/deuterium exchange, Mass spectrometry, Conformational change, Dynamics, Interaction, Membrane protein

1 Introduction

Hydrogen/deuterium exchange associated with mass spectrometry (HDX-MS) plays a major role in the fine characterization of proteins and helps to investigate the interplay of structure, function, and dynamics [1]. HDX-MS can locate conformational changes linked to mutations, interaction with a ligand (e.g., protein or membrane), or occurring in functioning mechanisms (activation, transport, etc.). The studies take advantage of the mass spectrometry features: sensitivity (tens to hundreds of μg are necessary), possibility to study large proteins or complexes (hundreds of kDa) in close to physiological conditions (buffers or detergents), with a few (or even single) amino acid [2] resolution, in a short time scale (days to weeks).

In HDX-MS experiments, the protein is diluted in the buffer of choice where H_2O is replaced by D_2O to let the amide hydrogens exchange against deuteriums. After variable incubation times, the exchange is quenched by decreasing the pH and the temperature, where the back-exchange is minimum [3]. The protein is then quickly digested with acidic proteases working in these conditions. The masses of the deuterated peptides are measured online after their separation by HPLC or UPLC. One mass unit increase of a peptide

Isabelle Mus-Veteau (ed.), *Heterologous Expression of Membrane Proteins: Methods and Protocols*, Methods in Molecular Biology, vol. 1432, DOI 10.1007/978-1-4939-3637-3_17, © Springer Science+Business Media New York 2016

indicates the exchange of one amide hydrogen by deuterium. The exchange depends on the accessibility of the amide to the solvent and its implication in hydrogen bonds (secondary and tertiary structure). For instance, when a peptide in a specific state of the protein shows a different exchange than in another state, it indicates that this peptide belongs to a region whose dynamics is influenced by the state change.

Membrane proteins have been studied only rather recently using HDX-MS. This is due to their hydrophobic character which makes them difficult to characterize by biophysical techniques in general and MS in particular. With the peripheral and the amphitropic proteins, phospholipid vesicles have been generally used to mimic the cell membrane and enable the partial insertion of the protein. HDX-MS has already been used since some years to characterize their conformation and their interaction with the membrane [4–6], in spite of the high amount of lipids giving intense MS signals. In contrast, the integral membrane proteins are usually supplemented with detergents to replace their natural membrane environments. The first reports describing the use of HDX-MS for this type of membrane proteins are more recent [7–10]. However the use of lipid membrane mimetics for the integral membrane proteins has also been described [11, 12].

Lipids or detergents are covering the hydrophobic parts of the protein, which render them inaccessible for the protease. Therefore, digestion may not be trivial and larger and highly hydrophobic fragments are often produced. It is advisable to monitor the digestion trials using SDS-PAGE and/or by MALDI-TOF MS to ensure that the digestion is quantitative. Full sequence coverage, optimal length of the peptides (less than ten amino acids), and overlapping peptides represent the major goals of the first experimental step detailed below, after the instrumental setup.

In the following sections, description of the HDX-MS protocol including automated labeling, MS analysis, and data processing is provided.

2 Materials

1. A PAL HTC-xt autosampler (CTC Analytics) equipped with an Injection Unit (*see* **Note 1**), a Stack Cooler (Peltier cooled), a Fast Wash Station, and a Needle Port (*see* **Note 2**).
2. A 54-vial (2 mL) tray (tray 1) and a PCR cooler with a 96-well PCR plate (tray 2) (Eppendorf) into the stack cooler tuned at 4 °C (Fig. 1).
3. An electronically actuated 6-port valve (injection valve, ***Inj V)*** (Idex Health & Science) placed in a polystyrene box cooled at 4 °C with a Peltier cooling device (UWE electronic).
4. A protection 0.2 μm filter (Agilent), an ACQUITY UPLC Protein BEH C_{18} VanGuard pre-column, 300 Å, 1.7 μm,

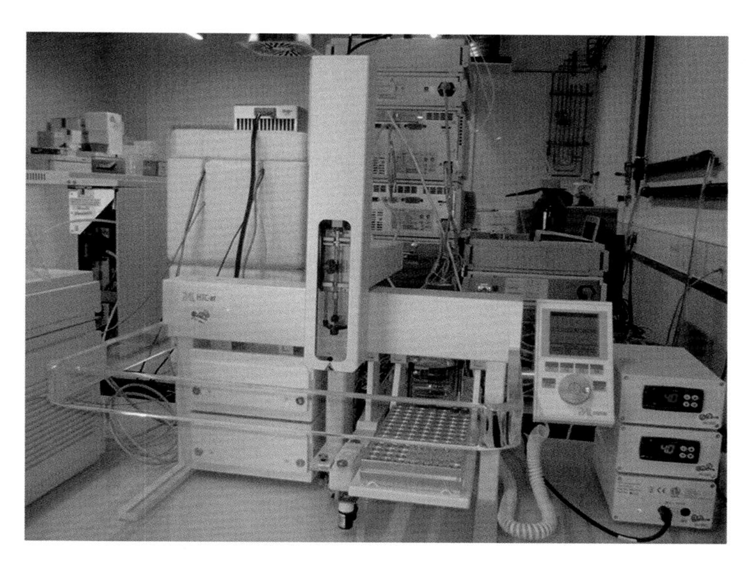

Fig. 1 Photograph of the PAL autosampler

2.1 mm × 5 mm (trap) (Waters), an ACQUITY UPLC Peptide BEH C_{18} column, 300 Å, 1.7 μm, 1 mm × 100 mm (Waters), and an electronically actuated 10-port valve (Switching Valve, ***Sw V)*** (Idex Health & Science).

5. A 1200 HPLC pump (Agilent), a 1290 Infinity UPLC pump (Agilent), and an electrospray mass spectrometer (*see* **Note 3**).
6. 1/16″ External and 0.005″ or 0.002″ internal tubing, in stainless steel or PEEK.
7. 1 mL Conical bottom vials.
8. 1 M Glycine-HCl buffer: 3.75 g of glycine in 50 mL ultrapure water, set pH to 2.3 by the addition of concentrated hydrochloric acid.
9. Hydrochloric acid (HCl).
10. Acetonitrile.
11. Formic acid.
12. Protein buffer (*see* **Note 4**).
13. Deuterium oxide (D_2O).
14. Porcine pepsin A (Sigma).
15. Rizhopuspepsin (Sigma or prepared recombinantly) [13, 14].
16. Protease from *Aspergillus saitoi* (Sigma) [13].
17. Nepenthesins (concentrated pitcher fluid or recombinant preparation) [15–18].
18. Membrane protein to be studied.
19. HDExaminer software (Sierra Analytics).

3 Methods

3.1 Instrumental Setup (See Note 5)

1. Connect the Needle Port of the autosampler to ***Inj V*** placed in the box cooled at 4 °C. Connect a 200 μL loop to Inj V as shown in Figs. 2 and 3.
2. In the cooled box, connect the protection 0.2 μm filter, the trap (pre-column), the UPLC column, and ***Sw V***; *see* Figs. 2 and 3.
3. Connect the HPLC pump, the UPLC pump, and the electrospray source of the mass spectrometer to the valves as indicated in Figs. 2 and 3.
4. Use 1/16″ external and 0.005″ internal tubing, except for the connections downstream UPLC column-***Sw V*** and Sw V-electrospray source (0.002″). Use stainless steel to connect pumps and valves with 1 m coiled in the cooled box to cool the solvents. Also use stainless steel for the connections trap-***Sw V*** and Sw V-upstream UPLC column to stand the high pressure. Use PEEK for all other connections.
5. Use the HPLC pump at 200 μL/min, with 0.4 % formic acid in water, to inject the peptides loaded in the loop and to trap and desalt them with the trap column.
6. Use the UPLC pump at 50 μL/min to elute the trapped peptides and separate them with the UPLC column with gradient 15–70 % B in 10 min (A: 0.4 % formic acid in water; B: 95 % acetonitrile, 5 % water, 0.4 % formic acid).
7. Electronically connect the autosampler with both valves, the mass spectrometer, and both pumps.
8. Program the mass spectrometry acquisition(s) with external start (given by the autosampler) and monitoring of both pumps.

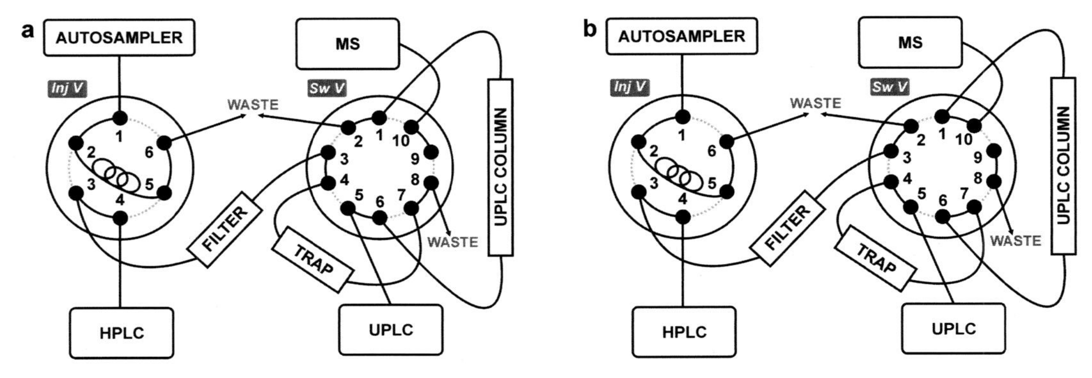

Fig. 2 Scheme of the fluidic connections between the different instruments used in the HDX-MS experiments, with injection valve in position load; switching valve in position wash (**a**) and elution (**b**)

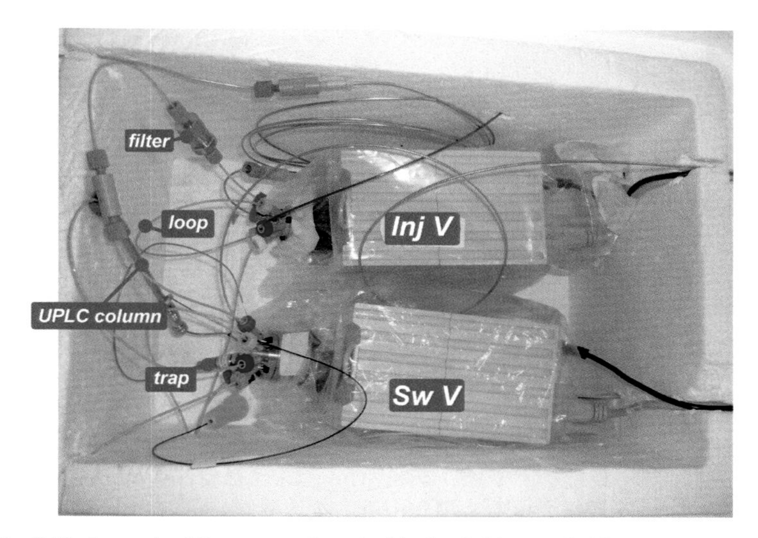

Fig. 3 Photograph of the connections inside the Peltier-cooled box

3.2 Optimizing Digestion Conditions

1. Dilute the protein of interest in the buffer suitable for the protein (*see* **Note 4**) to make 10–20 μM solution.
2. Prepare 50 mL of 1 M glycine-HCl buffer (quench solution) by dissolving 3.75 g of glycine in ultrapure water and set pH to 2.3 by the addition of concentrated hydrochloric acid.
3. Prepare 100 mM, 250 mM, and 500 mM glycine-HCl buffer by diluting the 1 M stock solution. Using these solutions test quenching conditions by mixing equal volumes of the glycine buffer and the buffer in which the protein is supplied. The goal is to stop (quench) the HDX reaction by setting pH between 2.3 and 2.5, where the HDX is minimal. Check the final pH using pH meter. Eventually try other dilutions. For further experiments use the lowest concentration of glycine-HCl required to shift the pH to 2.3–2.5 (*see* **Note 6**).
4. Prepare pepsin solution by dissolving 10 mg of porcine pepsin A in 1 mL of glycine-HCl buffer. Keep the solution on ice. Calculate the amount of pepsin for in-solution digestion. Starting protein:protease ratio 1:1 (w:w). Mix required amount of pepsin with glycine-HCl buffer to reach final volume of 1 mL (*see* **Note 7**).
5. Fill a 1 mL conical bottom vial with 10 μL protein sample per planned experiment and put it in position 1 of tray 1. Fill a 1 mL vial with 1 mL protein buffer and put it in position 10 of tray 1. Fill a 1 mL vial with 1 mL quench solution with protease and put it in position 12 of tray 1.
6. Program a method with the following steps, using the PAL Cycle Composer software (CTC Analytics).

7. Set ***Sw V*** in the wash position (Fig. 2a).
8. In the sample list, indicate the mixing well in tray 2: n (between 1 and 96).
9. Wash the 100 μL syringe with 0.4% formic acid in the wash station.
10. Aspirate 10 μL of protein sample in vial 1 of tray 1 and dispense 10 μL in well n of tray 2 (PCR plate) (*see* **Note 8**).
11. Wash the 100 μL syringe with 0.4% formic acid in the wash station.
12. Aspirate 40 μL of protein buffer in vial 10 of tray 1 and dispense 40 μL in well n of tray 2. Aspirate and dispense twice in well n of tray 2 to mix.
13. Wash the 100 μL syringe with 0.4% formic acid in the wash station.
14. Aspirate 50 μL of quench buffer in position 12 of tray 1 and dispense 50 μL in position n of tray 2.
15. Aspirate and dispense twice in position n of tray 2 to mix.
16. Wait 1 min (digestion) (*see* **Note 9**).
17. Set ***Inj V*** in the load position (Fig. 2a and 2b).
18. Aspirate 100 μL of the digested protein in position n of tray 2 and inject to the port.
19. Aspirate 100 μL of 0.4% formic acid in the wash station and inject to the port (washing of the transfer tubing).
20. Set ***Inj V*** in the inject position (ports 2–3, 4–5, and 6-1 connected in Fig. 2a and 2b).
21. Wash the 100 μL syringe with 0.4% formic acid in the wash station.
22. Wait for 3 min (desalting).
23. Set ***Sw V*** in the elution position (Fig. 2b).
24. Send signal Start to the mass spectrometer (beginning of the mass spectra acquisition and of the UPLC gradient).
25. Wait for 10 min (*see* **Note 10**).
26. Set ***Sw V*** in the wash position to send the detergent to waste (disconnection of the UPLC column from the mass spectrometer).
27. Run LC-MS/MS analysis using the standard setting (*see* **Note 11**). In optimal case, use identical gradient as will be further used for the analysis of partially deuterated samples. For slower scanning mass spectrometers and/or highly complex peptide mixtures, longer gradient may be used for the purpose of peptide identification. However, in such case, retention times of the individual peptides must be corrected to fit the elution in the LC-MS runs.

28. Search the LC-MS/MS data using suitable program (e.g., MASCOT) against two protein database containing sequence of the protein of interest and sequence of porcine pepsin A. Set enzymatic cleavage to "none" and apply no taxonomic restriction. Transfer the MASCOT result to an Excel spreadsheet and create a list of identified peptides. Plot the results using DrawMap script, part of MSTools (http://ms.biomed.cas.cz/MSTools/DrawMap/DrawMap.php [19], to check sequence coverage, peptide length, and redundancy.
29. Redo the experiments with other proteases working in acidic conditions, such as rhizopuspepsin [13, 14], protease from *Aspergillus saitoi* (also known as protease type XIII or aspergillopepsin) [13], or nepenthesins [15–18].
30. Select the protease and the digestion conditions giving the best sequence coverage with suitably sized (e.g., less than ten amino acids in average) and overlapping peptides, which will result in the best spatial resolution.

3.3 HDX-MS Experiments

1. Fill a 1 mL vial with 1 mL deuterated buffer and put it in position 11 of tray 1.
2. Program a slightly different method, using the PAL Cycle Composer software (CTC Analytics), adding or changing the following steps.
3. Replace previous **step 12** by aspirating 40 μL of deuteration buffer in vial 11 of tray 1 and dispensing 40 μL in well n of tray 2. Aspirate and dispense twice in well n of tray 2 to mix.
4. Add this step before previous **step 14**: wait m min (from $m = 0$ for 1-min deuteration to $m = 19$ for 20-min deuteration) (*see* **Note 12**).
5. With different n values in the sample list, redo the experiment at different deuteration times (to plot kinetics) and in different states of the protein (at different conformations or with a partner) (*see* **Note 13**).
6. Set the mass spectrometer to acquire data in the full-scan mode only (LC-MS mode).

3.4 HDX-MS Data Processing Using HDExaminer Software (Sierra Analytics)

1. Do at least one nondeuterated experiment and a series of deuterated experiment.
2. Use Excel "import file" with three columns: amino acid sequence of the peptide, charge state (one line per charge state), retention time created in Subheading 3.2, **step 26**.
3. In HDExaminer software, open the FASTA file of the protein, and add peptide source (import file) and mass spectrometry files for nondeuterated and deuterated experiments which may be classified in different states of the protein.

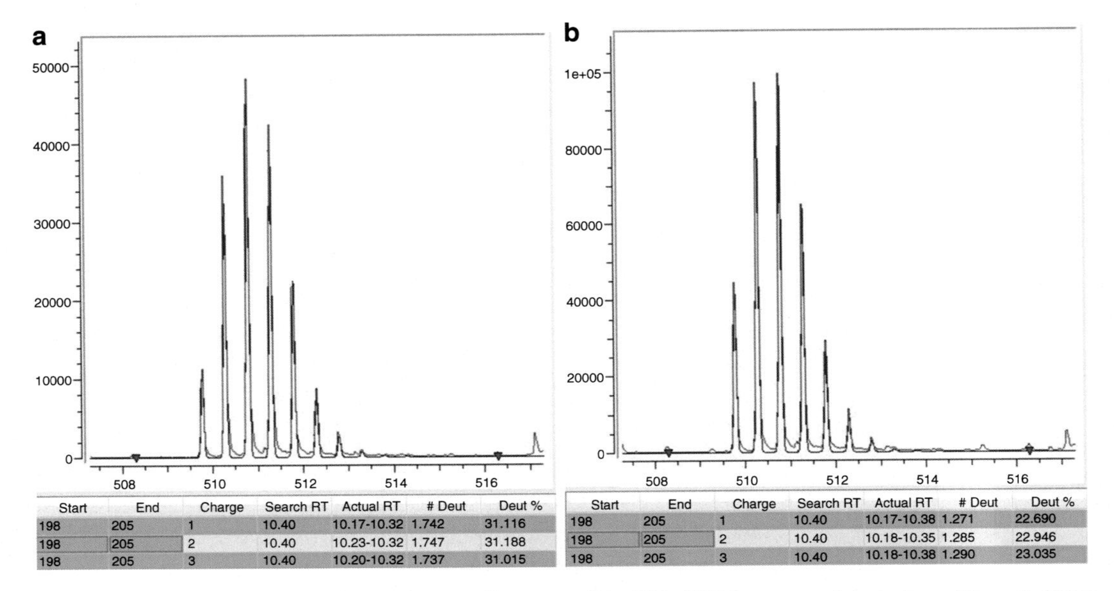

Fig. 4 Extract of HDExaminer deuteration results on peptide 198–205 from a protein in two different states (state 1 (**a**) and state 2 (**b**)). The *upper panels* show the superimposition of the experimental spectra with spectra calculated by the software. The *lower panels* indicate the first and last amino acids (Start and End), the charge states shown in the *upper panels* (2), the given (in the import file) retention time and the found one (Search RT and Actual RT), the number and percentage of deuteriums (# Deut and Deut %). A significant difference of percentage of deuterium is measured for this peptide: 31.1 % in state 1 and 22.9 % in state 2 (average of the three charge states)

4. Export the number of deuterium and the percentage of deuteration (*see* Fig. 4) for each peptide in each experiment into an Excel file. Export the kinetics of deuteration of each peptide for the different states of the protein. Export the heat map (rainbow color code) indicating the percentage of deuteration of each peptide on the primary structure or on the 3D structure, using Pymol software (if PDB file is available).
5. Locate the regions of interest, either showing a conformational change or a protection effect induced by the binding of a partner.

4 Notes

1. Either 100 μL or 250 μL syringes may be installed in the Injection Unit.
2. This model of autosampler is the simplest to fully automate the whole procedure. More sophisticated ones, equipped with two syringes, may also be used.
3. Alternatively, other HPLC or UPLC pumps may also be used.

4. H/D exchange is usually carried out in the buffer which is used for the other experiments. The method has very high tolerance to the buffer composition and concentration of the individual components. However if possible the buffer composition should be kept as simple as possible and the concentration of the individual components should also be kept at the lowest concentrations. Alkyl glycoside detergents (beta-octylglycoside, dodecyl maltoside, etc.) are well tolerated. Other nonionic detergents (Triton X100, polyoxyethylene-based, PEG, PPG) may also be used, with a more complex procedure using a chlorinated solvent [20].
5. Alternatively, the procedure described below in an automated way may be done manually, using pipettes and Eppendorf tubes (tubes stored in an ice bath); the Peltier-cooled box may be replaced by an ice-water bath and the electronic actuated valves by manual ones.
6. Lowering the pH can also be performed by, e.g., phosphoric or hydrochloric acid. Their concentration as well as ratio must be optimized with the protein buffer used in the study. Commonly used concentration range is between 50 mM and 250 mM and the buffer:quench solution ratio usually varies between 10:1 and 1:10.
7. Pepsin is quite tolerant to denaturing and reducing agents, so these can be used to enhance the efficiency of the digestion. However, high ionic strength (caused by the high concentration of guanidium hydrochloride) may cause precipitation of the hydrophobic membrane proteins. Therefore the optimal concentration of TCEP and guanidinium chloride must be found. In some cases, urea, thiourea, or their mixtures may also prove helpful. Also, keep in mind that phosphines (e.g., TCEP) are the only reducing agents working at low pH.
8. In case of diluted protein sample (below 5 μM), a 250 μL syringe may be used instead of a 100 μL one. The different aspirated volumes should be then multiplied by 2.5, except in **step 18**, and the injection loop should be changed to a 350 μL one instead of 200 μL.
9. It is possible to tune the digestion by changing the digestion time or to do online digestion: insert an immobilized protease column (e.g., commercial Poroszyme®, Applied Biosystems) between the both valves.
10. After this 10-min waiting time, the UPLC column is disconnected from the mass spectrometer. It thus avoids saturating the electrospray source with the detergent DDM that elutes at 10.5 min using the gradient indicated above. This time must be adjusted if the gradient is modified or if different alkyl glycoside detergent is used.

11. In this particular case an ESI-q-FT-ICR MS with 15 T superconducting magnet (SolariX XR, Bruker Daltonics) was used. The mass spectrometer was operated in a positive ion mode and data collected in data-dependent mode. Spectra were acquired over the mass range of 150–2000 *m/z*. Each full scan was followed by MS/MS of the six most intense ions. Dynamic exclusion with duration of 0.5 min was enabled. Data were processed in DataAnalysis 4.2 (Bruker Daltonics) using "Find Auto MSn compounds" feature and SNAP peak picking. Processed data were exported to a Mascot Generic File which was further used for MASCOT searches.
12. Shorter deuteration times than 1 min (down to 15 s) may be obtained with a slightly different method: dispense 50 μL of quench buffer in well n; aspirate 10 μL of protein, 10 μL of air, and 40 μL of deuteration buffer; dispense and mix in well $n+12$; aspirate in well $n+12$, dispense, and mix in well n. Longer incubation times, up to hours, can also be followed. However the length of the experiment is given by the stability of the studied protein.
13. It is possible to program a full series of experiments (nondeuterated experiment, replicates, and different deuteration times) by defining the methods and the different *n* values in a sample list in the Cycle Composer software and by programming them in a sample list in the mass spectrometry acquisition software. Longer incubation times, up to hours, can also be followed. However the length of the experiment is given by the stability of the studied protein.

Acknowledgements

PM acknowledges, LQ1604 NPU II provided by MEYS and CZ.1.05/1.1.00/02.0109 BIOCEV provided by ERDF and MEYS and grant from Charles University in Prague (GAUK 389115).

References

1. Konermann L, Pan JX, Liu YH (2011) Hydrogen exchange mass spectrometry for studying protein structure and dynamics. Chem Soc Rev 40:1224–1234
2. Rand KD, Zehl M, Jensen ON, Jorgensen TJD (2009) Protein hydrogen exchange measured at single-residue resolution by electron transfer dissociation mass spectrometry. Anal Chem 81:5577–5584
3. Bai YW, Milne JS, Mayne L, Englander SW (1993) Primary structure effects on peptide group hydrogen-exchange. Proteins 17:75–86
4. Man P, Montagner C, Vernier G, Dublet B, Chenal A, Forest E, Forge V (2007) Defining the interacting regions between apomyoglobin and lipid membrane by hydrogen/deuterium exchange coupled to mass spectrometry. J Mol Biol 368:464–472
5. Hsu YH, Burke JE, Li S, Woods VL, Dennis EA (2009) Localizing the membrane binding region of group VIA Ca2+-independent phospholipase a(2) using peptide amide hydrogen/deuterium exchange mass spectrometry. J Biol Chem 284:23652–23661

6. Man P, Montagner C, Vitrac H, Kavan D, Pichard S, Gillet D, Forest E, Forge V (2011) Accessibility changes within diphtheria toxin T domain upon membrane penetration probed by hydrogen exchange and mass spectrometry. J Mol Biol 414:123–134
7. Rey M, Man P, Clemencon B, Trezeguet V, Brandolin G, Forest E, Pelosi L (2010) Conformational dynamics of the bovine mitochondrial ADP/ATP carrier isoform 1 revealed by hydrogen/deuterium exchange coupled to mass spectrometry. J Biol Chem 285:34981–34990
8. West GM, Chien EYT, Katritch V, Gatchalian J, Chalmers MJ, Stevens RC, Griffin PR (2011) Ligand-dependent perturbation of the conformational ensemble for the GPCR beta(2) adrenergic receptor revealed by HDX. Structure 19:1424–1432
9. Clemencon B, Rey M, Trezeguet V, Forest E, Pelosi L (2011) Yeast ADP/ATP carrier isoform 2 conformational dynamics and role of the RRRMMM signature sequence methionines. J Biol Chem 286:36119–36131
10. Mehmood S, Domene C, Forest E, Jault J-M (2012) Dynamics of a bacterial multidrug ABC transporter in the inward and outward facing conformations. Proc Natl Acad Sci U S A 109: 10832–10836
11. Hebling CM, Morgan CR, Stafford DW, Jorgenson JW, Rand KD, Engen JR (2010) Conformational analysis of membrane proteins in phospholipid bilayer nanodiscs by hydrogen exchange mass spectrometry. Anal Chem 82: 5415–5419
12. Duc NM, Du Y, Thorsen TS, Lee SY, Zhang C, Kato H, Kobilka BK, Chung KY (2015) Effective application of bicelles for conformational analysis of G protein-coupled receptors by hydrogen/deuterium exchange mass spectrometry. J Am Soc Mass Spectrom 26:808–817
13. Cravello L, Lascoux D, Forest E (2003) Use of different proteases working in acidic conditions to improve sequence coverage and resolution in hydrogen/deuterium exchange of large proteins. Rapid Commun Mass Spectrom 17:2387–2393
14. Rey M, Man P, Brandolin G, Forest E, Pelosi L (2009) Recombinant immobilized rhizopuspepsin as a new tool for protein digestion in hydrogen/deuterium exchange mass spectrometry. Rapid Commun Mass Spectrom 23: 3431–3438
15. Rey M, Yang ML, Burns KM, Yu YP, Lees-Miller SP, Schriemer DC (2013) Nepenthesin from monkey cups for hydrogen/deuterium exchange mass spectrometry. Mol Cell Proteomics 12: 464–472
16. Kadek A, Mrazek H, Halada P, Rey M, Schriemer DC, Man P (2014) Aspartic protease nepenthesin-1 as a tool for digestion in hydrogen/deuterium exchange mass spectrometry. Anal Chem 86:4287–4294
17. Kadek A, Tretyachenko V, Mrazek H, Ivanova L, Halada P, Rey M, Schriemer DC, Man P (2014) Expression and characterization of plant aspartic protease nepenthesin-1 from Nepenthes gracilis. Protein Expr Purif 95:121–128
18. Yang M, Hoeppner M, Rey M, Kadek A, Man P, Schriemer DC (2015) Recombinant nepenthesin II for hydrogen/deuterium exchange mass spectrometry. Anal Chem 87: 6681–6687
19. Kavan D, Man P (2011) MSTools-Web based application for visualization and presentation of HXMS data. Int J Mass Spectrom 302:53–58
20. Rey M, Mrazek H, Pompach P, Novak P, Pelosi L, Brandolin G, Forest E, Havlicek V, Man P (2010) Effective removal of nonionic detergents in protein mass spectrometry, hydrogen/deuterium exchange, and proteomics. Anal Chem 82:5107–5116

Chapter 18

Lessons from an α-Helical Membrane Enzyme: Expression, Purification, and Detergent Optimization for Biophysical and Structural Characterization

Jennifer L. Johnson, Sibel Kalyoncu, and Raquel L. Lieberman

Abstract

This chapter outlines the protocol developed in our lab to produce a multipass α-helical membrane protein. We present our work flow, from ortholog selection to protein purification, including molecular biology for plasmid construction, protein expression in *E. coli*, membrane isolation and detergent solubilization, protein purification and tag removal, biophysical assessment of protein stability in different detergents, and detergent concentration determination using thin-layer chromatography. We focus on results from our ongoing work with intramembrane aspartyl proteases from archaeal organisms.

Key words Membrane protein, Cloning, Expression, Purification, Detergent screening, Signal peptide peptidase, Intramembrane aspartyl protease, Circular dichroism, Thin-layer chromatography

1 Introduction

Poor expression level is a major contributor to the reason that membrane protein structures lag far behind their soluble counterparts [1]. Mammalian membrane proteins often need to be expressed in eukaryotic expression systems because the proteins of interest require chaperones, specific lipids, and/or posttranslational modification for proper folding [1, 2]. Eukaryotic expression systems are often costly, time consuming, and relatively low yielding. In contrast, prokaryotic membrane proteins, which do not have posttranslational modifications, can be expressed in high abundance using bacterial expression systems that require simple media and grow rapidly [3]. Thus, bacterial and archaeal orthologs of mammalian proteins are attractive to pursue for structural studies, a strategy that has been met with considerable success (*see* refs. 4–7, for example).

The protocol presented in this chapter was developed in our lab for the production of orthologs of signal peptide peptidase (SPP) and can be easily adapted for other α-helical membrane

Isabelle Mus-Veteau (ed.), *Heterologous Expression of Membrane Proteins: Methods and Protocols,* Methods in Molecular Biology, vol. 1432, DOI 10.1007/978-1-4939-3637-3_18, © Springer Science+Business Media New York 2016

proteins. SPP is an intramembrane aspartyl protease (IAP) with orthologs found in a range of organisms from humans to extremophilic archaea [8]. IAP family members are multipass membrane proteins that share a conserved, membrane-embedded signature motif, YD…GXGD, where X is any amino acid [9]. In humans, SPP uses two aspartate residues to cleave type-2 signal peptides from the endoplasmic reticulum (ER) membrane, and the remnant short peptides act as signaling molecules for cell–cell communication, among other activities [10]. Orthologs of human SPP have similar inhibition profiles [11–14] and cleavage patterns [11], all of which strongly suggest that these enzymes share a similar structure and utilize a similar chemical mechanism. Unlike human SPP, archaeal SPPs are not glycosylated and are active after overexpression and purification from heterologous bacterial hosts [15–17].

Our approach to the molecular characterization of SPP has been to study orthologs recombinantly expressed in *E. coli* (Subheading 3.1). We use a commercial vector containing a *pelB* leader sequence for insertion into the periplasmic membrane and a C-terminal hexahistidine tag (Subheading 3.2), express the protein in *E. coli* (Subheading 3.3), isolate membrane from cells containing the protein of interest (Subheading 3.4), solubilize the membrane in detergent, purify protein of interest using Ni^{2+}-affinity chromatography (Subheading 3.5), and further polish the sample using size-exclusion chromatography (SEC, Subheading 3.7). Our characterization protocol includes monitoring SEC chromatograph profiles in different detergents and corresponding thermal stability using circular dichroism (CD) melts (Subheading 3.8) because protein stability correlates with crystallizability and activity [18], and protein in a stabilizing detergent solution will maintain monodispersity longer [19], allowing for more time to perform activity assays and crystallization trials. We are also concerned with the amount of detergent in our final purified sample (Subheading 3.9), as excess detergent can lead to ligand and subunit dissociation and phase separation in the crystallization drop [20]. Our methods were developed for our small lab with limited resources and thus can be at best considered as low/medium throughput. Wherever possible, we have included notes for troubleshooting and alternative methods.

2 Materials

2.1 Molecular Biology for Target Membrane Protein

1. Organism genome (ATCC).
2. Primers for gene amplification.
3. pET-22b(+) vector.
4. Polymerase Chain Reaction (PCR) mix or kit and thermocycler.

5. PCR and gel cleanup kit.
6. Restriction enzymes (suggested: *Sal*I-HF and *Nco*I-HF).
7. Agarose gel and electrophoresis equipment.
8. Dephosphorylation and ligation kit.
9. *E. coli* cells for plasmid preparation.

2.2 Membrane Protein Expression

1. *E. coli* Rosetta 2 (DE3) competent cells (EMD Millipore #71400).
2. pET-22b(+) plasmid with gene of interest (*see* Subheading 3.2).
3. LB agar culture plate with appropriate antibiotics (ampicillin and chloramphenicol when using pET-22b(+) plasmid with Rosetta 2 (DE3) cells).
4. LB agar plate with colonies containing plasmid of interest.
5. LB broth.
6. Appropriate antibiotics (ampicillin and chloramphenicol when using pET-22b(+) plasmid with Rosetta 2 (DE3) cells).
7. 6 × 2 L baffled flasks.
8. 1 × 500 mL Erlenmeyer flask.
9. Temperature controlled shaking incubator.
10. Isopropyl β-D-1-thiogalactopyranoside (IPTG).
11. Centrifuge and rotor.
12. Liquid nitrogen.

2.3 Membrane Isolation from Harvested Cells

1. Cell paste containing protein of interest (*see* Subheading 3.3).
2. French press and 35 mL French press cell.
3. Ultracentrifuge and rotor.
4. Dounce homogenizer, 7 mL (for example, Sigma Aldrich).
5. Cell lysis buffer: 50 mM HEPES pH 7.5 (or an alternative buffer with the appropriate pH range), 200 mM NaCl, complete EDTA Free Protease Inhibitor (Roche, 1 tablet to 50 mL buffer).

2.4 Solubilization from Membrane and Purification

1. Isolated membrane containing the protein of interest (*see* Subheading 3.4).
2. Dounce homogenizer, 7 mL.
3. Membrane resuspension buffer: 50 mM HEPES pH 7.5, 500 mM NaCl, 20 mM imidazole.
4. Desired detergent for protein solubilization (suggested: *n*-dodecyl β-D-maltoside (DDM, Anatrace), *see* Subheading 2.6).

5. Ultracentrifuge and tubes.
6. Superloop (e.g., from GE Healthcare).
7. AKTA FPLC instrument (GE Healthcare) with detector capable of measuring absorbance at 280 nm.
8. 1 mL Ni^{2+}-affinity chromatography column.
9. Ni^{2+}-affinity chromatography Buffer A: 50 mM HEPES pH 7.5, 500 mM NaCl, 20 mM imidazole, + detergent (*see* subheading 2.6, **step 2**).
10. Ni^{2+}-affinity chromatography Buffer B: 50 mM HEPES pH 7.5, 500 mM NaCl, 500 mM imidazole, + detergent.

2.5 Optional: Cleavage of the Hexahistidine Tag and Purification of Cleaved Protein

1. Gel filtration buffer: 50 mM HEPES pH 7.5, 200 mM NaCl, + starting detergent.
2. 1.5 mg/ml of purified tobacco etch virus (TEV) protease in a stabilizing buffer containing 10 mM phosphate buffer pH 7.2, 200 mM NaCl, 20% glycerol [21].
3. Ni^{2+}-affinity chromatography Buffer A.
4. Ni^{2+}-affinity chromatography Buffer B.

2.6 Detergent Screening by Gel Filtration

1. Sample purified by Ni^{2+}-affinity chromatography (*see* Subheading 3.5).
2. Gel filtration buffers: 50 mM HEPES pH 7.5, 200 mM NaCl, + detergent, each containing 2× the critical micelle concentration (CMC) of *n* different detergents. Suggested detergents include those that have been successful for membrane protein crystallography: N,N-dimethyldodecylamine-N-oxide (LDAO), *n*-octyl-β-D-glucopyranoside (OG), octyltetraoxyethylene (C8E4), *n*-decyl-β-D-maltopyranoside (DM), and *n*-dodecyl-β-D-maltopyranoside (DDM) [22].
3. AKTA FPLC instrument with detector for absorbance at 280 nm.
4. Superose 12 10/300 or 3.2/300 gel filtration column.

2.7 Protein Stability in Different Detergents Measured by Circular Dichroism (CD)

1. Purified membrane protein sample in *n* different detergents (*see* Subheading 3.6).
2. Amicon Ultra centrifugal filtration device of the appropriate molecular weight cutoff (MWCO) for concentration of protein (e.g., 3× smaller than protein mass, sizes 10K, 30K, 50K, 100K available from EMD Millipore).
3. CD spectropolarimeter.
4. CD cell (e.g., 0.1 cm diameter).
5. Graphing and analysis software (for example, GraphPad Prism).

2.8 Determination of Detergent Amount in Protein Sample by Thin-Layer Chromatography (TLC) and Optimal Method for Concentrating Protein (Protocol Adapted from Ref. 29)

1. Purified protein in gel filtration buffer
2. 0.5 mL Amicon Ultra centrifugal filtration devices of different MWCO (10–100 kDa suggested).
3. Standards containing known concentrations of detergent in gel filtration buffer (*see* Subheading 2.6, **item 2**).
4. Silica 60 TLC plates.
5. TLC chamber.
6. TLC solvent (63:35:5 chloroform:methanol:ammonium hydroxide).
7. Iodine and iodine chamber.
8. ImageQuant, ImageJ, Photoshop, or similar program for image analysis.

3 Methods

3.1 Target Protein Ortholog Selection

1. Select your target protein based on laboratory interests. We selected signal peptide peptidase (SPP, accession number CAD13132) as our target protein.
2. Run Basic Local Alignment Search Tool for proteins (BLASTp [23], http://blast.ncbi.nlm.nih.gov/) and search using the Domain Enhanced Lookup Time Accelerated (DELTA) feature on the target protein to find more suitable targets (*see* **Note 1**). The search can be limited to a certain organism or taxonomic group, or can exclude a taxonomic group. We excluded eukaryotes (taxid: 2759) to select for non-eukaryotic orthologs. Figure 1 shows the final Clustal Omega [24] alignment between human SPP and our selected targets, SPP from archaeal *Haloarcula marismortui* (*H. mar*), *Halobacterium salinarum* (*H. sal*), and *Methanoculleus marisnigri* (*M. mar*) (*see* **Note 2**) rendered in ESPript [25] with secondary structure information from the reported crystal structure of *M. mar* SPP (PDB code 4HYC) [6].

3.2 Molecular Biology for Target Membrane Protein (See Note 3)

1. Order the organism genome from ATCC (http://www.atcc.org/). We ordered the genomes of the three archaeal organisms mentioned above.
2. Use online signal sequence prediction software to predict if the target protein contains a signal sequence (*see* **Note 4**). Check several signal sequence prediction web servers for comparison, so the likelihood of a signal sequence being present can be assessed. Figure 2 shows the signal sequence prediction output from Signal-3L [26] and SignalP 4.0 [27]; the former predicts that the first 23 amino acids of *M. mar* SPP might be a signal sequence, and the latter predicts that a signal sequence is not

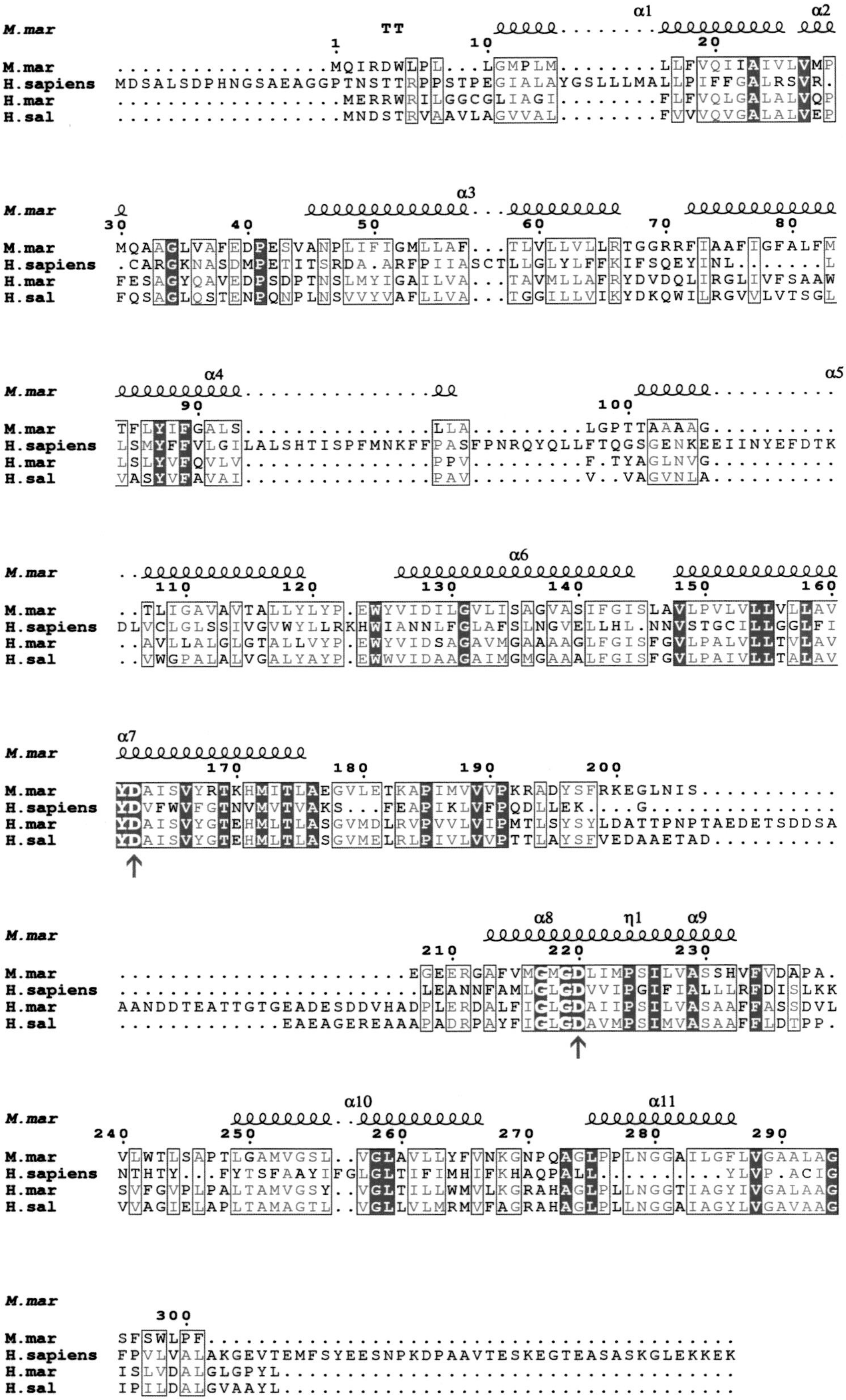

```
                                      TT               α1                 α2
M.mar                                  1         10         20
M.mar      ..................MQIRDWLPL...LGMPLM........LLFVQIIAIVLVMP
H.sapiens  MDSALSDPHNGSAEAGGPTNSTTRPPSTPEGIALAYGSLLLMALLPIFFGALRSVR.
H.mar      ..................MERRWRILGGCGLIAGI........FLFVQLGALALVQP
H.sal      ..................MNDSTRVAAVLAGVVAL........FVVVQVGALALVEP

                                      α3
M.mar       30        40        50          60        70        80
M.mar      MQAAGLVAFEDPESVANPLIFIGMLLAF...TLVLLVLLRTGGRRFIAAFIGFALFM
H.sapiens  .CARGKNASDMPETITSRDA.ARFPIIASCTLLGLYLFFKIFSQEYINL.......L
H.mar      FESAGYQAVEDPSDPTNSLMYIGAILVA...TAVMLLAFRYDVDQLIRGLIVFSAAW
H.sal      FQSAGLQSTENPQNPLNSVVYVAFLLVA...TGGILLVIKYDKQWILRGVVLVTSGL

               α4                                            α5
M.mar           90                               100
M.mar      TFLYIFGALS...............LLA.........LGPTTAAAAG..........
H.sapiens  LSMYFFVLGILALSHTISPFMNKFFPASFPNRQYQLLFTQGSGENKEEIINYEFDTK
H.mar      LSLYVFQVLV...............PPV.........F.TYAGLNVG..........
H.sal      VASYVFAVAI...............PAV.........V..VAGVNLA..........

                                      α6
M.mar          110       120        130       140       150       160
M.mar      ..TLIGAVAVTALLYLYP.EWYVIDILGVLISAGVASIFGISLAVLPVLVLLVLLAV
H.sapiens  DLVCLGLSSIVGVWYLLRKHWIANNLFGLAFSLNGVELLHL.NNVSTGCILLGGLFI
H.mar      ..AVLLALGLGTALLVYP.EWYVIDSAGAVMGAAAAGLFGISFGVLPALVLLTVLAV
H.sal      ..VWGPALALVGALYAYP.EWWVIDAAGAIMGMGAAALFGISFGVLPAIVLLTALAV

           α7
M.mar              170       180       190       200
M.mar      YDAISVYRTKHMITLAEGVLETKAPIMVVVPKRADYSFRKEGLNIS...........
H.sapiens  YDVFWVFGTNVMVTVAKS...FEAPIKLVFPQDLLEK....G...............
H.mar      YDAISVYGTEHMLTLASGVMDLRVPVVLVIPMTLSYSYLDATTPNPTAEDETSDDSA
H.sal      YDAISVYGTEHMLTLASGVMELRLPIVLVVPTTLAYSFVEDAAETAD..........
           ↑

                                          α8        η1     α9
M.mar                                  210       220       230
M.mar      ........................EGEERGAFVMGMGDLIMPSILVASSHVFVDAPA.
H.sapiens  .........................LEANNFAMLGLGDVVIPGIFIALLLRFDISLKK
H.mar      AANDDTEATTGTGEADESDDVHADPLERDALFIGLGDAIIPSILVASAAFFASSDVL
H.sal      .............EAEAGEREAAAPADRPAYFIGLGDAVMPSIMVASAAFFLDTPP.
                                               ↑

                                α10                   α11
M.mar      240        250       260       270       280       290
M.mar      VLWTLSAPTLGAMVGSL..VGLAVLLYFVNKGNPQAGLPPLNGGAILGFLVGAALAG
H.sapiens  NTHTY...FYTSFAAYIFGLGLTIFIMHIFKHAQPALL..........YLVP.ACIG
H.mar      SVFGVPLPALTAMVGSY..VGLTILLWMVLKGRAHAGLPLLNGGTIAGYIVGALAAG
H.sal      VVAGIELAPLTAMAGTL..VGLLVLMRMVFAGRAHAGLPLLNGGAIAGYLVGAVAAG

M.mar
             300
M.mar      SFSWLPF...........................................
H.sapiens  FPVLVALAKGEVTEMFSYEESNPKDPAAVTESKEGTEASASKGLEKKKEK
H.mar      ISLVDALGLGPYL.....................................
H.sal      IPILDALGVAAYL.....................................
```

a ------------------------------Prediction Results------------------------------

According to Signal-3L engine for your selected species, the signal peptide is:1-23

MQIRDWLPLLGMPLMLLFVQIIAIVLVMPMQAAGLVAFEDPESVANPLIFIGMLLAFTLV
LLVLLRTGGR

b

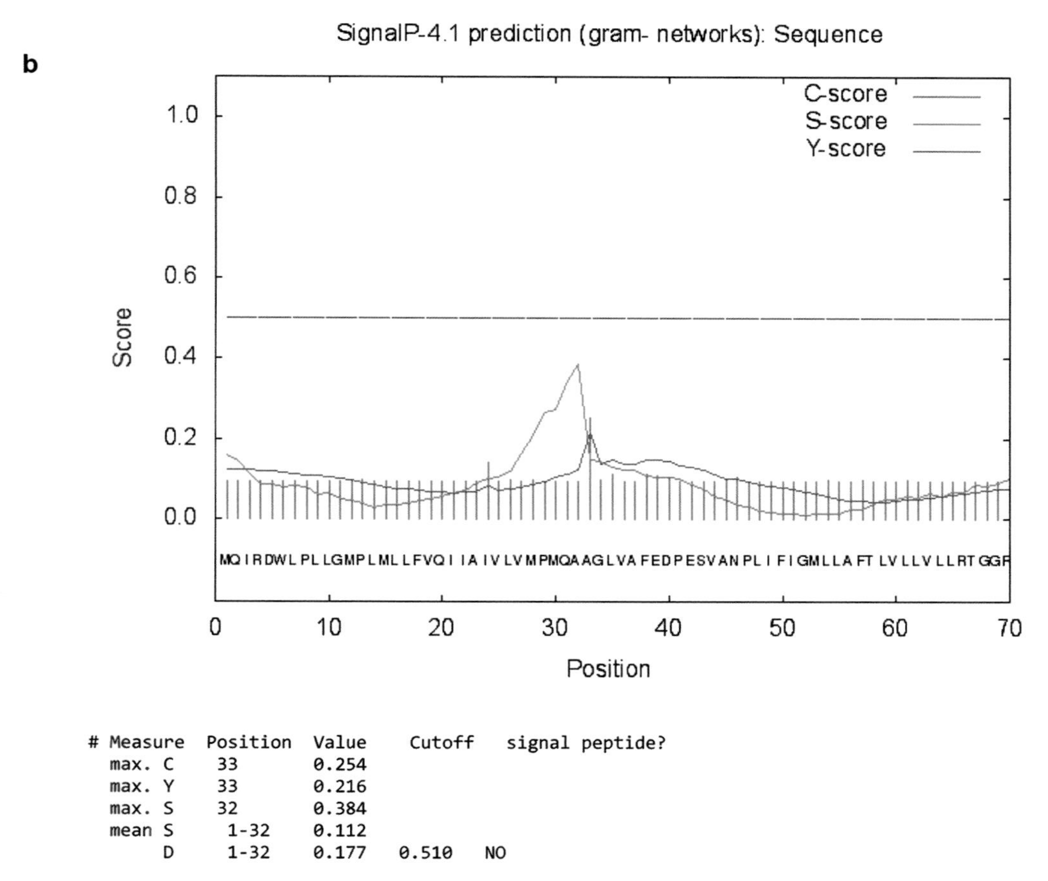

# Measure	Position	Value	Cutoff	signal peptide?
max. C	33	0.254		
max. Y	33	0.216		
max. S	32	0.384		
mean S	1-32	0.112		
D	1-32	0.177	0.510	NO

Fig. 2 Signal sequence prediction. (**a**) Output from Signal-3L predicting that the first 23 amino acids (*red*) from *M. mar* SPP are a signal sequence. (**b**) Output from SignalP predicting the same protein sequence does not contain a signal peptide

Fig. 1 Alignment of human SPP with three archaeal SPP orthologs. Alignment was performed using Clustal Omega and rendered in ESPript [43]. Identical residues in all four sequences are *white with red background*. Similar residues are in *red* and conserved patches are *boxed in blue*. The conserved motif of all IAP family members is in *bold white*, with the catalytic aspartate residues denoted with a *red arrow*. α-Helices are marked by spiral along the top of each row and are based on the crystal structure of *M. mar* SPP (PDB code 4HYC)

```
               T7 promoter         lac operator
AGATCTCGATCCCGCGAAATTAATACGACTCACTATAGGGGAATTGTGAGCGGATAACAATTCCCCTCTAGAAATAATTTTGTTTAACTTTAAGAAGGAGA

                                                                                     NcoI
                                 pelB leader
TATACATATGAAATACCTGCTGCCGACCGCTGCTGCTGGTCTGCTGCTCCTCGCTGCCCAGCCGGCGATGGCCATGGATATCGGAATTAATTCGGATCCGAATTCGAGCTCC
      MetLysTyrLeuLeuProThrAlaAlaAlaGlyLeuLeuLeuLeuAlaAlaGlnProAlaMetAlaMetAspIleGlyIleAsnSerAspProAsnSerSerSer
                                                                        |Signal peptidase

SalI                His Tag
GTCGACAAGCTTGCGGCCGCACTCGAGCACCACCACCACCACCACTGAGATCCGGCTGCTAACAAAGCCCGAAAGGAAGCTGAGTTGGCTGCTGCCACCGCTGAGCAATAAC
ValAspLysLeuAlaAlaAlaLeuGluHisHisHisHisHisHisEnd

              T7 terminator
TAGCATAACCCCTTGGGGCCTCTAAACGGGTCTTGAGGGGTTTTTTG
```

Fig. 3 pET-22b(+) DNA and amino acid sequence in the area where the ortholog DNA is inserted. Areas of interest are marked. The *red* DNA and amino acid sequence were replaced with the target DNA. The *orange* amino acids were removed and replaced with the DNA sequence for TEV protease cleavage sequence ENLYFQS. Image was adapted from that available from Novagen website

present. Consider making constructs including and omitting the predicted signal sequence residues.

3. Design primers to amplify the gene by PCR with the addition of restriction sites on each side (*see* **Note 5**). We selected restriction sites *Nco*I and *Sal*I of pET-22b(+) vector because they were suitable for cloning of all three genes of interest in parallel (*see* Fig. 3) with an N-terminal pelB leader sequence for periplasmic membrane insertion and a C-terminal hexahistidine tag for purification Ni^{2+}-affinity chromatography.
4. Amplify the gene of interest by PCR.
5. Clean up PCR products.
6. Perform restriction digest on both the plasmid and the gene of interest to form complementary ends.
7. Run agarose gel on digested reaction samples.
8. Cut the appropriate bands from the agarose gel and purify using appropriate kit.
9. Dephosphorylate the vector and ligate the DNA.
10. Transform the plasmid into a cell line for plasmid preparation; e.g. *E. coli* GigaSingles or XL1-Blue.
11. Sequence plasmid to confirm correct insertion of desired gene containing no errors. MWG Operon (www.operon.com) was used for our plasmid sequencing.
12. Optional: Insert DNA sequence corresponding to TEV protease cleavage site (ENLYFQS) between the DNA of the target protein and the hexahistidine tag (*see* Fig. 3, **Note 6**) using site-directed mutagenesis. Confirm correct cleavage site insertion by sequencing.

3.3 Membrane Protein Expression (See Note 7)

1. Perform standard heat-shock plasmid transformation into Rosetta 2 (DE3) cells (*see* **Note 8**); use a single colony to express immediately or wrap edges of the agar plate with Parafilm and place in the 4 °C refrigerator for storage up to 1 week.
2. Autoclave 1 × 200 mL LB broth in a 500 mL Erlenmeyer flask and 6 × 1 L LB broth in a 2 L baffled flask.
3. Add appropriate antibiotics to each flask when cool. When using pET-22b(+) plasmid with *E. coli* Rosetta 2 (DE3) cells, ampicillin (plasmid resistance, 60 μg/mL recommended) and chloramphenicol (cell resistance, 34 μg/mL recommended) antibiotics are used.
4. Add a single colony from transformation plate into the 200 mL flask (starter culture) and incubate the flask at 37 °C overnight with shaking at 225 RPM.
5. After the starter culture is incubated for 12–16 h, inoculate each of the six 1 L cultures with 10 mL of the starter culture. Incubate at 37 °C with shaking at 225 RPM, checking the optical density at 600 nm ($O.D._{600\ nm}$) after 2 h and every 30 min thereafter.
6. When the $O.D._{600\ nm}$ reaches 0.6–0.8, reduce the incubation temperature to 18 °C. Continue shaking for 1 hour to allow the temperature to equilibrate.
7. Induce protein expression by the addition of 0.5 mM IPTG to each flask. Continue shaking at 225 RPM at 18 °C for 16–20 h.
8. After expression for 16–20 h, harvest the cells by centrifugation at 5000 × *g* for 10 min. Distribute the cell pellet in small plastic bags and flash freeze in liquid nitrogen. Store at −80 °C until cell paste is needed.

3.4 Membrane Isolation from Harvested Cells

1. Thaw and resuspend 7–8 g of frozen cell paste in a 50 mL tube on ice by adding 25–30 mL lysis buffer and pipetting gently up and down with a 25 mL serological pipette, until the mixture is fully resuspended and homogeneous.
2. Lyse resuspended cells using a chilled French press cell maintained at 1200 psi pressure. Pass the cells through the French press at least twice to ensure complete lysis.
3. Pellet cellular debris by centrifuging lysate at 5000 × *g* for 15 min at 4 °C. Place the supernatant in a new centrifuge tube and centrifuge again. Repeat centrifugation step until no further pellet is discernible.
4. Transfer the supernatant to an ultracentrifuge tube and centrifuge at 120,000 × *g* for 45 min at 4 °C.
5. Discard the supernatant. Place the pellet in a 7 mL Dounce homogenizer with 7 mL cell lysis buffer (without protease inhibitor). Resuspend the membrane pellet with the loose plunger first, and then the tight plunger.

6. Ultracentrifuge the resuspended membrane pellet again at 120,000 × *g* for 45 min at 4 °C.
7. Discard the supernatant. Transfer the pelleted membrane into a tared microcentrifuge tube and record its mass. We typically obtain around 0.9–1.2 g of membrane from 7 to 8 g cells.
8. Flash cool membranes with liquid nitrogen in ~0.3 g aliquots in a microcentrifuge tube, and store in −80 °C freezer until ready for protein purification.

3.5 Solubilization from Membrane and Purification Using Ni^{2+}-Affinity Chromatography

1. Place 0.3–1.0 g frozen membrane in Dounce homogenizer and add 7 mL membrane resuspension buffer.
2. Resuspend the membrane using first the loose plunger, and then the tight plunger.
3. Weigh out an amount of the desired detergent (*see* **Note 9**) equal to the mass of membrane. Dissolve the detergent in enough membrane resuspension buffer to make a final 1 % solution. For 1 g membrane, add 1 g detergent to 93 mL membrane resuspension buffer, where the last 7 mL will be the membrane suspension.
4. Add the resuspended membrane to the detergent solution. Stir or rock gently at 4 °C for at least 1 h, minimizing bubble formation. Once solubilization is complete the solution will appear translucent.
5. Centrifuge at 120,000 × *g* at 4 °C for 45 min to remove any unsolubilized material. Discard any remaining pellet and store the supernatant on ice for purification. *See* Fig. 4, *lane 1*, for sodium dodecyl sulfate polyacrylamide-gel electrophoresis (SDS-PAGE) after membrane solubilization (*see* **Note 10**).
6. Load protein-detergent solution into a superloop appropriate for the total volume. Purify the protein over a 1 mL Ni^{2+}-affinity chromatography column using Buffer A for the wash and Buffer B for protein elution, using a suitable gradient. Track the absorbance at 280 nm to identify protein elution. Figure 5 shows Ni^{2+}-affinity chromatograms of two SPP constructs. Figure 4, *lane 2*, shows SDS-PAGE analysis after Ni^{2+}-affinity chromatography.

3.6 Optional: Cleavage of the Hexahistidine Tag and Purification of Cleaved Protein (See Note 11)

1. Buffer exchange protein using appropriate-sized Amicon Ultra centrifugal filter into gel filtration buffer to remove excess imidazole.
2. Add TEV protease (1.5 mg/ml in stabilizing buffer) directly to protein sample in a 1:1 TEV:target membrane protein mass ratio.
3. Allow cleavage reaction to proceed at 4 °C for 16–20 h.
4. Repurify sample over a Ni^{2+}-affinity chromatography column, collecting the flow through and discarding protein that bound

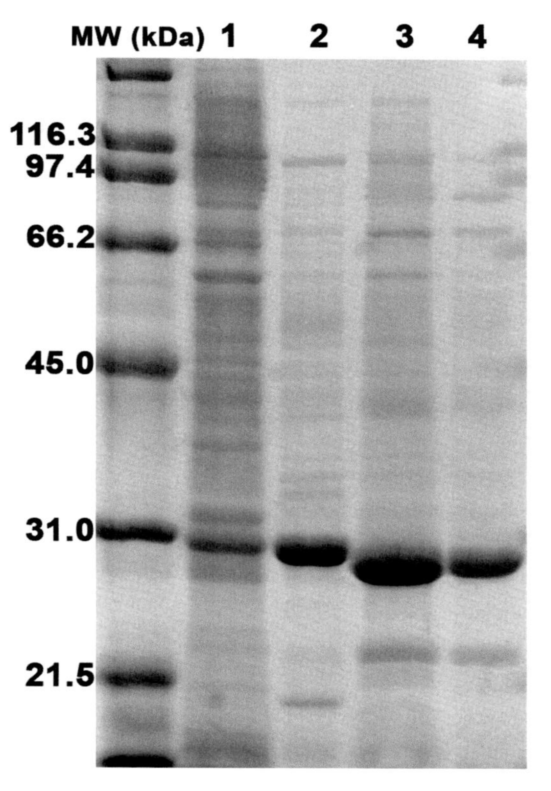

Fig. 4 SDS-PAGE of *M. mar* SPP samples during each purification step. A broad-range molecular weight marker with band sizes is shown on the left. *M. mar* SPP is the prominent band just under 30 kDa in each lane. *Lane 1* is the sample after membrane solubilization (Subheading 3.5, **step 5**). *Lane 2* is the protein sample after Ni^{2+}-affinity chromatography purification (Subheading 3.5, **step 6**). *Lane 3* is the protein purified over Ni^{2+}-affinity chromatography purification a second time after TEV protease cleavage of the hexahistidine tag (Subheading 3.6, **step 4**). *Lane 4* is the purified protein sample after size-exclusion chromatography on the Superose 12 column in gel filtration buffer with 0.0174 % DDM (Subheading 3.8, **step 4**)

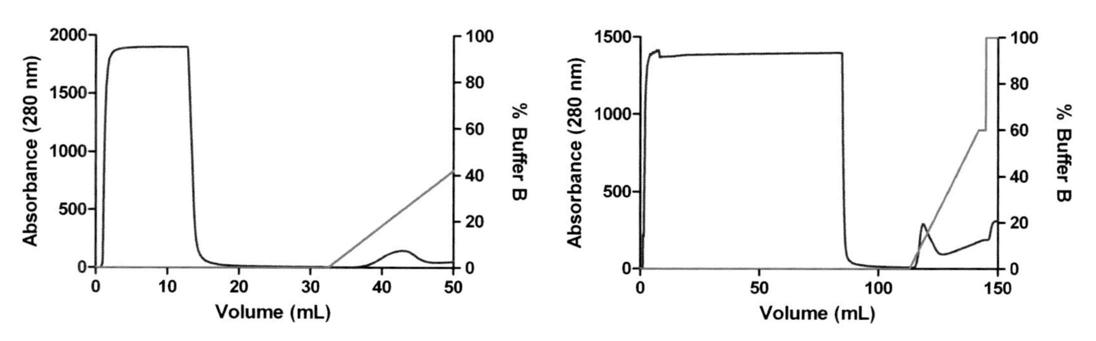

Fig. 5 Ni^{2+}-affinity chromatograms obtained using Unicorn software of *H. mar* SPP (*left*) and *M. mar* SPP (*right*). *Blue trace* is absorbance at 280 nm and *green trace* is % buffer B

to the column. *See* Fig. 4, *lane 3*, for SDS-PAGE analysis of sample after Ni^{2+}-affinity chromatography following TEV protease cleavage step.

3.7 Detergent Screening by Gel Filtration (See Fig. 6 for Overview of Workflow)

1. After protein purification by Ni^{2+}-affinity chromatography, divide the sample into *n* equal volumes, approx. 250 μL each, to test *n* different detergents by gel filtration.
2. Equilibrate the Superose 12 10/300 gel filtration column on an AKTA instrument with at least two column volumes of gel filtration buffer (*see* **Note 12**).
3. After equilibrating with two column volumes of buffer, inject the 250 μl sample onto the column. Track the absorbance at 280 nm.
4. Run each sample with a different detergent with a concentration of 2× CMC in the gel filtration buffer, each with 2 column volumes (CV) equilibration prior to sample injection onto the column. This extensive wash step removes the prior detergent. *See* Fig. 4, *lane 4*, for SDS-PAGE of a sample after gel filtration.
5. Compare the elution peak shape and intensity. High-intensity, Gaussian peaks are desirable. The size of the protein compared to the expected mass based on elution volume must also be considered. The column retention is rarely an accurate measure of membrane protein mass due to the fact that it is a protein-detergent complex (PDC), whose size also depends on detergent characteristics. Figure 7 shows the results of gel filtration of two SPP constructs, each in six different detergents.
6. If an activity assay is available, conduct the assay on protein samples in different detergent-containing buffers to ensure

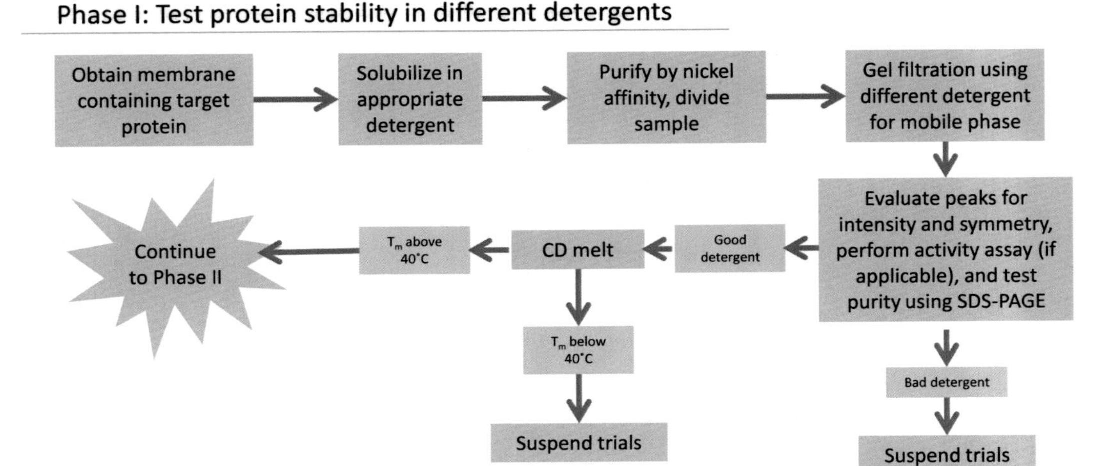

Fig. 6 Workflow to determine protein stability in different detergents using circular dichroism. *See* Fig. 9 for Phase II

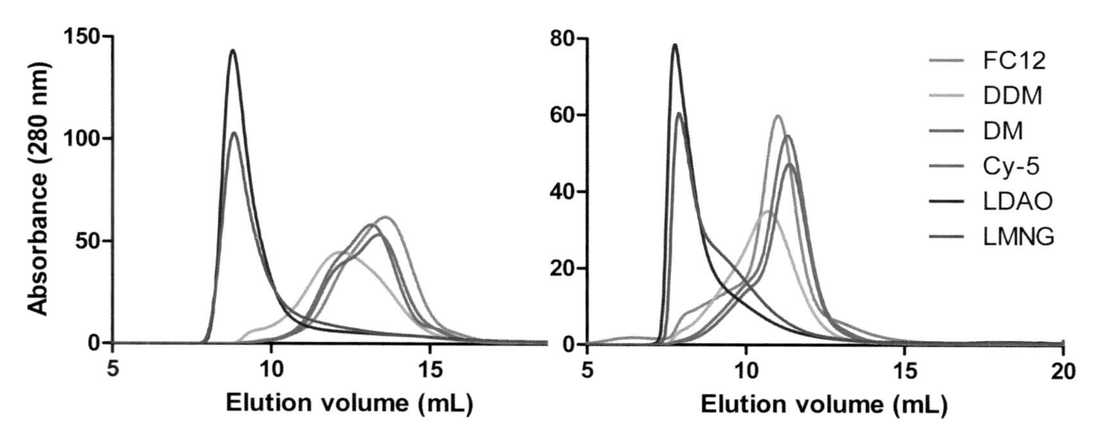

Fig. 7 Gel filtration chromatograms of *H. sal* SPP in six detergents (*left*) and *H. mar* SPP in six detergents (*right*). Detergent abbreviations are as follows: *FC12* fos-choline-12, *DDM* *n*-dodecyl β-D-maltoside, *DM* decyl β-D-maltoside, *Cy-5* 5-cyclohexyl-1-pentyl-β-D-maltoside, *LDAO* *N,N*-dimethyldodecylamine-*N*-oxide, *LMNG* 2,2-didecylpropane-1,3-bis-β-D-maltopyranoside or lauryl maltose neopentyl glycol

active protein. Samples in detergents that yielded favorable gel filtration elution profiles and active protein are then further studied by CD.

3.8 Protein Stability in Different Detergents Measured by Circular Dichroism

1. Concentrate each protein sample using an Amicon Ultra centrifugal filtration device with an appropriate molecular mass cutoff to 8–10 μM, e.g., as measured by absorbance at 280 nm using a calculated extinction coefficient and molecular weight.
2. Run a CD melt on each sample from 4 to 90 °C.
3. For each sample, plot the temperature versus the normalized molar ellipticity at the minimum wavelength (222 nm or 208 nm for alpha helical proteins [28]).
4. Find the melting temperature (T_m) by Boltzmann sigmoid or first derivative numerical analysis of the melt curves using software such as GraphPad Prism, Igor, or Matlab. High protein stability is desired; detergents that maintain stability to the protein of interest can be used for further study. Figure 8 shows the normalized ellipticity versus temperature graph for one of our SPP orthologs. *See* **Note 13** for alternative methods for determining the stability of proteins in different conditions.

3.9 Determination of Detergent Amount in Protein Sample by TLC and Optimal Method for Concentrating Protein

Protocol is adapted from ref. 29; *see* Fig. 9 for overview of workflow, and *see* **Note 14** for alternative methods for determining the concentration of detergent in the protein sample.

1. For each purified protein in each detergent to be tested, pool gel filtration elution peak, and measure the protein concentration.
2. Add 500 μL of the sample into each Amicon Ultra 0.5 mL centrifugal filter to be tested. Testing the MWCO 10 K, 30 K, 50 K, and 100 K filters is suggested.

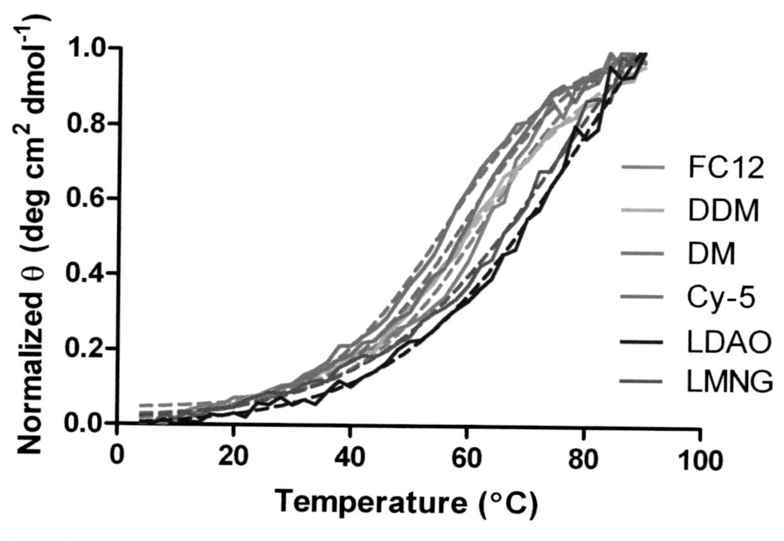

Fig. 8 Normalized CD thermal unfolding experiment of *H. mar* SPP in six detergents. The *solid trace* is the normalized CD data, the *dashed trace* is the Boltzmann sigmoidal fit

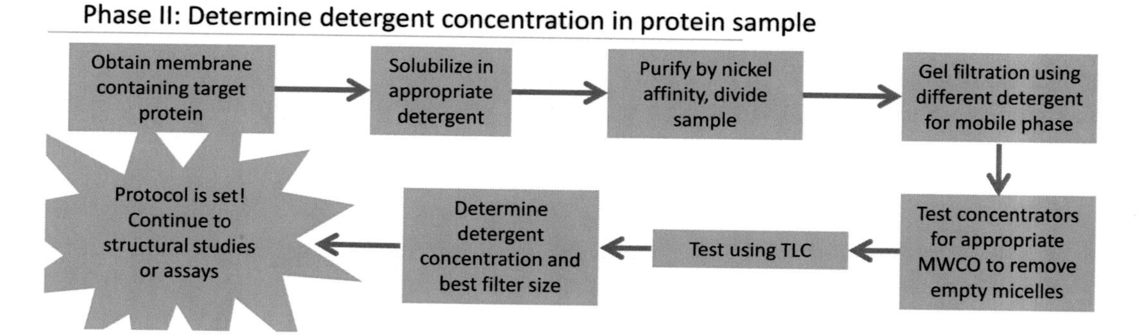

Fig. 9 Workflow to determine optimal filter MWCO for concentrating protein samples. TLC is used to determine detergent concentration in the samples

3. Centrifuge the samples at 8500 × *g* and 4 °C until less than 50 μL remains above the filter. Measure the total volume and protein concentration of each concentrated sample.
4. Determine any protein loss for each filter by comparing the expected concentration based on **steps 1** and **3** to the actual final protein concentration.
5. Place the filtrate from **step 3** into a new 10 K Amicon Ultra 0.5 mL filter. Centrifuge the samples at 8500 × *g* and 4 °C until less than 50 μL remains in the filter. This will concentrate all the detergent that passed through the original filter so that it can be visualized on a TLC plate.

6. Prepare detergent standards for a standard curve in the gel filtration buffer, including points above and below the expected detergent concentration. For example, prepare standards of 0.25 %, 0.5 %, 1.0 %, and 2.0 % (w/v) for an expected concentration of 0.5–1.0 %.
7. Spot 5 μL of each of the detergent standards, concentrate (retained on top), and filtrate (filtered to bottom) samples onto the baseline of a silica 60 TLC plate. Allow all samples to dry for at least 30 min.
8. Place the TLC plate in a TLC chamber containing about 0.5 in. solvent (chloroform:methanol:ammonium hydroxide, 63:35:5, v/v/v).
9. Allow the solvent to run at least half way up the plate. Remove the plate from the chamber and allow the solvent to evaporate in open air for approximately 5 min.
10. Place the plate in an iodine chamber and allow staining for at least 5 min. Remove the TLC plate and image immediately.
11. Quantify the detergent by densitometry of the detergent spots using ImageQuant, ImageJ, or Photoshop. Determine the unknown detergent concentrations using a prepared standard curve.
12. Establish the optimal filter for protein concentration by comparing the protein loss with the detergent loss. The best choice balances lowest detergent concentration in the sample and the least protein loss (*see* **Note 15**).

4 Notes

1. This protocol is for *E. coli* expression. Human membrane proteins are often very difficult to express in *E. coli*, so an alternative ortholog may be needed. In our lab, we have chosen to work with archaeal orthologs of human SPP (*see* Subheading 1). If crystallization is the goal, pay attention to the length of predicted loop regions as long, disordered loops may be difficult to crystallize.
2. We started with ~10 orthologs because the ability to express and purify these proteins is highly protein specific; starting with more targets will give a better chance of success in the pipeline.
3. It may be more cost effective and/or timely to order a codon-optimized gene already inserted into a vector from a synthesis company such as DNA2.0, GenScript, or MWG Operon.
4. The pET-22b(+) vector has the pelB leader sequence at the N-terminus of the protein. The addition of the pelB leader

sequence with a natural signal sequence may cause incorrect membrane insertion and result in poor expression. It may be best to make constructs with and without the predicted signal sequence residues and test expression and activity (if applicable) of each construct.

5. For primer design, *in silico* PCR web servers can be used to confirm the specific amplification of gene of interest (http://insilico.ehu.es/PCR/). Another approach that has had success in our lab [17] is restriction-free cloning [30].
6. Cleaving off the hexahistidine tag can prevent co-purification of an *E. coli* membrane protein contaminant (*see* **Note 11**). To the best of our knowledge, no commercial plasmid includes the TEV protease cleavage site before the hexahistidine tag.
7. Protein expression is a step that will need to be optimized for each protein. We test media, cell line, O.D. at induction, induction temperature, and induction length. *See* ref. 2 for a review of methods to monitor membrane protein overexpression. Also consider protocols in [31] and [1]. If the C-terminus of the membrane protein is in the cytosol (C_{in} topology), a membrane protein-green fluorescent protein (GFP) fusion can be used to rapidly screen homologues and expression conditions by whole-cell and in-gel fluorescence [32, 33].
8. We use *E. coli* Rosetta 2 (DE3) cells for membrane protein expression, which contains plasmids to correct for rare codon usage. Other options include standard *E. coli* BL21 (DE3) or *E. coli* C43 (DE3), which is reputed to produce higher levels of membranes. Try one of the other cell lines if your first choice does not work, or conduct small-scale expression trials to determine the best cell line.
9. Buffers and starting detergent must be optimized. Start with a detergent that works, which may take some trial. LDAO, OG, C8E4, DM, and DDM have been very successful for the crystallization of membrane proteins [22]. You may need to do a small-scale detergent screen, for example, with batch purification using Ni^{2+}-affinity resin to determine a detergent that is optimal for membrane solubilization. Another approach is using fluorescent size-exclusion chromatography (FSEC) on lysed and solubilized cells in different detergents containing a membrane protein-green-fluorescent-protein (GFP) fusion [32–34].
10. SDS-PAGE samples containing membrane proteins should not be boiled prior to loading sample onto the gel.
11. *E. coli* membrane protein acriflavine resistance protein B (AcrB) is, unfortunately, often co-purified on Ni^{2+}-affinity chromatography columns, is too large to pass through concentration devices, and is apparently hypercrystallizable even at low levels not detected by SDS-PAGE. Several labs [35], including ours, have crystallized and solved the structure of AcrB instead of our

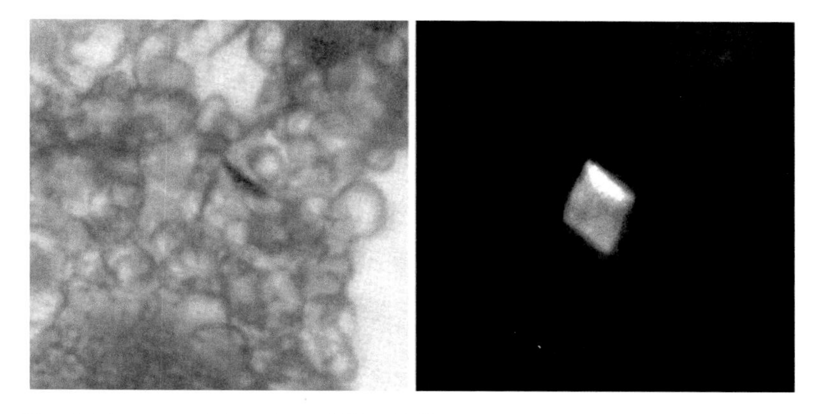

Fig. 10 AcrB crystals using visible (*left*) and UV (*right*) light. Crystals were obtained using crystallization condition containing 20 mM sodium citrate pH 5.6, 0.1 M NaCl, and 17 % polyethylene glycol 3350. The crystals formed within 6 days at room temperature. Space group is R32 and cell dimensions (Å) are $a=144$ $b=144$ $c=518.36$ with angles (°) of $\alpha=90$ $\beta=90$ $\gamma=120$

intended membrane protein. It is best to be wary of crystals resembling AcrB (*see* Fig. 10) and to search the Protein Data Bank for unit cell dimensions of any membrane protein crystals grown to make sure that the dimensions differ from those of AcrB. To prevent co-purification of AcrB or any other protein contamination, we have added a TEV cleavage site between the protein and the hexahistidine tag (*see* **Note 5**). After purifying SPPs over the Ni^{2+}-affinity chromatography column, we cleave the hexahistidine tag using TEV protease. Though our TEV cleavage protocol can likely be optimized for each protein, we have found that TEV protease can be readily prepared in the laboratory in high yield [21] and is insensitive to the specialized detergents used for membrane protein solubilization and purification. TEV itself contains a C-terminal histidine tag, so a second purification over the Ni^{2+}-affinity chromatography column traps TEV, uncleaved SPP, the tags, and AcrB; cleaved SPP is collected in the flow through fractions.

12. The Superose 12 10/300 column is a 24 mL column. There are smaller columns that would help make this step higher throughput. If a smaller column is used, the sample volume recommendations for the column should be followed.
13. Denaturing SDS-PAGE and size-exclusion high-pressure liquid chromatography (SE-HPLC) has also been used to determine the effect of detergents, pH, additives, and lipids on the proportion of monomeric protein [19, 36, 37]. FSEC is a good way to determine stabilizing detergents using a membrane protein-GFP fusion without the need to purify the protein first (*see* **Note 9**) [32–34]. If buried cysteines are present, effects of detergents and additives on purified pro-

tein can be measured in higher throughput by a thermal unfolding assay using *N*-[4-(7-diethylamino-4-methyl-3-coumarinyl)phenyl]maleimide (CPM), a dye that fluoresces upon reaction with a thiol [38–40].

14. Several other methods for determining detergent concentration have been developed. Colorimetric assays, molybdate assay for total phosphate, phenol/sulfuric acid assay for sugar content, contact angle measurements, NMR, gas chromatography, and other methods are covered in ref. 41.

15. We have found that actual detergent concentrations at the completion of the purification protocol are higher than expected. Though our buffer used for gel filtration contained 0.0174 % DDM, our protein sample contained about 0.0275 % as measured by TLC (*see* Fig. 11 and Table 1). The 0.0275 % DDM estimation was made as follows. We assumed that the 10 K MWCO Amicon Ultra centrifugal filter concentrates all

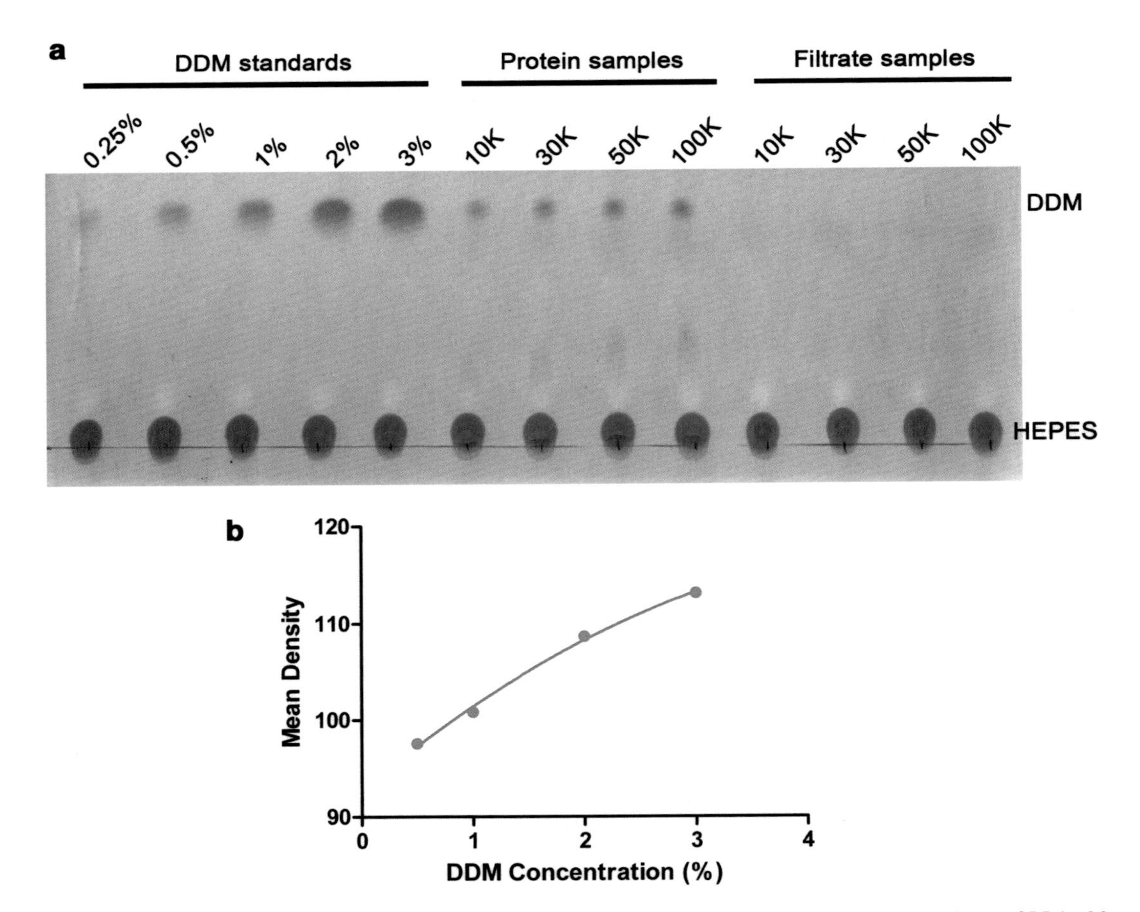

Fig. 11 Detergent concentration assessment using TLC. (**a**) TLC from detergent filter test of *M. mar* SPP in 20 mM HEPES pH 7.5, 200 mM NaCl, 0.0174 % DDM. (**b**) Standard curve prepared using mean density values from Photoshop for each standard DDM concentration. $R^2 = 0.9960$

Table 1
Data table for TLC detergent test

Filter MWCO	10 K	30 K	50 K	100 K
Starting concentration (mg/mL)[a]	0.054	0.054	0.054	0.054
Volume after concentrating (μL)	26	17.5	14	9
Concentration factor[b]	19.2	28.6	35.7	55.6
Theoretical concentration (mg/mL)[c]	1.038	1.543	1.929	3.000
Measured concentration (mg/mL)[d]	0.995	1.296	1.388	1.848
Protein recovery (%)[e]	95.8	84.0	72.0	61.6
Theoretical [detergent][f]	0.335	0.497	0.621	0.967
Actual [detergent][g]	0.529	0.687	0.771	0.885
Detergent recovery (%)[h]	1.58	1.38	1.24	0.92
Detergent recovery (%) using 0.0275 % as starting detergent concentration[i]	100.0	87.5	78.5	57.9

[a]Starting protein concentration was measured using a nanodrop and the protein extinction coefficient calculated by ExPASyProtParam [44]
[b]Concentration factor = 500 μL/(volume after concentrating)
[c]Theoretical concentration = (Starting concentration) × (Concentration factor)
[d]Measured concentration was measured using a nanodrop and the protein extinction coefficient calculated by ExPASyProtParam [44]
[e]Protein recovery = (Measured concentration)/(Theoretical concentration)
[f]Theoretical detergent concentration = 0.0174 × (Concentration factor)
[g]Actual detergent concentration was estimated using TLC (Fig. 11a) and the standard curve prepared using mean density calculation in Photoshop from DDM standards on TLC (Fig. 11b)
[h]Detergent recovery = (Theoretical detergent concentration)/(Actual detergent concentration)
[i]It is assumed that the 10 K filter concentrates all the detergent (100 %). Since the actual detergent concentration was larger than theoretical detergent concentration in most cases, the starting detergent concentration was recalculated using the actual detergent concentration and the concentration factor for the 10 K filter, and then applied to the other filters. That detergent concentration (0.0275 %) was used to calculate detergent recovery

the DDM in the sample. We then used the TLC standard curve to calculate the DDM concentration in each of the concentrated samples, then divided by the protein concentration factor. Notably, the excess detergent micelles and our protein of interest apparently have similar elution volumes on the gel filtration column, which could be confirmed by the use of an in-line refractive index detector [42].

References

1. Mancia F, Love J (2010) High-throughput expression and purification of membrane proteins. J Struct Biol 172(1):85–93
2. Wagner S, Bader ML, Drew D, de Gier J-W (2006) Rationalizing membrane protein overexpression. Trends Biotechnol 24(8):364–371
3. Fernandez JM, Hoeffler JP (1999) Introduction. In: Fernandez JM, Hoeffler JP (eds) Gene expression systems. Academic, San Diego, pp 1–5
4. Yamashita A, Singh SK, Kawate T, Jin Y, Gouaux E (2005) Crystal structure of a bacterial homologue of Na+/Cl—dependent neurotransmitter transporters. Nature 437(7056):215–223
5. Yernool D, Boudker O, Jin Y, Gouaux E (2004) Structure of a glutamate transporter homologue from Pyrococcus horikoshii. Nature 431(7010):811–818
6. Li X, Dang S, Yan C, Gong X, Wang J, Shi Y (2013) Structure of a presenilin family intramembrane aspartate protease. Nature 493(7430):56–61
7. Corringer PJ, Baaden M, Bocquet N, Delarue M, Dufresne V, Nury H, Prevost M, Van Renterghem C (2010) Atomic structure and dynamics of pentameric ligand-gated ion channels: new insight from bacterial homologues. J Physiol 588(Pt 4):565–572
8. Weihofen A, Binns K, Lemberg MK, Ashman K, Martoglio B (2002) Identification of signal peptide peptidase, a presenilin-type aspartic protease. Science 296(5576):2215–2218
9. Haass C, Steiner H (2002) Alzheimer disease gamma-secretase: a complex story of GxGD-type presenilin proteases. Trends Cell Biol 12(12):556–562
10. Martoglio B, Golde TE (2003) Intramembrane-cleaving aspartic proteases and disease: presenilins, signal peptide peptidase and their homologs. Hum Mol Genet 12(2):R201–206
11. Sato T, Nyborg AC, Iwata N, Diehl TS, Saido TC, Golde TE, Wolfe MS (2006) Signal peptide peptidase: biochemical properties and modulation by nonsteroidal antiinflammatory drugs. Biochemistry 45(28):8649–8656
12. Narayanan S, Sato T, Wolfe MS (2007) A C-terminal region of signal peptide peptidase defines a functional domain for intramembrane aspartic protease catalysis. J Biol Chem 282:20172–20179
13. Nyborg AC, Kornilova AY, Jansen K, Ladd TB, Wolfe MS, Golde TE (2004) Signal peptide peptidase forms a homodimer that is labeled by an active site-directed gamma-secretase inhibitor. J Biol Chem 279(15):15153–15160
14. Weihofen A, Lemberg MK, Friedmann E, Rueeger H, Schmitz A, Paganetti P, Rovelli G, Martoglio B (2003) Targeting presenilin-type aspartic protease signal peptide peptidase with gamma-secretase inhibitors. J Biol Chem 278(19):16528–16533
15. Torres-Arancivia C, Ross CM, Chavez J, Assur Z, Dolios G, Mancia F, Ubarretxena-Belandia I (2010) Identification of an archaeal presenilin-like intramembrane protease. PLoS One 5(9) e13072
16. Dang S, Wu S, Wang J, Li H, Huang M, He W, Li YM, Wong CC, Shi Y (2015) Cleavage of amyloid precursor protein by an archaeal presenilin homologue PSH. Proc Natl Acad Sci U S A 112(11):3344–3349
17. Naing SH, Vukoti KM, Drury JE, Johnson JL, Kalyoncu S, Hill SE, Torres MP, Lieberman RL (2015) Catalytic properties of intramembrane aspartyl protease substrate hydrolysis evaluated using a FRET peptide cleavage assay. ACS Chem Biol 10(9):2166–74
18. Serrano-Vega MJ, Magnani F, Shibata Y, Tate CG (2008) Conformational thermostabilization of the β1-adrenergic receptor in a detergent-resistant form. Proc Natl Acad Sci U S A 105(3):877–882
19. Boulter JM, Wang DN (2001) Purification and characterization of human erythrocyte glucose transporter in decylmaltoside detergent solution. Protein Expr Purif 22(2):337–348
20. Tribet C, Audebert R, Popot J-L (1996) Amphipols: polymers that keep membrane proteins soluble in aqueous solutions. Proc Natl Acad Sci U S A 93(26):15047–15050
21. Tropea J, Cherry S, Waugh D (2009) Expression and purification of soluble His6-tagged TEV protease. In: Doyle S (ed) High throughput protein expression and purification, vol 498. Humana Press, Totowa, NJ, pp 297–307
22. Privé GG (2007) Detergents for the stabilization and crystallization of membrane proteins. Methods 41(4):388–397
23. Altschul SF, Gish W, Miller W, Myers EW, Lipman DJ (1990) Basic local alignment search tool. J Mol Biol 215(3):403–410
24. Sievers F, Wilm A, Dineen D, Gibson TJ, Karplus K, Li W, Lopez R, McWilliam H, Remmert M, Söding J, Thompson JD, Higgins DG (2011) Fast, scalable generation of high-

quality protein multiple sequence alignments using Clustal Omega. Mol Syst Biol 7:539
25. Robert X, Gouet P (2014) Deciphering key features in protein structures with the new ENDscript server. Nucleic Acids Res 42(W1):W320–W324
26. Shen HB, Chou KC (2007) Signal-3L: a 3-layer approach for predicting signal peptides. Biochem Biophys Res Commun 363(2):297–303
27. Petersen TN, Brunak S, von Heijne G, Nielsen H (2011) SignalP 4.0: discriminating signal peptides from transmembrane regions. Nat Methods 8(10):785–786
28. Greenfield NJ (2006) Using circular dichroism spectra to estimate protein secondary structure. Nat Protoc 1(6):2876–2890
29. Eriks LR, Mayor JA, Kaplan RS (2003) A strategy for identification and quantification of detergents frequently used in the purification of membrane proteins. Anal Biochem 323(2):234–241
30. van den Ent F, Löwe J (2006) RF cloning: a restriction-free method for inserting target genes into plasmids. J Biochem Biophys Methods 67(1):67–74
31. Zerbs S, Giuliani S, Collart F (2014) Chapter eleven - small-scale expression of proteins in E. coli. In: Jon L (ed) Methods in enzymology, vol 536. Academic, New York, pp 117–131
32. Drew D, Lerch M, Kunji E, Slotboom D-J, de Gier J-W (2006) Optimization of membrane protein overexpression and purification using GFP fusions. Nat Methods 3(4):303–313
33. Newstead S, Kim H, von Heijne G, Iwata S, Drew D (2007) High-throughput fluorescent-based optimization of eukaryotic membrane protein overexpression and purification in *Saccharomyces cerevisiae*. Proc Natl Acad Sci U S A 104(35):13936–13941
34. Kawate T, Gouaux E (2006) Fluorescence-detection size-exclusion chromatography for precrystallization screening of integral membrane proteins. Structure 14(4):673–681
35. Veesler D, Blangy S, Cambillau C, Sciara G (2008) There is a baby in the bath water: AcrB contamination is a major problem in membrane-protein crystallization. Acta Crystallogr Sect F: Struct Biol Cryst Commun 64(Pt 10):880–885
36. Engel CK, Chen L, Prive GG (2002) Stability of the lactose permease in detergent solutions. Biochim Biophys Acta 1564(1):47–56
37. Lemieux MJ, Reithmeier RA, Wang DN (2002) Importance of detergent and phospholipid in the crystallization of the human erythrocyte anion-exchanger membrane domain. J Struct Biol 137(3):322–332
38. Tomasiak TM, Pedersen BP, Chaudhary S, Rodriguez A, Colmanares YR, Roe-Zurz Z, Thamminana S, Tessema M, Stroud RM (2014) General qPCR and plate reader methods for rapid optimization of membrane protein purification and crystallization using thermostability assays. Curr Protoc Protein Sci 77:29.11.21–29.11.14
39. Sonoda Y, Newstead S, Hu N-J, Alguel Y, Nji E, Beis K, Yashiro S, Lee C, Leung J, Cameron AD, Byrne B, Iwata S, Drew D (2011) Benchmarking Membrane Protein Detergent Stability for Improving Throughput of High-Resolution X-ray Structures. Structure 19(1):17–25
40. Alexandrov AI, Mileni M, Chien EY, Hanson MA, Stevens RC (2008) Microscale fluorescent thermal stability assay for membrane proteins. Structure 16(3):351–359
41. Prince C, Jia Z (2013) Detergent quantification in membrane protein samples and its application to crystallization experiments. Amino Acids 45(6):1293–1302
42. Miercke LJW, Robbins RA, Stroud RM (2001) Tetra detector analysis of membrane proteins. Curr Protoc Protein Sci 77(29):1–29
43. Gouet P, Robert X, Courcelle E (2003) ESPript/ENDscript: extracting and rendering sequence and 3D information from atomic structures of proteins. Nucleic Acids Res 31(13):3320–3323
44. Gasteiger E, Hoogland C, Gattiker A, Duvaud S, Wilkins MR, Appel RD, Bairoch A (2005) Protein identification and analysis tools on the ExPASy server. In: Walker JM (ed) The proteomics protocols handbook. Humana Press, New York, pp 571–607

Chapter 19

Method to Screen Multidrug Transport Inhibitors Using Yeast Overexpressing a Human MDR Transporter

Laura Fiorini and Isabelle Mus-Veteau

Abstract

Multidrug resistance has appeared to mitigate the efficiency of anticancer drugs and the possibility of successful cancer chemotherapy. The Hedgehog receptor Patched is a multidrug transporter expressed in several cancers and as such it represents a new target to circumvent chemotherapy resistance. In this chapter, we describe the screening test developed to identify molecules able to inhibit the drug efflux activity of Patched. This screening test uses yeast overexpressing functional human Patched that have been shown to resist to chemotherapeutic agents. This test can be adapted to other MDR transporters.

Key words Membrane protein, Overexpression in yeast, Patched, Multidrug transporters, Screening test, Chemotherapy resistance

1 Introduction

Multidrug resistance (MDR) is a phenomenon of resistance of tumors to chemically unrelated anticancer drugs, and is one of the most formidable challenges in the field of cancer chemotherapy [1, 2]. MDR can have many causes such as alterations in DNA repair, defective regulation of apoptotic gene expression, enhanced intracellular drug detoxification, but the most common mechanism is the efflux of cytotoxic drugs by membrane transporters. In human, most of the multidrug transporters belong to the large ATP-binding cassette (ABC) transporter super family of membrane proteins from which P-glycoprotein (P-gp) was the first member to be identified [3–5]. Many inhibitors of P-gp have been tested in clinical trials to assess their pharmacological potential. Unfortunately, most of them have failed because they displayed nonspecific toxicity [6].

Emerging data from many human tumors have shown that the chemotherapy-resistant phenotype of cancer cells correlates with the activation of Hedgehog (Hh) signaling, and that the Hh pathway regulates cancer stem cells or tumor initiating cells [7–10].

Isabelle Mus-Veteau (ed.), *Heterologous Expression of Membrane Proteins: Methods and Protocols,* Methods in Molecular Biology, vol. 1432, DOI 10.1007/978-1-4939-3637-3_19, © Springer Science+Business Media New York 2016

The Hh receptor Patched being an Hh target gene, this 12 transmembrane domains protein is overexpressed in many recurrent and metastatic tumors such as breast, lung, colorectal, ovarian, prostate cancers or melanoma. We recently discovered that Patched is a multidrug efflux pump that transports different chemotherapeutic agents out of cells, and particularly doxorubicin, using the proton motive force like the bacterial multidrug efflux pumps from the RND family [11]. This is a real breakthrough which suggests that the Hh receptor Patched participates to the resistance to chemotherapy of cancer cells, and allows proposing Patched as a new target for anti-cancer therapy. We then designed innovative screening tests to identify molecules able to inhibit the drug efflux activity of Patched, and we recently showed that a natural compound isolated from the Mediterranean sponge *Haliclona mucosa* inhibits the resistance to the chemotherapeutic agent doxorubicin (dxr) conferred to yeast by the expression of human Patched [12].

In this chapter, we describe the method used to express functional human Patched in the yeast *Saccharomyces cerevisiae* [13], and the screening test developed to identify inhibitors of the drug efflux activity of Patched which uses yeast expressing human Patched that have been shown to resist to chemotherapeutic agents [11]. This test can be adapted to other MDR transporters.

2 Materials

2.1 Strains

1. Competent *E. coli* DH5α for vectors preparation.
2. *S. cerevisiae* strain K699 (Mata, ura3, and leu 2–3) for heterologous expression.

2.2 Vectors Preparation

1. A yeast expression vector allowing a functional expression of your MDR transporter. We use the Yep-PMA-MAP vector containing the plasma membrane proton ATPase (PMA) promoter, a sequence giving autotrophy for an amino acid for leucine and a Multitag Affinity Purification (MAP) sequence encoding (1) a factor Xa, a TEV, and a thrombin cleavage site to eliminate the MAP sequence; (2) a calmodulin binding domain (CBD), a streptavidin tag, and an hexahistidine tag for affinity chromatography; and (3) an hemagglutinin peptide (HA) for anti-HA western-blot analysis [13].
2. cDNAs from the target membrane protein.
3. ProofStart polymerase.
4. Polymerase chain reaction (PCR).
5. pCR™2.1 plasmid.
6. Restriction enzymes.
7. Ligase.

2.3 Yeast Transformation

1. Plate mixture: 1 M lithium acetate, 90% (w/v) polyethylene glycol (PEG) 4000, 1 M UltraPure™ Tris–HCl pH 7.5, 0.5 M ethylene diamine tetraacetic acid (EDTA) disodium salt dihydrate.
2. 10 mg/mL salmon sperm DNA.
3. 1 M 1,4 dithio-DL-threitol (DTT).
4. Water bath set at 42 °C.
5. Expression vector.
6. 25 mL sterile flasks with vented caps.

2.4 Yeast Culture

1. Minimal medium (composition for 1 L): dissolve 8 g yeast nitrogen base without amino acids, 55 mg tyrosin, 55 mg uracil, and 55 mg adenosine hemisulfate in 900 ml of Milli-Q water, and autoclave at 120 °C. Store at room temperature. For solid plates, add 20% (w/v) agar.
2. 20% d-(+)-glucose (10×): dissolve progressively 200 g of d-glucose (minimum 99.5%) in 800 mL of Milli-Q water, and then adjust to 1 L with Milli-Q water (*see* **Note 1**), and sterilize on 0.22 μm filter and store at +4 °C. The shelf life of this solution is approximately 1 year.
3. Drop-out (without leucine) (100×): dissolve 500 mg each of l-arginine, l-histidine, l-iso-leucine, l-lysine, l-methionine, l-phenylalanine, l-threonine, l-tryptophan in 100 ml of water, sterilize on 0.22 μm filter, and store in aliquots at −20 °C. The shelf life of this solution is approximately 1 year.
4. Rich medium (composition for 1 L): dissolve 11 g yeast extract, 22 g bacteriological peptone, 55 mg adenine hemisulfate in 900 mL of Milli-Q water and autoclave at 120 °C. Store at room temperature.
5. Erlenmeyer flasks or sterile flasks with vented caps.
6. Spectrophotometer (Eppendorf) and cuvettes.
7. Incubator shaker at 30 °C and 18 °C.

2.5 Yeast Membrane Fraction Preparation

1. Glass beads, acid-washed, 425–600 μm.
2. 10% glycerol.
3. Heidolph™ Multi Reax vortexer.
4. Buffer A (2×): dissolve 12.114 g Tris, 17.53 g NaCl in 900 mL Milli-Q water. Adjust pH to 7.4 by adding HCl. Adjust to 1 L with Milli-Q water. This is the grinding buffer.
5. Buffer A (1×): 50 mM Tris–HCl, 500 mM NaCl, pH 7.4.
6. Buffer B (2×): dissolve 12.114 g Tris, 17.53 g NaCl in 700 mL Milli-Q water. Add 200 mL glycerol. Adjust pH to 7.4 by adding HCl. Adjust volume to 1 L with Milli-Q water.

7. PMSF (200×) (to inhibit serine proteases): 0.35 g in 10 mL absolute ethanol. 200 mM stock solution is stable for months at +4 °C (*see* **Note 2**).
8. Benzamidine hydrochloride 4 mM (200×) (to inhibit serine proteases): 1.25 g of benzamidine hydrochloride hydrate minimum 98 % in 10 mL Milli-Q water. Store at +4 °C.
9. EDTA (200×) (to inactivate metalloproteases by depleting ions): 8.18 g in 50 mL of water. EDTA solution will not go into solution until the solution is adjusted to approximately pH 8.0 by adding NaOH pellets. Store at +4 °C.

2.6 SDS-Polyacrylamide Gel Electrophoresis (SDS-PAGE)

1. Separating buffer: 375 mM UltraPure ™ Tris–HCl (pH 8.7).
2. 30 % acrylamide–bisacrylamide solution (37.5:1).
3. Stacking buffer: 125 mM Tris–HCl (pH 6.8).
4. 0.1 % (w/v) UltraPure ™ sodium dodecyl sulfate (SDS).
5. 0.1 % (w/v) ammonium persulfate (APS).
6. *N,N,N,N′*-tetramethyl-ethylenediamine (TEMED).
7. Running buffer (5×): 125 mM UltraPure ™ Tris, 960 mM glycine, 0.5 % (w/v) SDS. Store at room temperature.
8. Loading buffer (4×): 250 mM UltraPure ™ Tris–HCl pH 6.8, 8 % (w/v) SDS, 0.04 % (w/v) bromophenol blue, 40 % (w/v) glycerol, 20 % (v/v) 2-β-mercaptoethanol (*see* **Note 3**).
9. Prestained protein ladder.
10. Bio-Rad Mini-PROTEAN® 3 cell.

2.7 Western Blot

1. Transfer buffer (10×): Dissolve 58 g UltraPure™ Tris (do not adjust pH), 29 g glycine, and 3.7 g SDS in 1 L of water. Store at room temperature (with cooling during use).
2. 20 % ethanol.
3. Nitrocellulose membrane.
4. Tris-Buffered Saline with Tween 20 (TBS-T): 20 mM UltraPure ™ Tris (pH 7.4), 450 mM NaCl, and 0.1 % (v/v) Tween 20 (polyoxyethylene sorbitan monolaurate Polysorbate 20, cell culture and bacteriology grade).
5. Primary antibody directed against your MDR transporter or a tag (6His or HA).
6. Secondary polyclonal immunoglobulins conjugated to horseradish peroxidase.
7. Chemiluminescent HRP substrate (HRP substrate peroxide solution and HRP substrate Luminol reagent).
8. 0.1 % (w/v) amido black dissolved in 25 % (v/v) acetic acid.
9. Imager.

10. Plastic bags and plastic films.
11. Rotating wheel.
12. Blocking buffer: 4% (w/v) nonfat dry milk in TBS-T.
13. Amido black 0.1% solution: Dissolve 0.1 g of amido black in 25 mL acetic acid, 10 mL isopropanol, and 65 mL of Milli-Q water.

2.8 Screening Test

1. 50 mL sterile tubes.
2. Sterile 96 deep wells of 1 or 2 mL.
3. Sterile 96-well plates.
4. Multichannel pipette (12 channels) 20–200 μL.
5. Microtitre plate shaker in a thermostat cabinet at 18 °C.
6. Rich medium.
7. 20% d-glucose.
8. Sterile DMSO.
9. 10 mM doxorubicin (dissolved in sterile water).
10. Molecules to be tested at 10 mM in DMSO in 96-well plates.

3 Methods

3.1 Construction of Expression Vectors

1. PCR with the ProofStart polymerase was carried out on the target cDNA using the adequate primers to introduce the restriction sites allowing the subcloning of the cDNA into the expression vector. In our case, *Xba*I, *Spe*I restriction sites and a sequence of six adenosines were added at the 5′ end, and a *Nhe*I restriction site at the 3′ end of *Ptch1* cDNA.
2. Sequence the PCR product.
3. Digest and subclone target cDNA in the expression vector. In our case, *Ptch1* cDNA was digested by *Spe*I and *Nhe*I and subcloned in MAP *Xba*I/*Nhe*I sites, giving YEpPMA-*Ptch1*-MAP expression vector.
4. Transform competent *E. coli* DH5α with YEpPMA-*Ptch1*-MAP expression vector for vector amplification.

3.2 Yeast Transformation

Day 0

1. Start an overnight (O/N) culture of *S. cerevisiae K699* cells to be transformed. In a 25 mL sterile flask with vented cap, inoculate 5 mL of rich medium containing 2% d-glucose with *S. cerevisiae K699* cells. Incubate O/N at 30 °C in an incubator shaker at 200 rpm.

Day 1

2. Four hours before transformation, dilute the saturated O/N culture to the third with rich medium containing 2% d-glucose. Incubate for 2–4 h at 30 °C at 200 rpm in the incubator shaker.
3. In a 1.5 mL Eppendorf tube, spin down 0.5 mL of the culture (*see* **Note 4**)
4. Discard supernatant. This leaves 50–100 μL of medium.
5. Add 10 μL of 10 mg/mL carrier salmon sperm DNA and resuspend cells with the pipette.
6. Add 1 μg of your plasmid (*see* **Note 5**). Vortex.
7. Add 0.5 mL of sterile plate mixture. Vortex
8. Add 20 μL of 1 M DTT (*see* **Note 6**). Vortex.
9. Incubate O/N at room temperature on your bench top.

Day 2

10. Cells will have settled to the bottom of the tube.
11. Heat-shock cells by incubating for 10 min at 42 °C.
12. Use a pipette to suck up the bottom 50 μL of cells and plate mixture (*see* **Note 7**).
13. Plate cells out on selective medium agar dishes. In our case: minimal medium containing 2% d-glucose and 1% drop-out without leucine and agar.
14. Incubate for 2 days at 30 °C. Clones growing onto the dishes should express you protein.

3.3 Yeast Culture for Expression of Your Target MDR Protein

1. Inoculate 5 mL of minimal medium complemented with 2% of d-glucose and 1% of drop-out without leucine in a 25 mL sterile flask (with vented cap) with an isolated colony of transformed *Saccharomyces cerevisiae*.
2. Grow overnight at 30 °C in the incubator under vigorous shaking at 200 rpm (*see* **Note 8**).
3. In the morning, check absorbance at 600 nm (Abs_{600}) making a 1:10 dilution in a spectrophotometer cuvette. Yeast should have reached stationary phase ($Abs_{600} > 2$).
4. Dilute preculture in 10 mL of minimal medium complemented with 2% of d-glucose and 1% of drop-out without leucine in a 25 mL sterile flask (with vented cap) in order to reach an Abs_{600} of 2 in the afternoon. Yeasts double every 2 h at 30 °C.
5. Keep at 30 °C under shaking at 200 rpm.
6. Check the absorbance of your preculture and verify that it is free from any bacterial contamination using a microscope.
7. Inoculate with a volume of preculture 100 mL of rich medium complemented with 2% d-glucose in a sterile Erlenmeyer of

500 mL that will allow obtaining an Abs_{600} of 5 after an ON culture at 18 °C knowing that yeast divide approximately every 4 h at 18 °C (*see* **Note 9**).

8. Put the culture in an incubator at 18 °C under shaking at 200 rpm O/N.
9. 24 h post-incubation, yeasts should have reached an Abs_{600} between 6 and 7.
10. Check that your culture has not been contaminated by bacteria using a microscope.
11. Harvest the cells by centrifugation 10 min at 2000 × *g* at +4 °C.
12. Resuspend the pellet first with cold water.
13. Centrifuge for 10 min at 2000 × *g* at 4 °C. Discard the supernatant.
14. Wash with ice-cold buffer A (1×) extemporary supplemented with protease inhibitors: 1 mM PMSF, 4 mM benzamidine hydrochloride, 2.5 mM EDTA pH 8.
15. Centrifuge again 10 min at 2000 × *g* at 4 °C. Discard the supernatant.
16. Freeze the pellet at −80 °C until next use.

3.4 Membrane Preparation

1. Prepare 200 mL of buffer A (1×) extemporary supplemented with 1 mL PMSF 200×, 1 mL EDTA 200×, 1 mL benzamidine hydrochloride 200× and keep cold.
2. Thaw yeast pellet by adding three volumes of this freshly prepared ice-cold buffer A supplemented with protease inhibitors.
3. In a 50-mL Falcon tube, put 17.5 mL of yeast suspension and 2.5 mL of glass beads (*see* **Note 10**).
4. Grind yeasts by vortexing at 2000 rpm at 4 °C using a Heidolph™ Multi Reax device. The optimal time for yeast expressing Patched is 15 min but it can be different for your MDR transporter. The optimal time can be determined by tacking samples at increasing times.
5. With a pipette, separate the supernatant from the beads and centrifuge at 3000 × *g* for 10 min to eliminate unbroken cells, debris, and nuclei. Here you can estimate the grinding's efficiency by measuring the ratio between the starting and the remaining volume of pellet (*see* **Note 11**).
6. The supernatant is then centrifuged at 18,000 × *g* for 1 h at 4 °C (*see* **Note 12**).
7. The supernatant is discarded.
8. The pellet containing yeast plasma membrane is resuspended in ice-cold buffer A supplemented with protease inhibitors and centrifuge at 18,000 × *g* for 1 h at +4 °C.

9. Repeat **steps 7** and **8**.
10. Finally, resuspended the pellet in buffer A.
11. Using DC Bio-Rad assay reagents and BSA at concentrations ranging from 2 to 20 mg/mL as a standard, quantify total membrane proteins.

3.5 Protein Expression Analysis by Western-Blotting

The following protocol assumes the use of Bio-Rad Mini-PROTEAN gel system.

1. Glass plates to be used should be very well cleaned and extensively rinsed with distilled water and ethanol.
2. Prepare a 1.5 mm thick 8 % resolving gel for a protein of 120–150 kDa. For 10 mL of resolving gel, mix 4.6 mL of H_2O, 2.7 mL of acrylamide, 2.5 mL of separating buffer, 100 μL of 10 % SDS, 100 μL of 10% ammonium persulfate, and 6 μL of TEMED. Wear gloves because acrylamide is a neurotoxic and carcinogenic when unpolymerized.
3. Pour the gel immediately leaving space (1 cm below comb teeth) for the stacking gel, and overlay with 100 μL of H_2O. Allow the gel to polymerize for at least 10 min. Polymerization time depends on room temperature.
4. Prepare 5 mL of stacking gel mixing 3.4 mL of H_2O, 0.83 mL of acrylamide, 0.63 mL of stacking buffer, 50 μL of 10 % SDS, 50 μL of 10 % ammonium persulfate, and 5 μL of TEMED.
5. Pour off water from the surface of the separating gel, pour the gel and immediately insert the comb. Wait few minutes to polymerize before removing the comb.
6. Once the stacking gel has set, remove carefully the comb (*see* **Note 13**).
7. Put the gel in the electrophoresis unit and add 1× running buffer and wash wells with running buffer using a syringe.
8. Mix 15 μL of your sample with 5 μL of the 4× sample buffer. Let the sample at room temperature 5–10 min before loading (*see* **Note 14**).
9. Load the samples and prestained molecular weight markers into the wells with a Hamilton syringe or a pipette using gel loading tips.
10. Place the lid on the Mini Tank, insert the electrical leads into a suitable power supply with the proper polarity. A constant 100 V is the recommended for SDS-PAGE. Run time is around 2 h depending on the molecular weight of your protein.
11. During electrophoresis, prepare the transfer buffer 1× by diluting 100 mL of the 10× transfer buffer in 700 mL Milli-Q

water and add 200 mL ethanol (to have a final concentration of 20 %).

12. Cut the membrane and the filter paper to the dimensions of the gel. Always wear gloves when handling membranes to prevent contamination. Soak the membrane, filter paper, and fiber pads in transfer buffer 15 min before blotting.
13. After electrophoresis is complete, remove the tank lid and carefully lift out the Inner Chamber Assembly. Pour off and discard the running buffer.
14. Remove the gels from the Gel Cassette Sandwich by gently separating the two plates of the gel cassette. The green, wedge shaped, plastic Gel Releaser may be used to help pry the glass plates apart.
15. Remove the gel by floating it off the glass plate by inverting the gel and plate under transfer solution, agitating gently until the gel separates from the plate.
16. Prepare the gel sandwich. The following instructions assume the use of Bio-Rad Mini Trans-Blot electrophoretic transfer cell.
17. Place the cassette, with the black side down, in a recipient containing 1× transfer buffer.
18. Place one pre-wetted fiber pad on the gray side of the cassette, a sheet of filter paper on the fiber pad, the gel on the filter paper and the pre-wetted membrane on the gel.
19. Remove any air bubbles between gel and membrane by rolling a 5-mL tube on the membrane.
20. Complete the sandwich by placing a piece of filter paper on the membrane and then the last fiber pad.
21. Close the cassette firmly, being careful not to move the gel and filter paper sandwich.
22. Place the cassette into the transfer tank such that the nitrocellulose membrane is between the gel and the anode. It is vitally important to ensure this orientation or the proteins will be lost from the gel into the buffer rather than transferred to the nitrocellulose.
23. Add the frozen cooling unit. Place in tank and completely fill the tank with 1× transfer buffer and run the blot under constant voltage of 100 V for at least 1 h (High intensity field ~ 350 mA). For Patched, we transfer 1h15.
24. Upon completion of the run, disassemble the blotting sandwich and remove the membrane. The colored molecular weight markers should be clearly visible on the membrane.
25. To check whether your transfer has been correctly done or not, incubate the membrane for 2 min with 0.1 % of amido black solution to stain the transferred bands and wash twice with TBS-T buffer to remove amido black.

26. Incubate the membrane in 50 mL blocking buffer for 1 h at room temperature on a rocking platform.
27. Discard blocking buffer and place the membrane in a plastic bag containing 5 mL of blocking buffer supplemented with the primary antibody (here 1:20 dilution of anti-HA) overnight at 4 °C on a rotating wheel.
28. Remove primary antibody and wash the membrane three times for 10 min each with 50 mL PBS-T.
29. Place the membrane in a plastic bag containing 5 mL of blocking buffer supplemented with the convenient secondary antibody coupled to horseradish peroxidase (HRP) at the dilution given by the manufacturer (here it's the secondary polyclonal anti mouse immunoglobulin antibody used as 1:5000-fold dilution). Place the bag on a rotating wheel for 2 h at 4 °C or 45 min at room temperature.
30. Discard the secondary antibody and wash the membrane at least three times for 10 min each with TBS-T buffer.
31. Mix HRP substrate peroxide solution and HRP substrate Luminol reagent according to manufacturer's recommendations and proceed to membrane revelation using autoradiography films or a luminescent image analyzer.

3.6 Drug Resistance of Yeast Expressing Your MDR Protein

1. Inoculate 5 mL of minimal medium complemented with 2% of d-glucose and 1% of drop-out without leucine in a 25 mL sterile flask (with vented cap) with yeast overexpressing your MDR protein and 5 mL of rich medium supplemented with 2% glucose with control yeast, and proceed to preculture as described in Subheading 3.3, **steps 1–5**.
2. Check the absorbance at 600 nm of your preculture and verify that it is free from any bacterial contamination under a microscope.
3. Prepare two tubes with 50 mL of rich medium complemented with 2% d-glucose and add a volume of yeast expressing your MDR protein or of control yeast preculture in respective tubes to obtain an $Abs_{600\ nm}$ of 0.01 (Fig. 1a).
4. In a 2 mL 96-deep-well plate, put 1 ml of medium containing control yeast in wells from the first lane (A) and 1 mL of medium containing yeast expressing your MDR protein in wells from the second lane (B) (Fig. 1b).
5. Add increasing concentration of a drug known to be a substrate of your MDR protein in deep wells from lanes A and B
6. Transfer 150 μL of each deep well per well of a sterile 96-well plate using a 12-channel pipette as shown in Fig. 1c. Each condition will be done in quadruplicate.
7. Cover the plate with a transparent film and make two holes with a syringe on the top of each well to let the air flow.

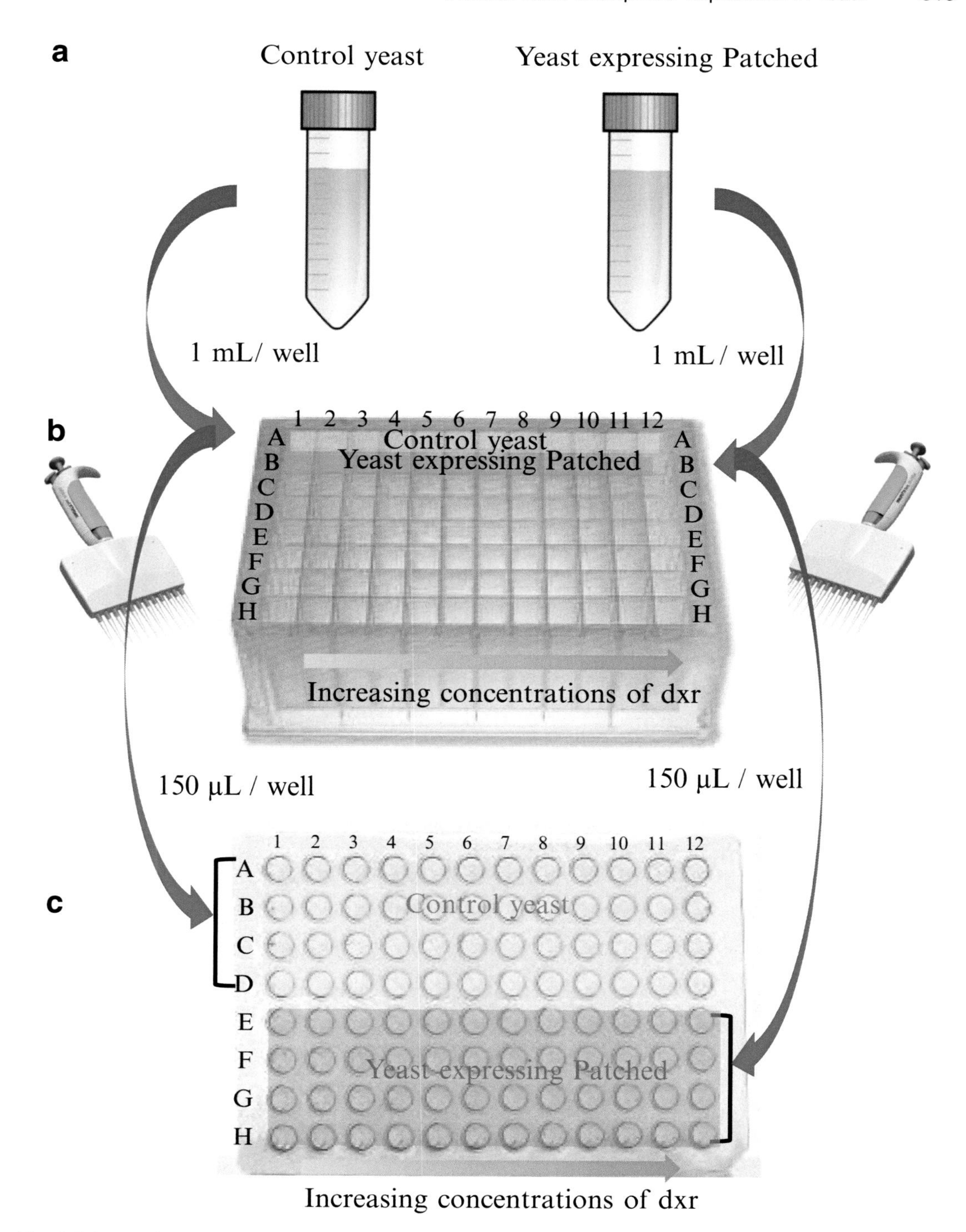

Fig. 1 Scheme describing the method developed to test the drug resistance of yeast expressing your Patched protein

8. Place the plate in an incubator at 18 °C under shaking at 1250 rpm.
9. Measure the absorbance at 600 nm twice a day (at the beginning and the end of the day) during 4 or 5 days.
10. Calculate the mean of the four absorbance values for each drug concentration for control yeast and yeast expressing your MDR protein, respectively, and draw the curve of the absorbance in function of time for each drug concentration using Excel software. The expression of your MDR protein should confer to yeast a resistance to the drug in comparison to control yeast as illustrated in Fig. 2a.

3.7 Efflux Inhibitor Screening

1. Inoculate 5 mL of minimal medium complemented with 2 % of d-glucose and 1 % of drop-out without leucine in a 25 mL sterile flask (with vented cap) with yeast overexpressing your MDR protein and proceed to preculture as described in Subheading 3.3, **steps 1–5**. The preculture must be done overnight in order to inoculate the 96-well plates the next day (*see* **Note 15**).
2. Check the absorbance at 600 nm of your preculture and verify that it is free from any bacterial contamination by a microscope observation.
3. Prepare 100 mL of rich medium complemented with 2 % d-glucose, add a volume of yeast preculture to obtain an $Abs_{600\ nm}$ of 0.01, and separate the medium containing yeast in two Falcon tubes of 50 mL (Fig. 3a).
4. Add in one of the tubes a concentration of drug that slows but does not completely inhibit the growth of yeast expressing your MDR protein (for yeast expressing Patched: 15 μM of doxorubicin).
5. In a 2 mL deep-well plate, put 1 ml of medium containing yeast at $Abs_{600\ nm}$ of 0.01 without drug in wells from the first lane (A) and 1 mL of medium containing yeast $Abs_{600\ nm}$ of 0.01 with the drug in wells from the second lane (B) (Fig. 3b).
6. Add 1 μL of the chemical compounds to be screened in deep wells from lanes A and B (from a 10 mM mother solution in DMSO in a 96-well plate). You can test 11 different compounds (Fig. 3c). Put 1 μL of DMSO in wells number 12 of the deep-well plate for the control.
7. Transfer 150 μL of each deep well per well of a sterile 96-well plate as shown in Fig. 3d. Each condition will be done in quadruplicate. Half of the 96-well plate will contain medium without the drug and half of the plate will contain medium with the drug.
8. Cover the plate with a transparent film and make two holes with a syringe on the top of each well to let the air flow.

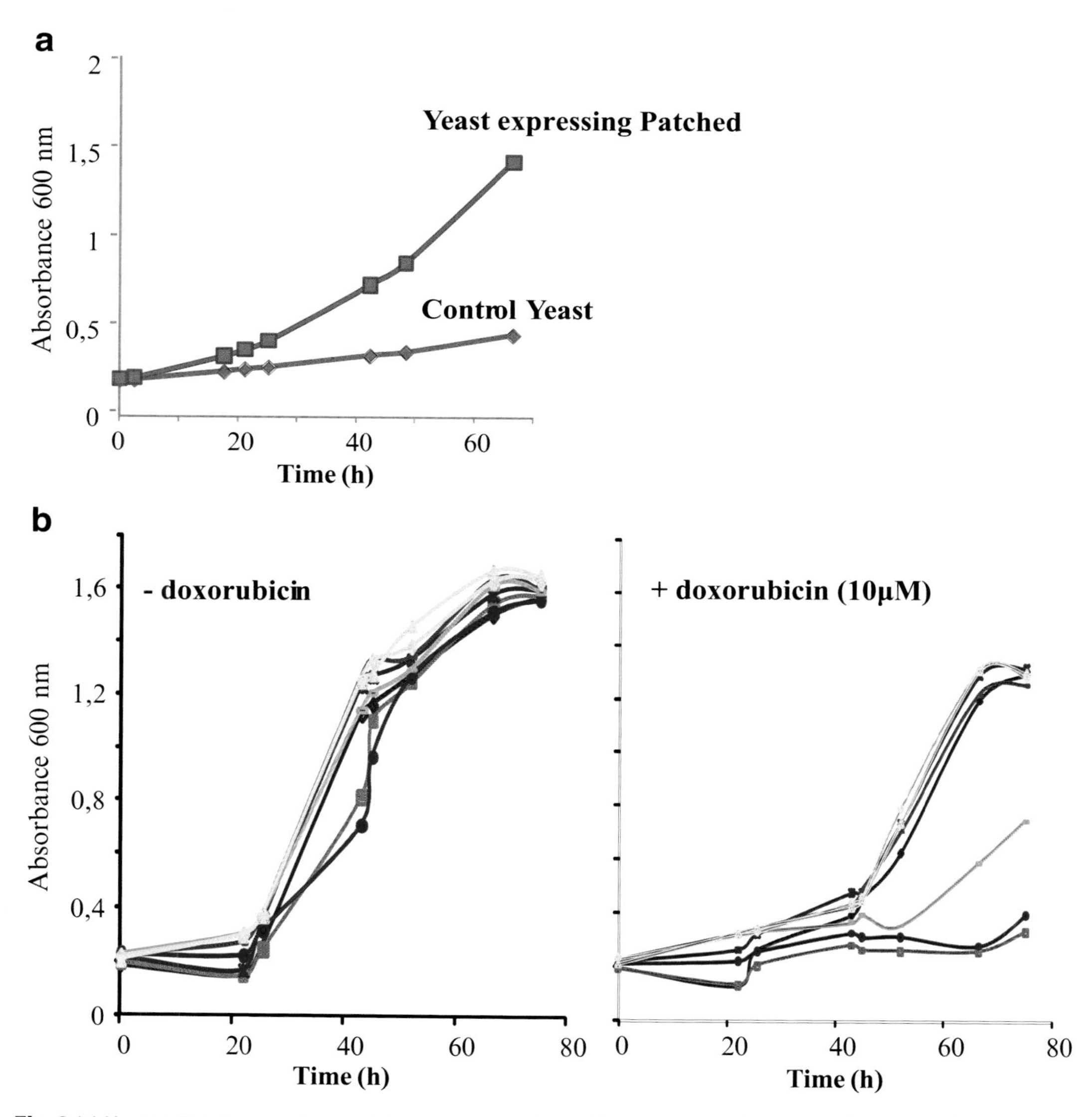

Fig. 2 (**a**) Human Patched confers resistance to doxorubicin. The presence of doxorubicin in the culture medium prevents control yeast growth (in *gray*) but not yeast expressing Patched growth (in *red*). (**b**) Screening of potential doxorubicin efflux inhibitors on yeast expressing Patched. Yeast expressing Patched were grown in presence of 10 μM of compounds to be tested with or without doxorubicin. While the "pink" and the "brown" compounds have no effect on yeast growth without doxorubicin, they completely inhibit yeast growth in presence of doxorubicin suggesting that these two compounds could be inhibitors of Patched drug efflux

9. Repeat the same procedure for the other compounds you want to test.
10. Measure the absorbance at 600 nm for each 96-well plate.
11. Place the plates in an incubator at 18 °C under shaking at 1250 rpm.
12. Measure the absorbance at 600 nm twice a day (at the beginning and the end of the day) during 4 or 5 days. Stop measure-

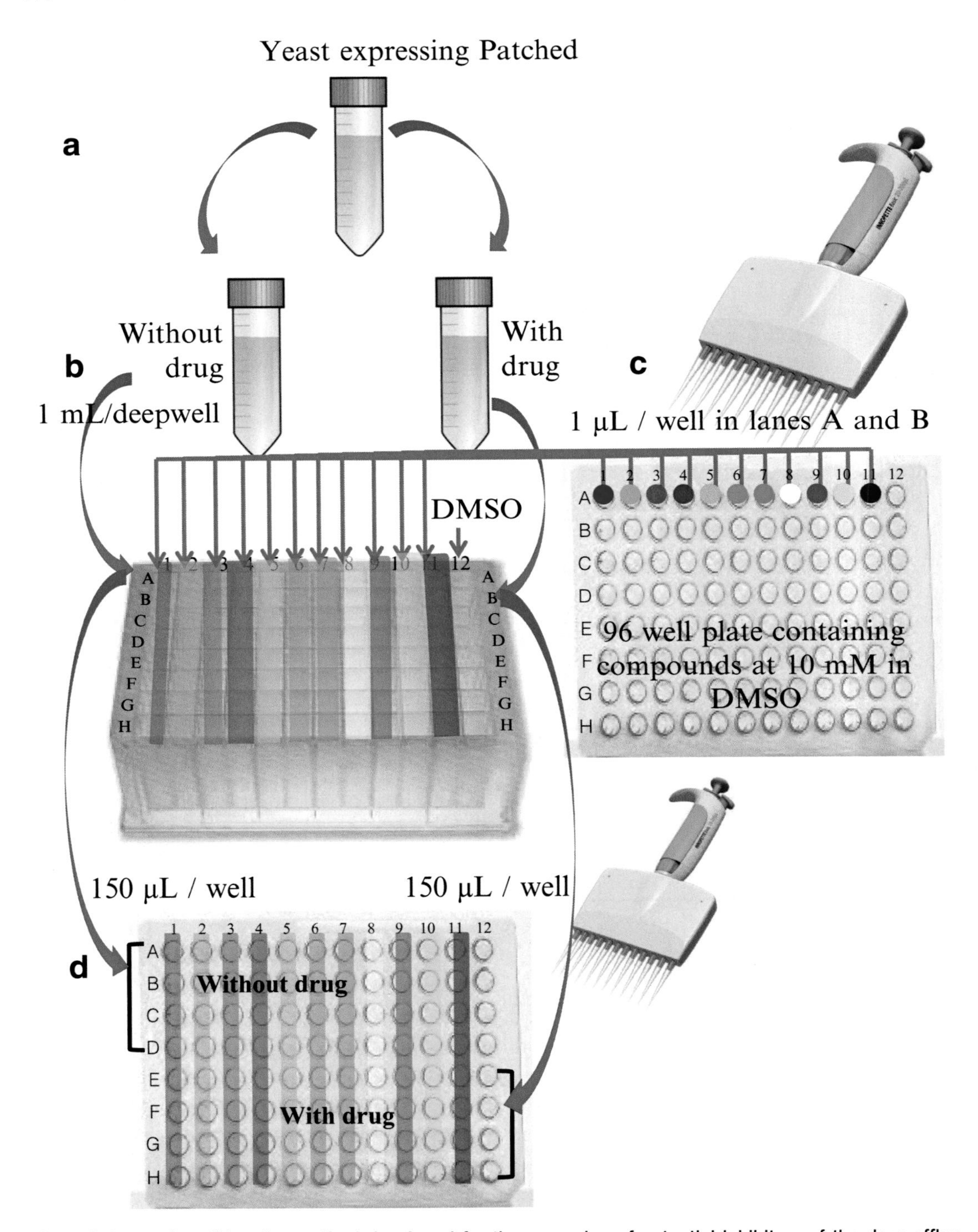

Fig. 3 Scheme describing the method developed for the screening of potential inhibitors of the drug efflux activity of Patched using drug resistance of yeast expressing Patched protein activity

ments when the control wells containing the drug and DMSO reach an $Abs_{600\ nm}$ of 1.

13. Calculate the mean of the four absorbance values for each compound and the control with and without drug, and draw the curve of the absorbance in function of time for each chemical compound tested using Excel software. An inhibitor of drug efflux should inhibit the growth of yeast in presence of the drug. If a compound also inhibits the growth of yeast in medium without drug, this compound is surely cytotoxic. Fig. 2b gives an example of the kind of results you can expect.

4 Notes

1. Add glucose progressively in water under agitation.
2. PMSF is poisonous and should be handled carefully. Do not dissolve PMSF in water. The half-life of PMSF in water is 1 h.
3. 2-β-mercaptoethanol should be added extemporaneously.
4. Five seconds are sufficient to pellet the cells.
5. Usually 1–3 μL of mini-prep.
6. DTT is very instable in solution. It should be prepared and immediately aliquoted and stored at −20 °C.
7. Try to get as many cells as possible but don't worry about keeping every cells. You should easily get 80–90 % of cells.
8. At this stage, the preculture can be done over 2 days or a week end.
9. We grow yeast at 18 °C to give time to the heterologous MDR protein to correctly insert into the plasma membrane.
10. Optimal repartition. Exceeding these volumes may alter vortexing and therefore grinding will not be efficient.
11. To estimate grinding's efficiency you can compare the volume of the yeast pellet before and after grinding if you are using the 50-mL Falcon tubes. If not weigh the pellet before and after grinding. Generally this ratio is around 50–60 %.
12. 18,000 × *g* for 1 h allows to mainly pelleting plasma membrane where the heterologous MDR protein should be addressed.
13. Remove the comb slowly and vertically to avoid snatching the wells.
14. Heating at 95 °C often induces aggregation of membrane proteins.
15. Inoculations of microplates can take several hours (count about 1 h per plate).

Acknowledgements

This work was supported by the European Community Specific Target Research Project grant FP6-2003-LifeSciHealth "Innovative Tools for Membrane Structural Proteomics", the CNRS (Soutien au transfert), the PACA Region (PatchedWork project APRF2013 DEB 14-495), and the France Cancer Association.

References

1. Ribacka C, Pesonen S, Hemminki A (2008) Cancer, stem cells, and oncolytic viruses. Ann Med 40:496–505
2. Zinzi L, Contino M, Cantore M, Capparelli E, Leopoldo M, Colabufo NA (2014) ABC transporters in CSCs membranes as a novel target for treating tumor relapse. Front Pharmacol 5:163
3. Ambudkar SV, Dey S, Hrycyna CA, Ramachandra M, Pastan I, Gottesman MM (1999) Biochemical, cellular, and pharmacological aspects of the multidrug transporter. Annu Rev Pharmacol Toxicol 39:361–398
4. Szakacs G, Paterson JK, Ludwig JA, Booth-Genthe C, Gottesman MM (2006) Targeting multidrug resistance in cancer. Nat Rev Drug Discov 5:219–234
5. Eckford PD, Sharom FJ (2009) ABC efflux pump-based resistance to chemotherapy drugs. Chem Rev 109:2989–3011
6. Lopez D, Martinez-Luis S (2014) Marine natural products with P-glycoprotein inhibitor properties. Mar Drugs 12:525–546
7. Zhao C, Chen A, Jamieson CH, Fereshteh M, Abrahamsson A, Blum J, Kwon HY, Kim J, Chute JP, Rizzieri D et al (2009) Hedgehog signalling is essential for maintenance of cancer stem cells in myeloid leukaemia. Nature 458:776–779
8. Queiroz KC, Ruela-de-Sousa RR, Fuhler GM, Aberson HL, Ferreira CV, Peppelenbosch MP, Spek CA (2010) Hedgehog signaling maintains chemoresistance in myeloid leukemic cells. Oncogene 29:6314–6322
9. Liu S, Dontu G, Mantle ID, Patel S, Ahn NS, Jackson KW, Suri P, Wicha MS (2006) Hedgehog signaling and Bmi-1 regulate self-renewal of normal and malignant human mammary stem cells. Cancer Res 66:6063–6071
10. Sims-Mourtada J, Opdenaker LM, Davis J, Arnold KM, Flynn D (2014) Taxane-induced hedgehog signaling is linked to expansion of breast cancer stem-like populations after chemotherapy. Mol Carcinog 54(11):1480–93. doi:10.1002/mc
11. Bidet M, Tomico A, Martin P, Guizouarn H, Mollat P, Mus-Veteau I (2012) The Hedgehog receptor patched functions in multidrug transport and chemotherapy resistance. Mol Cancer Res 10:1496–1508
12. Fiorini L, Tribalat M-A, Sauvard L, Cazareth J, Lalli E, Broutin I, Thomas OP, Mus-Veteau I (2015) Natural paniceins from Mediterranean sponge inhibit the multidrug resistance activity of Patched and increase chemotherapy efficiency on melanoma cells. Oncotarget 6(26):22282–97
13. Joubert O, Nehmé R, Fleury D, De Rivoyre M, Bidet M, Polidori A, Ruat M, Pucci B, Mollat P, Mus-Veteau I (2009) Functional studies of membrane-bound and purified human Hedgehog receptor Patched expressed in yeast. Biochim Biophys Acta 1788:1813–1821

Index

Isabelle Mus-Veteau (ed.), *Heterologous Expression of Membrane Proteins: Methods and Protocols*, Methods in Molecular Biology, vol. 1432, DOI 10.1007/978-1-4939-3637-3, © Springer Science+Business Media New York 2016

P

Q

R

S

T

V

W

X

Y